Exploring Occupant Behavior in Buildings

Andreas Wagner · William O'Brien
Bing Dong
Editors

Exploring Occupant Behavior in Buildings

Methods and Challenges

Springer

Editors
Andreas Wagner
Building Science Group
Karlsruhe Institute of Technology
Karlsruhe
Germany

Bing Dong
Department of Mechanical Engineering
The University of Texas at San Antonio
San Antonio, TX
USA

William O'Brien
Department of Civil and Environmental Engineering
Carleton University
Ottawa, ON
Canada

ISBN 978-3-319-61463-2 ISBN 978-3-319-61464-9 (eBook)
DOI 10.1007/978-3-319-61464-9

Library of Congress Control Number: 2017949493

Printed on acid-free paper

This Springer imprint is published by Springer Nature
The registered company is Springer International Publishing AG
The registered company address is: Gewerbestrasse 11, 6330 Cham, Switzerland

Foreword

As our knowledge of buildings and the built environment has increased, so has the complexity of optimizing the design and operation of buildings. It is now understood that as well as conforming to the needs of the users, buildings must also be sensitive to many other dynamic processes in the wider built environment. Thus, in addition to the growing understanding of occupant needs and comfort expectations, and the increasing awareness of the correlation between indoor environments and the health, well-being, and consequent productivity of a building's occupants, other significant factors that must be taken account of include: global climate change; fossil fuel depletion; variable prices for energy sources; and greater flexibility of organizations and the building stocks that they use.

The simultaneous management of these factors is highly complex and demands an integrated approach to both the design and operation of buildings. To deliver robust building designs and effective system solutions capable of conforming to future demands, it is essential to thoughtfully integrate cutting-edge practices from the fields of building construction and services, the available knowledge of ambient environment effects, and the requirements of a building's occupants and the activities they undertake.

One of the most powerful analysis techniques is computational modeling (i.e., creating a computer-based representation of a real system) and simulation (i.e., using a model to predict (future) behavior of a real system). This is now routinely used in a wide range of fields including astrophysics, climatology, chemistry, biology, economics, and engineering. However, it must be noted that (A) simulation aims to provide a greater understanding of a given system and does not claim to provide definitive answers to questions or solutions to problems and (B) ensuring the quality of simulation results is often difficult.

In terms of building performance simulation, various uncertainties (weather, occupant behavior, variable energy prices, etc.) hinder the reliability of the predictions made. Thus, the impact of such uncertainties in building performance predictions is a hot topic in current research. Improving the reliability of predictions of building performance is of prime importance if key concepts for the built environment are to be realized, most notably zero-carbon buildings and districts, energy performance contracting, and demand side management.

It is now generally accepted that the energy and comfort performance of buildings is often strongly influenced by the occupants and their behavior. This influence arises from both the presence of occupants and the control actions that they execute to adjust the conditions of the indoor environment (thermal, air quality, light, noise). While significant advances in efficiency have been made in many aspects of building design such as materials, equipment, and building envelopes, understanding of the impact of occupant behavior is less developed. Accordingly, research into how best to utilize building performance simulation software to model and predict occupant behavior related effects is getting a lot of attention.

Capably guided by the editors—Andreas Wagner, William O'Brien, and Bing Dong—leading researchers in occupant behavior in buildings describe their research methods and provide insights by discussing the outcomes thereof. Because of the interdisciplinary and challenging nature of the topic, the contributing authors were selected from a wide range of backgrounds—from building physics and sensor technologies to psychology and social sciences.

This book is intended to provide better insight into challenges and methodologies in the study of occupant behavior in buildings. Thus, the aim is to improve and increase experimental work in order to provide a better basis for the modeling of occupant behavior in the future. This in turn will help to improve building simulation programs, aid simulation users (professionals and researchers), and ultimately lead to better building performance.

This book is essential reading for researchers and practitioners aiming to understand and implement occupant behavior modeling for building performance simulation.

April 2017

Jan Hensen
Computational Building Performance Simulation group
Eindhoven University of Technology
Eindhoven, The Netherlands

Acknowledgements

The authors and editors of this book wish to acknowledge the following individuals for their important contributions.

The external expert reviewers of book chapters, including: Craig Brown, Dane Christensen, Andrea Gasparella, Lindsay Taylor Graham, Olivia Guerra Santin, Frederic Haldi, Jan Hensen, Mustafa Emre Ilal, Panagiota Karava, Roberto Lamberts, Jared Langevin, Adrian Leaman, Dennis Loveday, Alexander Maki, Manfred Schmitt, Marcel Schweiker, David Shipworth, Masanori Shukuya, Judy Theodorson, Jørn Toftum, Chris Tweed, Thanos Tzempelikos, Kevin Van Den Wymelenberg, Jennifer Veitch, Geun Young Yun, and John Zhai.

Chapter 2: Zsofia Belafi, Marilena de Simone, Burak Gunay, Astrid Roetzel, Dafni Mora, Jie Zhao for contributing to the literature review.

Chapter 5: Burcin Becerik-Gerber for photographs.

Chapter 6: Rune Korsholm Andersen and Fréderic Haldi for contributing comments, discussion, and photographs and Justin Berquist, Julia Day, Ardeshir Mahdavi, and Carolyn Wayne for contributing photographs and figures.

Chapter 7: Cornelia Moosmann and Michael Kleber for contributing comments, discussion, and photographs.

Chapter 8: Isis E. Bennet, Thomas Berker, Ian Beausoleil-Morrison, Cristina Carpino, Marilena De Simone, Xiaohang Feng, Burak Gunay, Dafni Mora, and William O'Brien for contributing example surveys.

Chapter 11: The authors are thankful to the following individuals who contributed to the literature review of this chapter: Xiaojing Xu and Alyssa Loveday.

Ongoing support of IEA EBC Annex 66 operating agents, Da Yan and Tianzhen Hong.

Financial support from their respective countries, with which participation in IEA EBC Annex 66 and their contributions to this book would not have been possible.

Special thanks to Nina Doré for her in-depth final reviewing and editing of the whole text.

Contents

Glossary

Accelerometer	An electromechanical sensor that provides information on acceleration due to gravity, movement, and acceleration, and can also be used to determine orientation
Accuracy	The metric to quantify the closeness between a measurement and the true value
Actor	The human initiator (agent) or control software that can trigger changes in the state of a control device, equipment, and associated settings
Actual Meteorological Year (AMY)	A dataset that consists of the twelve consecutive months of data that are not necessarily typical
Adjustments, environmental	Devices which help to alter the immediate environment of an occupant. These can be elements –also referred to as controls– that potentially cause a change in the physical conditions of the indoor environment such as windows, heating and cooling devices. They can also refer to systems such as electric equipment, which may have a minor effect on internal heat or moisture gains, but are primarily not used to control indoor environmental conditions
Adjustments, individual	Comfort-related physical adaptations of the human body, including for example the change of body posture, and the level of clothing worn
Adjustments, physiological	Internal biological comfort-related adaptations (e.g., the thermoregulation process) that cannot be actively influenced by thought alone
Adjustments, spatial	Actions that are related to the active movement within the room, the building or between the building and outside

Advanced Message Queuing Protocol (AMQP)	An application layer protocol for message oriented middleware
Annular	An object having the form of a ring. In the context of this book, it is the space between the inner and the outer panes of a window which can be conditioned with warm/cold air.
Anonymity	A characteristic of data or other records for which the identity of people, buildings, or other objects is not revealed
Application program interface (API)	A set of functions, code, and clearly defined methods that facilitate interfacing with computer software directly
Aural comfort	The satisfaction with the acoustic environment that refers to the actual aural perception of a subject, but also contains a psychological component which determines the acceptance of the acoustic signal(s)
Azimuth angle	The angle between a reference direction – normally North or South – and the direction of interest on a horizontal plane
Beneficence	A concept which means that researchers should seek to protect and improve the welfare of study participants
Bias	A form of systematic error whereby repeated measurements do not obtain the true value of the measurand
Binary data	A type of data that is represented by either zeros or ones
Bluetooth	A wireless technology standard for exchanging data over short distances (on the order of 10s to 100s of meters) using ultra-high frequency (UHF) radio waves
Boolean	A data type which has two values: true or false
Building Automation and Controls network (BACnet)	A common, open-source, manufacturer-independent building automation system (BAS) communication protocol that allows hardware systems to communicate with each other
Building automation system (BAS)	The hardware and software systems responsible for controlling -and often collecting data on- space heating, cooling, ventilation, lighting, access, and fire detection equipment
Building information modeling (BIM)	A process and system for digitally representing the functional and physical characteristics of a building in three or more dimensions

Bus system	A system for communication and data transfer between different devices/nodes following a specific protocol
Cassandra	A robust NoSQL database management system, which handles large amounts of data across multiple servers
Chilled beam air conditioning	A device for cooling which is normally integrated into a ceiling and internally circulates cold water to cause air around the beam to cool down, thus initiating convective air flow that cools the space. Active chilled beams increase the convective flow and the cooling capacity by ventilation, whereas passive beams rely on natural convection only
Close-ended question	A survey or interview question that consists of predetermined responses
Cluster sample	The result of a sampling technique whereby the researcher groups or clusters population units for a sample, which is often used to target participants in a specific geographic location or building
Coding	In the context of social sciences, a process to assign (pre-defined) codes to observed data in order to categorize (e.g., types of behaviors)
Computer vision	A computer-based method used to extract useful features of images or video and convert them into useful information
Concept	An abstract representation of a property of a process or system
Concrete core activation	The process of heating or cooling concrete walls, floors, or ceilings by passing a heated or cooled fluid through them in order to condition a space and provide a thermal buffer
Confidence interval	The range of values within which the parameter of interest (e.g. the mean) of the population from which the sample is drawn can be said to fall, based on that same parameter in the sample (e.g. the sample mean)
Confidentiality	The status of information whereby it is kept private and restricted
Construct	A concept in the context of a specific population, which has been defined as the reification of all actual or potential instances of a set of experiences in the in a specific population

Construct validity	A status obtained when the variables of interest being measured with a research instrument are logically connected with one another and align with the theoretical model being utilized by the researcher
Controller Area Network (CAN) bus	A serial bus system that facilitates communication between systems, such as microcontrollers without a centralized computer
Convector	A heating device with a high share of convective heat transfer, which can either be active, with forced convection, or passive, relying on buoyancy
Convenience sampling	A method to select study participants whereby the sample is drawn from a population based on convenience
Convergent parallel design	A study for which quantitative data collection and analysis and qualitative data collection and analysis are performed in parallel and then the results are compared and interpreted
Convergent validity	A state of data achieved when it is shown that the variables that should be related are in fact related, which can be ascertained by using related measures of the same construct
Coverage	The total percentage of the population that is included within the sampling frame
Coverage error	The error that occurs when the sample does not accurately represent the population
Covert	The state or strategy of a study, whereby it is kept hidden or secret from the participants
Data acquisition system	A hardware system or device which retrieves and stores data
Data mining	A technique for using software to systematically explore data to seek patterns and other useful information
Data point	A reference to a unit of information such as a measurement
Data repository	A resource for storage of a collection of data
Data source	The origin of the data, such as sensors, meters, or human agents
Deadband	An interval or band within which the given value of a control parameter requires no control action

Debriefing	A process that is a conducted after a study with human subjects, whereby subjects are informed about the real purpose of the study and sometimes about the results of the study
Dependent variable	The research outcome variable of interest that is hypothesized to be dependent on other variables
Descriptive statistics	A generic term for statistics that can be used to describe variables, e.g., frequencies or means
Device interaction	An action (e.g., button-pushes, knob-turns, etc.) that occurs with electronic/mechanical building interfaces (e.g., thermostats and light switches)
Digital Addressable Lighting Interface (DALI) bus system	A building automation protocol for controlling devices for lighting
Digital Subscriber Line (DSL)	A family of technologies which enables the transmission of digital data over telephone lines
Discriminant validity	A status that is accomplished when data reveal that unrelated variables do not have a relationship
Displacement ventilation	A mechanical ventilation approach whereby fresh air is provided close to the floor level at low air velocities and turbulence and exhaust air is extracted close to the ceiling where it concentrates due to the buoyancy effect
Distortion	The inaccurate reproduction or amplification of a signal, which can be caused by changes in the frequencies or unequal delay or amplitude of the components of the output wave
Ecological validity	The extent to which an experimental study's findings can be related and applied to real settings
Effect size	The magnitude of an effect expressed in percentage or absolute terms
Embedded database	A database management system within an application software that requires access to the stored data
Embedded design	A mixed methods research approach that includes applying a quantitative or qualitative method, but embedding a lesser amount of the opposite (e.g., a quantitative survey with a few qualitative questions within it)

Energy harvesting	A feature of building sensors that allows them to generate (and often store) energy on-board such that they do not require external energy sources to measure, record, and transmit data
Energy simulation tool	A modeling tool that conducts computer simulation to estimate energy consumption of a building or system
Entity-Relationship (ER) model	A diagram that illustrates entities with the relationships among them, together with the attributes of the entities and the relationships
Entrainment ventilation	A ventilation system for which air is introduced at a sufficiently high speed to cause significant mixing of supply air and room air
Ethernet	The most widely installed Local Area Network (LAN) technology
Event related data	A type of data that defines events (e.g., detection of a movement, opening of a window)
Explanatory sequential design	A mixed methods research approach that involves starting with quantitative data methods and only after using qualitative methods to help explain the quantitative data (opposite order of exploratory sequential design)
Exploratory sequential design	A mixed methods research approach that involves using qualitative methods to help inform quantitative methods (opposite order of explanatory sequential design)
External validity	The degree of generalizability of a study result with respect to populations, climate, culture, environment, treatment variables, and measurement variables
Extraneous variable	A factor not of interest to the researcher, but that needs to be controlled for as it can impact on the dependent variable. Also called a confounding factor.
Face validity	A subjective evaluation of the survey items to determine whether they appear logical and appropriate for the concepts being studied
Factor, contextual	A factor that can influence an adaptive or non-adaptive trigger on occupancy or behavior. They can be grouped into physical environmental factors, psychological factors, social factors, and physiological factors

Factor, physical environmental	A set of contextual factors related to the physical environment, which are not immediate triggers and generally remain unchanged over a period of time. These can be the quality of the building envelope or ease of using a specific device.
Factor, physiological	A set of contextual factors related to the characteristics of the occupant in a building, such as age, sex, state of health, etc.
Factor, psychological	A set of contextual factors pertaining to occupants' thought processes, that help explain occupants' interactions with building envelope, services and equipment through expectation, awareness, preference, etc.
Factor, social	A set of contextual factors that affect occupants' lifestyles and their attitude in relating to others, such as group interaction, education, social status, etc.
Focus group	A qualitative research approach in the form of a group interview format
Gateway	A hardware device in building automation systems that is used to translate between multiple communications protocols so that components can communicate with each other
Global Positioning System (GPS)	A United States-developed, space-based radionavigation system that helps pinpoint a three-dimensional position to about a meter of accuracy and provides nanosecond precise time anywhere on Earth
Granularity	A property of data describing the level of detail and depth
Ground truth	Data obtained by directly observing the phenomenon of interest, as opposed to data collected by sensors or otherwise inferred
Hawthorne effect	A phenomenon of a study whereby participants change their behaviors (knowingly or unknowingly) when observed
Heating, ventilation, and air-conditioning (HVAC) equipment	A broad term referring to all hardware and software that is responsible for providing mechanical heating, cooling, humidification, dehumidification, and fresh (e.g., filtered and outdoor air). HVAC encompasses the plant (hardware that converts a fuel or electrical power into heating and cooling), distribution (infrastructure to carry air and energy to occupied spaces), terminal units (equipment for conditioning rooms), and building automation system that controls the hardware.

High Dynamic Range (HDR) image	An image with a large ratio (dynamic range) between the lowest and highest luminance levels
Honorarium	A payment given to study participants to recognize the value of their time and effort
Human-in-the-loop	Occupancy and/or behavior data that are collected with humans involved in measurement and recording –knowingly or unknowingly– that are comprised of studies where a researcher manually records occupants as well as studies that use active engagement of occupants in their own recording (e.g., using thermostat interactions to collect data)
Hypertext Transfer Protocol (HTTP)	The fundamental basis for the communication protocol used by the World Wide Web
Illuminance	A photometric quantity giving the total luminous flux on a surface per unit area that quantifies the illumination of a surface by a light source. The unit for illuminance in SI is lux (lumens per square meter) and in IP units is foot-candle (lumens per square foot).
Image-based sensing	A technique that uses a camera or other sensor to detect electromagnetic waves and converts data into useful information using computer vision
In situ study	A study that is performed in a real environment (e.g., an operating office or home) without disturbing the occupants
In-memory database	A database management system that mainly relies on main memory for data storage
Independent variable	A variable measured in the research that is hypothesized to affect the dependent variable. In experimental research, the independent variable is under direct control of the researcher and used for creating experimental conditions - also called the treatment or intervention.
Indoor environmental quality (IEQ)	A broad measure of the comfort and healthiness of an indoor environment for occupants, that includes thermal comfort, visual comfort, aural comfort, and indoor air quality
Inferential statistics	A branch of statistics in which conclusions about the population are made on the basis of data from a sample drawn from that population. Also called 'inductive' statistics.

Informed consent	A step required for studies involving human subjects that includes notifying potential subjects about among others: the steps, risks, data handling procedures, and benefits of a specific study. It commonly includes getting an informed consent form signed by the subjects.
InnoDB	A storage engine for MySQL. See also MySQL.
Instrument	A tool or other means of measuring variables of interest in the data-collection process
Internal validity	The extent to which the findings from the study can be correctly attributed to the interventions being experimentally tested
Internet of Things (IoT)	A network of connected electronic devices (sensors, computers, meters, etc.) with unique identifiers that communicate via Internet protocol
Internet-enabled Sensors	Sensors that can directly transmit data using Internet protocol
Interval scale	A scale used in questionnaire items with the following criteria: the rank ordering and the distances among objects on the attribute are known, but the absolute magnitudes are unknown
Intervention program	A set of activities or strategies aiming to change behavior or attitudes
Interview	A qualitative research tactic in the form of a conversation (face to face, phone, or email), often between the researcher and a study participant
Interview, fully structured	An interview whereby questions are predetermined and the same questions are asked to all participants
Interview, semi-structured	An interview that consists of questions that are prepared before-hand, but with flexibility to add additional questions as new topics emerge during the interview
Java Message Service (JMS) API	A Java MOM API for sending messages. See also Message Oriented Middleware (MOM) and Application Programming Interface (API).
Laboratory study	A research project that is conducted in an instrumented and purpose-built facility that may or may not appear as a real evironment
Latency	The time between measurement sampling and availability on the data storage platform for further processing

Latent cooling	The thermal process of removing energy from a volume of air, by dehumidifying it and decreasing the wet bulb temperature
Linear diffuser	A ventilation system device that supplies fresh air to a room through a straight, elongated air outlet, located in the ceiling, wall, or floor
Local area network (LAN)	A network that connects computers at local scales (e.g. within a building or cluster of buildings)
Longitudinal	A characteristic for a study whereby data collection occurs over a period of time with the same sample of subjects so that individual changes and/or variation of a sample or individual can be measured
Luminance	A direction-specific photometric metric that quantifies the luminous intensity per unit area of light source, and is often used to describe the light intensity perceived by the human eye. The unit is candela per m^2 (cd/m^2).
Measurand	The construct that is to be measured
Measured value	The value of a variable provided by a data source
Measurement	The assignment of scores to objects or individuals according to rules so that the scores represent some characteristic of the objects or individuals.
Measurement error	In the context of surveys, a discrepancy encountered when responses to survey items are inaccurate or cannot be effectively compared to the responses of other survey participants
Measurement uncertainty	A parameter (e.g., standard deviation) that describes the variability of measured data
Message Oriented Middleware (MOM)	A software or hardware infrastructure that supports sending and receiving of messages among systems
Mixed methods research design	A study methodology that uses multiple research approaches (i.e., qualitative and quantitative using one or more of: in situ, laboratory, or survey methods) in parallel or series to better understand the study subject
Mixed sensing	A combination of multi-infrared, image-based and acoustic sensors to measure occupant position, action, orientation, etc.

Mixed-mode conditioning	A building space conditioning approach that combines passive cooling design features (facade design, thermal capacity, etc.), natural ventilation strategies, concrete core activation, and mechanical systems for (peak) cooling
Mixing ventilation, mixed flow ventila tion, or entrainment ventilation	A ventilation strategy that aims to supply fresh air to a room by a high degree of mixing thus achieving an evenly distributed temperature and contaminant level in the space
MongoDB	A cross platform and document oriented database
Monitoring System Toolkit (MOST)	A vendor/technology-independent building monitoring toolkit, developed to simplify measuring, processing, and visualizing different building data streams
Multiphase design	A mixed methods research approach that involves a combination of sequential and concurrent elements, and often includes three or more phases
MySQL	An open-source relational database management system
NewSQL	A modern relational database management system that provides the same performance of NoSQL systems for online transaction processing read-write workloads and maintains the atomicity, consistency, isolation, and durability assurances of traditional systems
Nominal data	A data type which represents categorically discrete data (e.g., gender, country, profession).
Non-intrusive load monitoring	A method to distinguish individual loads (e.g., appliances) from an aggregated load dataset (e.g., from a building-level meter)
Nonresponse error	A source of error resulting from missing data or an insufficient number of responses that renders the collected data insufficient for statistical analysis.
NoSQL	A non-relational database management system
Occupancy (also known as occupant presence)	The Boolean value of the state of an occupant being in a space; it could also refer to the number of occupants in a space
Occupant action	An act initiated by an occupant that affects the occupant, other occupants, the indoor environment, and/or building systems
Occupant attitudes	A broad range of information obtained from (reported by) an occupant, including subjective sensations, perceptions, and evaluations

Occupant attributes	The data describing occupants state, including clothing, activity, and physiological parameters
Occupant behaviour, adaptive	A category of adaptive behaviors with the goal of adapting (to) the indoor environment and are activated by adaptive triggers (e.g., opening windows due to increasing carbon dioxide concentration)
Occupant behaviour, non-adaptive	A category of adaptive behaviors that do not have the goal of adapting (to) the indoor environment and are activated by non-adaptive triggers (e.g., events based on schedules)
Oculography	A scientific method used to measure and record eye position and movement
Olfactory	A term referring to the sense of smell and the perception of air quality with regards to odor
Ontology	A formal structure to define classes, relations, functions, properties, and other objects and the relationships between them
OPC Unified Architecture (OPC UA)	A machine to machine communication industry standard protocol
Open Building Information Exchange (oBIX)	A standard forRESTful-based interfaces to building control systems. See also RESTful.
Open-ended question	A survey question that allows participants to answer in a text box or essay format such that it can be qualitatively analyzed.
Operationalization	The process of determining how best to measure constructs
Ordinal scale	A scale used in closed-ended survey questions, whereby the list of allowable responses can be ranked in some meaningful way but for which the absolute magnitude between variables is unknown (e.g., strongly disagree, disagree, neutral, agree, and strongly agree)
p-value	The probability of obtaining a result equal to or more extreme than what was actually observed, when the null hypothesis is true
Passive Infrared (PIR) motion sensor	A sensor that detects the infrared radiation from objects in its view field, often for the purpose of detecting occupants
Peel and stick	A class of wireless sensors that can be easily configured and adhered to surfaces

Periodic data	A property of data, whereby it is recorded and/or stored at regular time intervals
Personalized ventilation	A ventilation strategy that provides fresh air close to a person through special air terminals and enables individual control of air supply
Piezoelectric	A physical phenomenon whereby certain materials convert mechanical energy into an electric signal when they are compressed
Population	In the context of survey research, a population is representative of the entire group of people that the researcher hopes to better understand, from which a sample is drawn
Post Occupancy Evaluation (POE)	A type of survey that is performed after building completion, during the occupancy phase, for which the primary objective is to learn if building systems are working properly or not
Power	In the context of research methods, the probability of correctly rejecting the null hypothesis, i.e., to detect an effect when there is one
Pre-Analysis Plan (PAP)	A document laying out the research, and in particular, the proposed analysis of the data, in detail before the research is conducted. A PAP is usually uploaded to an online repository, and helps to avoid data mining and cherry-picking of results.
Precision	The closeness of results obtained from two or more repeated tests that are performed under identical conditions
Predicted mean vote (PMV)	The average thermal comfort vote predicted by a theoretical index for a group of subjects when exposed to a particular set of environmental conditions
Pseudonymity	An identification method used in research whereby a concealed identifier is used to maintain the relationship between a single person and its data
Pygmalion	The psychological theory that posits that people often conform to the expectations that other people have of them
Quantitative data	A data type which can be measured and written with numbers
Questionnaire	A survey that consists of one or more questions and normally comes in written form (including by paper and computer or mobile telephone)

Radar (Radio Detection and Ranging)	A technique using radio waves to detect the presence of objects in the atmosphere
Radiant heating and cooling	Systems that deliver/extract heat primarily through radiant heat transfer
Radio frequency Identification (RFID)	A generic term for technologies that use radio waves to automatically identify people or objects
Random error	A type of error caused by limitations of precision of the measurement device
Random sampling	Randomly selecting elements from the sample frame, with each element of the sample frame having an equal probability of being chosen
Ranked data	A data type which displays a certain order
Ratio scale	Interval scales with a true zero, for example, zero height or weight
Recruitment	The process of obtaining study participants, which includes activities such as promotion, communication with potential participants, and participant selection
Reed switch	An electromechanical sensor that can be used to detect state (e.g., whether a window is closed) using magnetic fields
Reliability	A measurement of the ability of a particular research method to achieve repeatable results after multiple trials
Representational State Transfer (REST)	Web services, which provide interoperability between computer systems on the Internet
RESTful	See "Representational State Transfer (REST)"
Reverberation time	A metric to define the frequency-dependent acoustic quality of a room; it is defined as the time required for the sound pressure level to decrease by 60 dB in a space
Robustness, data	In the context of occupant research, the probability that occupant data is delivered to the storage platform
Saliva	A fluid that primarily consists of water and is secreted by the salivary glands in the mouth
Sample	A subset of a given population, for which there are numerous methods to obtain it
Sample frame	A list containing (ideally all) members of a population (e.g. people, buildings) from which the sample can be drawn

Sampling error	Standard error plus any bias arising from the sampling process
Satisfaction, Occupant	The fulfillment of an occupant's expectations related to indoor environmental quality (i.e., thermal, visual, and aural comfort and air quality) and other (e.g., privacy, furniture) aspects of a space
Security, data	A measure of the degree to which data communication and storage cannot be interfered with or manipulated
Sensible cooling	The decrease of dry-bulb temperature which can be measured (or perceived) if heat is extracted from a space
Sensor fusion	The combination of data from multiple sensors to reveal new information or identify faults and inconsistencies, much like virtual sensors
Sick building syndrome (SBS)	A wide variety of health impacts experienced as a result of indoor environmental conditions, that are often a result of insufficient indoor air quality and have a significant impact on productivity, cogitative ability, and absenteeism
Skin perfusion	Passage of blood into human tissues
Skip logic	A strategy used in survey development whereby if a participant selects a particular answer, he/she will be advanced ahead in the survey depending upon their answer
Social desirability response (SDR) bias	The tendency for occupants to present a positive image of themselves on questionnaires, or in a way that is consistent with societal norms or beliefs
Sorption	A physical and/or chemical process in which a fluid is either dissolved by a liquid or a solid material (absorption) or in which one substance is attached (on an atom or molecule level) to the surface of another substance (adsorption)
Spatial attribute	The geometrical aspect of a variable's value, which can include a point, plane, or a three-dimensional volume information
Standard deviation	A measure for the dispersion of a dataset with respect to the mean value
Standard error	The standard deviation of a sampling distribution of statistics, such as the sample mean
Statistical confidence	A measure of the probability of avoiding a Type I error

Stratified sample	Stratified samples divide a given population into groups (i.e., strata) and then random samples can be taken for each strata (e.g., ASHRAE climate zones)
Structured Query Language (SQL) database	A programming language for a relational database management system
Survey	A broad category of research instruments whereby individual or groups of human participants are asked a series of questions in order to extract data
Swirl type diffuser	A ventilation system device that supplies fresh air to a room through a circular or square air outlet with (mostly radial) flow directing devices in order to create a swirling turbulent flow that mixes with surrounding air without providing drafts
System architecture	A conceptual representation of the configuration of systems and subsystems, such as sensors and data transmission paths
Systematic error	A type of error resulting from improper research methods that lead to consistent bias
Temporal attribute	A time-related aspect (or extension) of a variable's value which can include time stamp and sampling interval entries
Test bed	A comprehensive array of sensors and other monitoring equipment that is deployed in a laboratory or real building environment
Theory of Planned Behavior	One of the most well-known and tested theories in psychology to understand the antecedents of behavior; it postulates that the attitude toward a behavior, subjective norms, and perceived behavioral control, shape an individual's behavioral intentions and, ultimately, their behavior
Thermal comfort	A state of satisfaction with thermal conditions, that is frequently defined as "That condition of mind that expresses satisfaction with the thermal environment and is assessed by subjective evaluation." according to ANSI/ASHRAE Standard 55-2013)
Thermistor	A temperature sensor whose resistance is proportional to the temperature (Negative Temperature Coefficient (NTC): resistance decreases with temperature, Positive Temperature Coefficient (PTC): resistance increases with temperature).

Thermocouple	A sensor comprised of two conductors of different materials that form an electrical junction and result in a temperature-dependent voltage so that temperature can be measured
Thermoregulation process	The process that occurs in an organism that aims to keep its body temperature within a certain range, despite varying environmental temperatures
Total error	The difference between the measured value and the true value (reference value)
Trigger, adaptive	A category of triggers involving physiological or physical environmental conditions that motivate or influence occupants to initiate adaptive behaviors
Trigger, non-adaptive	A category of triggers involving time, schedule, or other contextual factor-related conditions that motivate or influence occupants to initiate non-adaptive behaviors
Trigger, physical environmental	A category of triggers involving physical properties that describe the indoor and outdoor environments and have been proven to stimulate occupant action in response to a thermal, olfactory, visual and aural stimulus, such as air and mean radiant temperatures, solar radiation, air pollutants, illuminance, acoustic sound pressure, etc. in order to voluntarily modify the surrounding built environment to restore or improve the comfort conditions
Trigger, physiological	A signal from the body causing it to activate an adaptive behavior
True random sample	A sample that is selected randomly, in which every person in population has a truly equal chance of being selected
Trueness	The closeness between measured data and true results
Tympanic membrane	The membrane that separates the outer and the inner ear and serves for reception of sound pressure and its transmission to the inner aural sensor system
Type 1 error	The error of concluding something is true when it is not
Type 2 error	The error of concluding that something is not true when it is true
Typical Meteorological Year (TMY)	A 8760-hour (one-year) weather data file that is designed to be representative of long-term historical weather data, while comprising of real weather data lasting shorter periods (e.g., months) of time that are combined
Ultra-wideband (UWB)	Very high-bandwidth communications over a large portion of the radio spectrum

Underfloor air distribution	A ventilation concept that utilizes the cavity below a raised floor (plenum) to directly distribute fresh air into a room without an additional duct system connected to the floor diffusers
Unified Modeling Language (UML)	A general modeling language that provides a standard way of visualizing the design of a structured system
Unordered categories	A type of closed-ended question whereby the provided answers have no particular order
Validation of measurement method	A process used to ensure that a measurement technique is valid (i.e., accurately obtaining the intended results)
Validity	A measure of the agreement between two different ways of measuring the same construct
Variable	A property that is subject to change
Variable air volume (VAV) system	A class of ventilation systems that is able to supply variable air flows (at constant temperature) according to the actual demand of fresh air and cooling load
Variance	A measure of the spread in a variable, defined as the average of the squared differences from the mean
Verification of measurement method	A set of procedures used to test the extent to which the performance data obtained by manufactures during method validation can be reproduced in the environments of end-users
Virtual data point	See "Virtual sensor".
Virtual reality	An electronic system for producing synthetic images and sounds such that the user perceives an environment and experience
Virtual sensor	Software- and/or model-based sensors that infer measurements from one or more other sensors
Visual comfort	The satisfaction with the visual environment, referring to the actual visual perception of a subject but also a psychological component, which determines the acceptance of the visual impressions
Volatile organic compounds (VOC)	Organic compounds with a boiling point under 250°C.
Wearable device	A type of sensing device that can be used to sense occupant comfort, metabolic rate actions, posture, location, orientation, etc.
WiFi	A communications technology that uses radio waves to provide network connectivity

Chapter 1
Introduction

Andreas Wagner, William O'Brien and Bing Dong

Abstract Occupant behavior has a major influence on building energy consumption. With tightening requirements towards building energy performance and sustainability, awareness of the importance of understanding building occupants' behavioral and presence patterns has risen, yet the latter are often treated in a highly simplistic manner. There is a lack of a consistent research framework with regard to data collection on occupant behavior from experiments. Therefore, this book which emerged from work in the IEA Energy in Buildings and Communities Annex 66 Definition and Simulation of Occupant Behavior in Buildings, provides guidance to organize research from conception and design to study phase, and then validation, data management, and ethics.

It is commonly known that people nowadays spend more than 80% of their time in buildings, which significantly impacts energy consumption in terms of space conditioning (heating, cooling, ventilation, and lighting) and the usage of appliances and other building services. According to the International Energy Agency (IEA), "buildings are the largest energy-consuming sector in the world, and account for over one-third of total final energy consumption". The level of this consumption differs widely between buildings, due in part to available technologies providing comfort and climatic, societal, cultural, and social contexts in different parts of the

---

A. Wagner (✉)
Building Science Group, Faculty of Architecture, Karlsruhe Institute of Technology, Englerstr. 7, 76131 Karlsruhe, Germany
e-mail: wagner@kit.edu

W. O'Brien
Department of Civil and Environmental Engineering, Carleton University, 3432 Mackenzie Building, 1125 Colonel By Drive, Ottawa, ON K1S 5B6, Canada
e-mail: liam.obrien@carleton.ca

B. Dong
Department of Mechanical Engineering, University of Texas at San Antonio, Engineering Building, One UTSA Circle, San Antonio, TX 78249-0670, USA
e-mail: bing.dong@utsa.edu

A. Wagner et al. (eds.), *Exploring Occupant Behavior in Buildings*,
DOI 10.1007/978-3-319-61464-9_1

world, but also to a significant extent to occupant behavior in buildings. Indeed, occupant behavior has been identified as one of the leading influences on building energy consumption in developed countries, particularly with the current trend to design and operate more energy-efficient buildings.

With tightening requirements regarding building energy performance and sustainability, researchers—and increasingly architects, planners, engineers, and building managers—have begun to recognize the importance of understanding building occupants' behavior and presence patterns. Studies have found that occupants can influence energy consumption by at least a factor of two in buildings with identical structures, and yet the field has only begun to understand why, let alone accurately predict this phenomenon. Occupant actions, such as adjusting a thermostat or opening a window, can be related to any number of drivers, including physical, physiological, psychological, and social. As buildings are designed to meet an array of increasingly ambitious design and operation performance objectives, the uncertainty of occupant behavior poses a major challenge.

A surge of interest in building occupants and their behavior has spurred considerable research, but aspects of occupant behavior are still treated in a highly simplistic manner. An essential step forward is the collection of representative occupant data, whether derived from real-life buildings (in situ measurements, surveys) or laboratory experiments. These data can, however, be difficult to obtain. Resolution, accuracy, and explanatory power of the data are highly dependent on the chosen research methodology. Monitoring occupant behavior is likely to require a substantial sensor infrastructure and other ongoing investments. Sensors may malfunction, certain behaviors may be difficult to measure, results may be counterintuitive and inexplicable, to name a few possible hiccups. Moreover, privacy and ethical issues can make accessing certain occupant data difficult, even impossible. Hence, occupant studies are challenging, costly, and time-consuming.

What can be done with occupant data once it has been successfully collected? The answer is plenty—at least in theory. Most applications for occupant data have been vastly underexploited, in particular using occupant data to develop models to support design and improve controls and operations. One application with significant potential is using occupant data to better predict occupancy and behavior in building performance simulation. Specifically, a more accurate representation (model) of occupants about which designers are confident would be invaluable in helping to design comfortable and energy efficient spaces and not greatly oversize equipment due to the uncertainty of occupants. The same model could later be used for performance optimization within the building management system. A second and equally important potential application is using occupant data to make inferences about the relationship between a building and its occupants in order to inform improvements to future building designs with regards to energy and comfort performance. For instance, if a recurring pattern is observed, e.g., open blinds on west-facing windows every afternoon, this could result in alternative design solutions with regard to floor plan (zoning), façade design, and choice of shading device. In existing buildings, occupant data could be applied to improve controls, operations, and renovations. For example, knowledge of the distribution of arrival times for occupants could help inform a more efficient operating schedule.

Overall, while progress is evident, there is a definite lack of a consistent research framework within which to tackle the complexities of the field. The result is a hodgepodge of research approaches and results where oftentimes studies are so inconsistent that external validation is impossible. To further complicate matters, the study of building occupants spans several disciplines, including engineering, architecture, information technology, and social sciences, to name the most relevant. Thus, individual researchers are rarely equipped with the tools and knowledge to comprehensively approach a new research project.

This book was written in response to this gap. It is the product of a collective effort by a multidisciplinary and international group of researchers who were brought together with the creation of International Energy Agency Energy in Buildings and Communities Annex 66 on Definition and Simulation of Occupant Behavior in Buildings. It is among the first to discuss research methods in the context of building occupant research. The objective of this book is to provide a concise guide for graduate students and other researchers who are preparing to embark on a building occupant research campaign. The chapters are organized roughly according to the chronology of the research process: from conception and design to study phase, and then validation, data management, and ethics (Fig. 1.1). Although a broad roster of expert authors from around the world has contributed to this book, the diligent researcher will undoubtedly read beyond these pages, as any of the chapters could quite easily be expanded to an entire book.

After reading this book, readers will be able to answer questions such as:

- How should research questions be framed in building occupant research?
- What variables should be measured to help predict occupancy and occupant actions?
- What sensing equipment and technologies are available to assist with research, and how have they been applied successfully (or unsuccessfully) in the past?
- What methods are most suitable to address the research questions of interest?
- How should the accuracy of research data be verified?
- How should data be structured and stored?
- What ethical and privacy concerns should be addressed prior to embarking on research involving human participants?

Following this Introduction, Chap. 2 begins by defining key terms in the field of building occupant research. It presents a taxonomy of relevant occupant behaviors and actions alongside a list of adaptive and non-adaptive triggers and contextual factors that could influence occupant behavior. The chapter closes with a comprehensive list of studies related to occupant behavior and corresponding predictors.

Chapter 3 proposes a systematic approach to occupant research design using a theoretical cause-effect model through concept mapping. It addresses the formulation of research questions and hypotheses, as well as the identification of relevant measurands and appropriate methods to measure them. The chapter introduces the

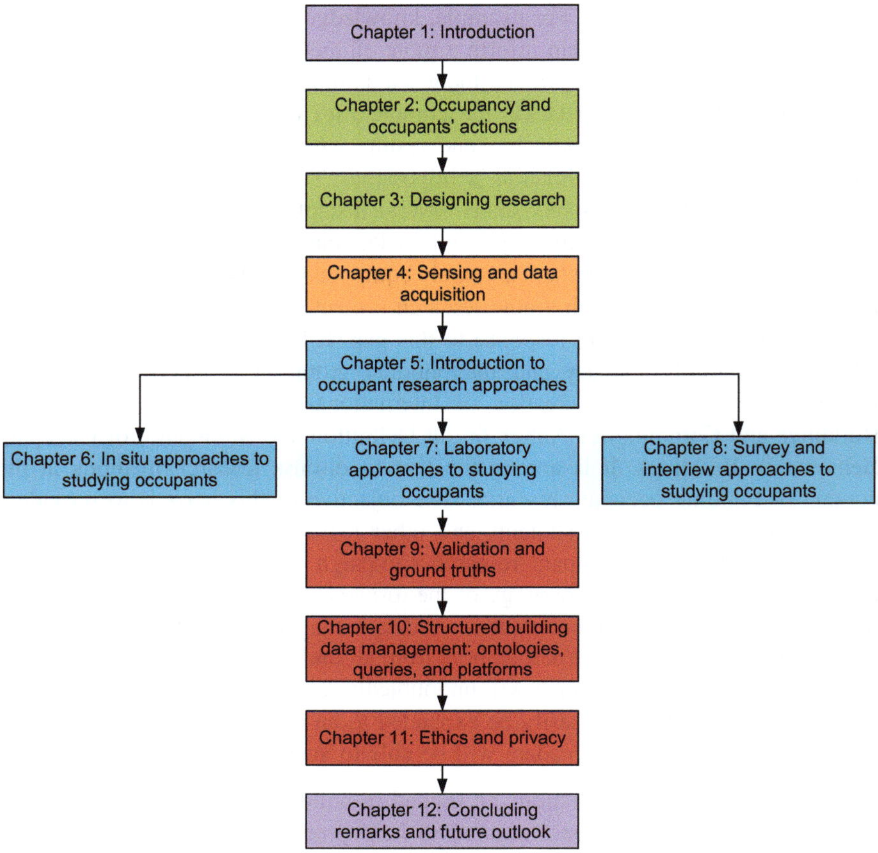

Fig. 1.1 Organization of the book chapters

concepts of reliability, validity, and uncertainty, and why knowledge about them is essential for any researcher.

Chapter 4 summarizes existing occupancy and occupant behavior sensing and data acquisition technologies. It describes and evaluates sensors for information on both occupants' presence and their interactions with the built environment according to nine outlined performance metrics. It also reviews different data acquisition systems in terms of type of data transmission, data storage, robustness, and security.

Chapter 5 lays the foundation for Chaps. 6–8. It introduces and briefly presents the advantages and disadvantages of four methods in occupant research: in situ, laboratory, survey, and virtual reality. It also explains mixed method research design, offering several illustrative examples from the literature. It closes with a comprehensive list of occupant-related phenomena of interest alongside a qualitative discussion of the merits of each research method in examining them.

Chapter 6 overviews in situ methods of studying occupant behavior and presence and recommends a systematic method for designing, conducting, and publishing this type of study. It includes a detailed discussion of in situ-specific sensor technologies and sensing strategies, and offers advice about the level of reporting required for in situ studies with attention to contextual factors. Finally, it discusses the potential of surveys to complement in situ monitoring.

Chapter 7 presents various types of laboratory facilities around the world and highlights their main features in terms of experimental opportunities. It describes the typical technical equipment and sensor technologies used in these lab environments and how they determine the range of indoor environmental scenarios that can be simulated under precisely controlled conditions. The chapter considers questions of appropriate lab design and experimental set-ups.

Chapter 8 introduces survey methods, including the conceptualization and articulation of survey questions and their intention, validity of surveys, selection of appropriate sample, and tools for data collection. It concludes with lessons learned from exemplary survey studies and a brief discussion of interview methods.

Chapter 9 defines the concepts of measurement validation and ground truth in occupancy and occupant behavior observations. It proposes methods for ground truth data collection and approaches to the verification and validation of collected data, with supporting examples. It suggests procedures for constructing an occupant behavior ground truth dataset based on the authors' experiences.

Chapter 10 opens with an ontology for the representation and incorporation of multiple layers of building monitoring data for applications like building performance simulation and building automation. It addresses common data processing requirements and provides examples of typical queries that building monitoring data repositories must support. Further, it outlines data repository specifications for structured collection, storage, processing, and multi-user exchange of monitored data.

Chapter 11 explains common concepts and applications of ethical standards, including informed consent, privacy, and confidentiality in order to help improve interactions with institutional ethics review boards and meet crucial requirements. It also points to additional ethical considerations in occupant research to ensure researchers understand how to conduct their research ethically.

Finally, Chap. 12 offers final thoughts and the future outlook of occupant behavior research.

As mentioned above, this book emerged from work in the IEA Energy in Buildings and Communities Annex 66 Definition and Simulation of Occupant Behavior in Buildings. The editors, authors, and contributors acknowledge the financial and other support from their respective countries, without which their participation in IEA EBC Annex 66 and their contribution to the book would not have been possible.

Chapter 2
Occupancy and Occupants' Actions

Marcel Schweiker, Salvatore Carlucci, Rune Korsholm Andersen, Bing Dong and William O'Brien

Abstract Occupants' presence and actions within the built environment are crucial aspects related to understanding variations in energy use. Within this chapter, first, a nomenclature for the field of research dealing with occupants in buildings is defined. This nomenclature distinguishes between occupants' presence and behavior, states and actions, adaptive triggers, non-adaptive triggers, and contextual factors. Second, an extensive list of occupant behaviors is provided and categorizations of occupants' actions are introduced. The list includes most of the possible phenomena that researchers may wish to study, measure, and ultimately model. The categories are physiological, individual, environmental, and spatial adjustments. Third, a list of adaptive and non-adaptive triggers together with contextual factors that could influence occupant behavior is presented. Individual elements are further grouped into physical environmental, physiological, psychological, and social aspects. Finally, a comprehensive table of studies related to occupant behavior and

M. Schweiker (✉)
Building Science Group, Faculty of Architecture, Karlsruhe Institute of Technology, Englerstr. 7, 76131 Karlsruhe, Germany
e-mail: marcel.schweiker@kit.edu

S. Carlucci
Department of Civil and Environmental Engineering, NTNU—Norwegian University of Science and Technology, Høgskoleringen 7A, 7491 Trondheim, Norway
e-mail: salvatore.carlucci@ntnu.no

R.K. Andersen
Department of Civil Engineering, Technical University of Denmark, Nils Koppels Allé, Building 402, Office 218, 2800 Kongens Lyngby, Denmark
e-mail: rva@byg.dtu.dk

B. Dong
Department of Mechanical Engineering, University of Texas at San Antonio, Engineering Building, One UTSA Circle, San Antonio, TX 78249-0670, USA
e-mail: bing.dong@utsa.edu

W. O'Brien
Department of Civil and Environmental Engineering, Carleton University, 3432 Mackenzie Building, 1125 Colonel By Drive, Ottawa, ON K1S 5B6, Canada
e-mail: liam.obrien@carleton.ca

A. Wagner et al. (eds.), *Exploring Occupant Behavior in Buildings*,
DOI 10.1007/978-3-319-61464-9_2

the corresponding significant and non-significant predictors, based on an extensive literature review, is shown. This table highlights areas of research where numerous studies have been conducted, as well as areas where hardly any research has been published. The conclusion highlights the importance of publishing future occupant monitoring campaigns with sufficient detail to inform future researchers and save redundant effort. Such detail is especially necessary in relation to the methodology, including, for example, a clear description of the type of variables monitored, and in relation to the results, where both the influencing factors that were found to be significant and insignificant should be documented.

2.1 Introduction

In most buildings, occupants have numerous options to interact with their built environment. This includes entering/leaving a room, adjusting thermal, visual, or aural indoor conditions (for example, via windows, doors, or blinds), and using other devices, such as electrical equipment. Thereby, an interaction which leads to a state change, and no interaction which leaves the current state unchanged are both aspects of occupants' behavior (see Fig. 2.1). With the opportunity to interact with the building controls and building envelope, occupants are empowered to operate these systems in ways that may be inconsistent with design intent (Deuble and de Dear 2012; Schakib-Ekbatan et al. 2015). At the same time, the real use of electrical equipment and occupancy patterns may also be very different from the assumptions made in the design phase.

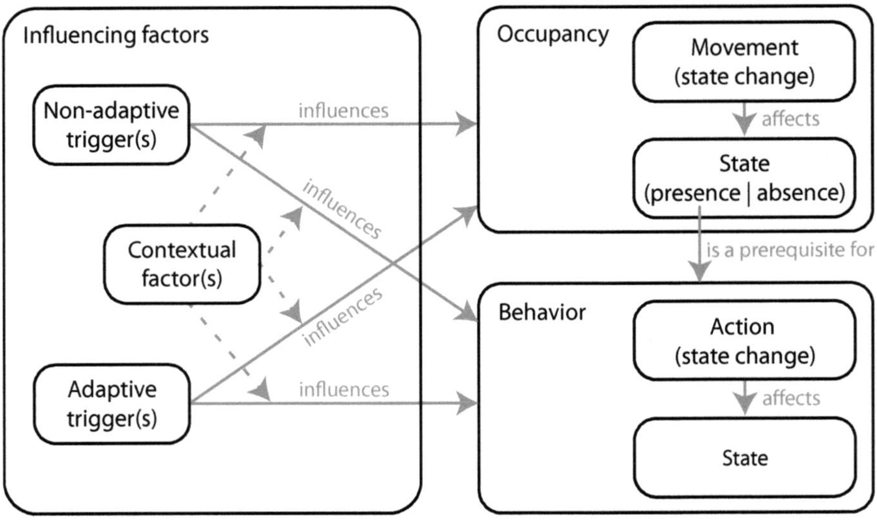

Fig. 2.1 Ontology of occupant-related phenomena

Uncertainties in the prediction of occupancy and occupants' actions increase the difficulty of predicting the performance of a building. Buildings' performance can be sensitive to the behavior and presence of occupants, and these aspects can result in large differences between actual performance and design (Bishop and Frey 1985; Macintosh and Steemers 2005; Majcen et al. 2013; Blight and Coley 2013; Andersen et al. 2016; Carlucci et al. 2016). Not surprisingly, identically-built homes have been found to vary in energy consumption by a factor of two or more as a result of occupants' behavior and occupancy patterns alone (Socolow 1978; Emery and Kippenhan 2006; Yun and Steemers 2008; Guerra Santin et al. 2009; Maier et al. 2009; Andersen 2012). The behavior of occupants may influence the variation in the actual energy use to a much larger degree than the thermal processes within the façade (Hoes et al. 2009) or the insulation measures (Schweiker and Shukuya 2010b). The influence may be asserted in many ways—for example, by opening/closing windows, adjusting thermostats, operating solar shading, light switching, etc. As insulation levels and airtightness of buildings increase as a consequence of stricter building regulations, occupants' control over window openings, thermostats, solar shading, etc. will have larger relative effects on the air change rate and temperature distribution in the buildings (Guerra Santin et al. 2009). For instance, occupant control of operable windows and window shading devices will significantly impact energy flows across façades (Hoes et al. 2009). Consequently, designers' inclusion of occupants' interactions with building controls and envelopes becomes increasingly important (Andersen 2012) and could lead to arguments for reducing opportunities for occupants to interact with their built environment.

On the other hand, occupants who have the possibility of controlling their own environment have been found, at least in office buildings, to be more content with the indoor environment and suffer fewer symptoms of sick building syndrome (SBS) than occupants who do not have the possibility of interacting with building controls and envelope devices (Toftum 2010). The same finding will likely be observed for other building types once investigated. Therefore, it is important to note that occupants should not be restricted in their behavior or forced to change their behavioral patterns, and that designers need to take more care to design buildings according to occupants' needs and likely to occupancy and behavioral patterns. However, leaving the control of a building and façade in the hands of its occupants demands that occupants' behavioral patterns be realistically represented in predictions of building performance.

In general, occupants can impact buildings in two ways: through the direct impact of their presence (heat production, carbon dioxide emission, moisture generation, etc.) and through their interactions with the building. In most circumstances, the presence of occupants is the prerequisite for any behavior, since occupants can only interact with the controls and the building envelope if they are present in the building. Occupants' presence and behavior can be modelled as states (e.g., state of light switch) and actions (e.g., the act of turning on or off a light switch) using different modeling approaches. In general, the action changes the state and the state remains constant until a new action is taken (e.g., the state of the window will remain closed until an occupant opens it). Note that the state may refer to two or more levels (e.g., a window can have several states such as open/closed or fully open/half open, X% open/closed), depending on the modeling approach.

As described in detail in Sect. 2.3, state changes can be triggered by either adaptive triggers (those which are rooted in occupant discomfort or expectations of discomfort) or non-adaptive triggers (those which are part of occupant tasks, like using a computer). These triggers, as well as the resulting action or non-action, are moderated by contextual factors. The adaptive triggers activate adaptive behaviors, such as opening windows due to increasing carbon dioxide (CO_2) concentration (indication of deteriorating indoor air quality) or closing blinds due to glare. Non-adaptive triggers activate behavior that is not performed with the goal of adapting to the indoor environment—for example, events based on schedules. Note that the same type of action, e.g., closing the window, could be triggered by an adaptive trigger (undesirable cold air entering the room) or a non-adaptive trigger (leaving the room and securing the building beforehand).

In the context of this book, the authors are interested in all occupant-related phenomena that impact building energy performance, comfort, and occupant satisfaction. These phenomena are presented in Fig. 2.1 and defined in the next section. The purpose of this chapter is to introduce readers—whether fundamental occupant behavior or comfort researchers, occupant modelers, or even model users —to the diverse occupancy and occupant behavior-related phenomena that should be considered prior to conducting experimental, modeling, or simulation-based research studies. First, a definition for the nomenclature used in the field of research dealing with occupants in buildings is presented. Second, an extensive list of occupant behaviors is provided and categorizations of occupants' actions are introduced. The list includes most of the possible phenomena that researchers may wish to study, measure, and ultimately model. Then, a list of triggers and contextual factors that could influence occupant behavior is provided. Finally, a comprehensive table of studies related to occupant behavior and the corresponding significant and non-significant predictors, based on a comprehensive literature review, is presented.

2.2 Categorization of Occupants' Actions

Energy-related occupant behavior was previously defined as "human being's unconscious and conscious actions to control the physical parameters of the surrounding built environment based on the comparison of the perceived environment to the sum of past experiences" (Schweiker 2010). As pointed out by Fabi et al. (2012), this definition is limited to actions related to the perceived environment, or as defined above to those actions caused by adaptive triggers. The aim of this section is to categorize all sorts of occupants' actions according to the distance between the occupant and the location the action takes place into physiological, individual, environmental, and spatial adjustments (Table 2.1).

Physiological adjustments are done unconsciously and cannot be actively influenced by thought alone. One group of physiological adjustments can be traced back to the human body's thermoregulation process. Thermoregulation processes occur within the human body in order to keep the core temperature within defined and narrow limits. The basic control of the thermoregulation system is done by the hypothalamus, which adjusts the body functions by discharging hormones into the bloodstream and activating the nervous system (e.g., for shivering). Human beings can develop a long-term adaptation to extreme environmental conditions. For example, people who grew up in a hot and humid climate show a higher number of sweat glands than people who grew up in a cold climate (Hori 1995). Thanks to a higher number of sweat glands, they sweat more effectively and can hence maintain their core temperature at a comfortable level in hot and humid conditions more easily compared to those with fewer sweat glands. It has been found that physiological adaptations do not only occur over a long time, but also over shorter periods, starting generally after four to five days of exposure to warm or cold conditions (Hori 1995; van der Lans et al. 2013). People who are exposed to severe climates adapt to the conditions prevailing in their environment and are thus able to accept a wider range of temperatures. As a consequence, people with more sweat glands may have a higher indoor temperature as triggering condition for opening a window or turning on a fan.

Physiological adaptations can also extend to visual aspects. For example, the human is able to adapt to a wide range of visual scenarios ranging from 1 lx at night with the moonlight to 100,000 lx during the daytime with sunlight. Physiological reactions to changes of visual conditions include optical (pupil dilation), neurological, and photochemical adaptations in order to regulate the amount of light passing through (Murdoch 2003; American Optometric Association 2016). With increasing age, the functionality of the eye is reduced (Lerman 1980; IJspeert et al. 1990; Winn et al. 1994; Van De Kraats and Van Norren 2007), so that visual requirements change (Moosmann 2014). This will also affect corresponding interactions with the building.

Individual adjustments include, for example, the change of body posture and the amount of clothes worn. Takahashi et al. (2000) found differences in individual adjustments pursued by subjects sitting in a natural ventilated room compared to those sitting in an air-conditioned room: those in the natural ventilated room were more active with respect to actions, such as "waving hands like a fan" or "making space under clothes with hands".

The definition for environmental adjustments used here is broad. Environmental adjustments include all actions related to elements of the room. These can be those elements—also referred to as "controls"—which potentially lead to a change in the physical conditions of the indoor environment, such as windows, heating and cooling devices, and elements such as electronic devices, which may have a minor effect on internal gains, but are primarily not used to control indoor environmental conditions.

Spatial adjustments consist of those actions related to the active movement within the room or building, or between the building and outside. It can be assumed that these actions are mainly driven by non-adaptive triggers, such as meetings in another room or the end of the workday. They could also include such adaptive actions as relocating away from daylight glare in a library. Related to adaptive triggers, a study found that the choice between sitting outside or inside a bar in Kassel, Germany, is very much dependent on the thermal outdoor conditions. Additionally, the expectation of weather and the activity planned have an influence on this decision (Katzschner et al. 2006).

Table 2.1 Categories of occupant actions (Schweiker 2010; Polinder et al. 2013)

Physiological adjustments	Individual adjustments	Environmental adjustments	Spatial adjustments
Sweating Relaxation of arrector pili muscles Arteriolar vasodilation (dilation of arteriole blood vessels) Piloerection (bristling of hairs due to cold) Shivering Arteriolar vasoconstriction (narrowing of arteriole blood vessels) Thermogenesis Pupillary light reflex Change in the sensitivity of the cones or rods of the eye Adjustment of ears' sensitivity to the noise level	Adjustment of clothing/bedding level Selection of clothing/bedding material Change of the body posture Change of activity level (siesta in summer afternoons) Use of earplugs or earphones to avoid excessive aural discomfort Drinking hot/cold beverages Eating high or low calorie food Sleeping in family group with the bodies pushed up against each other Fanning fresh air around the body Taking cold showers	Use of heating, ventilation, and air-conditioning systems Thermostat adjustment Window opening/closing Door opening/closing Solar shading devices adjustment Light switching/dimming Use of personal control devices for cooling effect (e.g., fan) Use of personal control devices for heating effect (e.g., small heaters) Seasonal adjustment of building envelope components (e.g., placement of storm windows and plastic film on windows) Use of electronic devices Use of household appliances for cooking or water boiling Use of work-related appliances Domestic hot water usage (e.g., shower usage)	Movement from one room to another Change of position or orientation in a room (e.g., to a place further away from window to reduce problems with glare or direct solar radiation) Relocation of activities to a different room (e.g., sleeping in a basement)

2.3 Potential Triggers and Contextual Factors Influencing Occupant Behavior in a Building

Whether the focus is on existing buildings or new buildings, the factors influencing the behavior patterns and patterns of occupancy together with influencing contextual factors are important for modeling purposes, supporting design decisions, and choosing operation modes. In order to study occupancy and occupants' behavior in buildings, a researcher needs to be aware of which variables are most critical to measure. This knowledge helps inform the researcher of how best to allocate resources towards sensors and labor for installing sensors and collecting and analyzing data.

People live and work in buildings and use them by interacting with their envelope components, energy systems, and equipment, as outlined above. Potential influencing factors that drive occupant behavior were first classified as internal and external factors (Schweiker and Shukuya 2009). Internal factors are aspects related to the social sciences, such as preference, attitudes, cultural background, etc.; external factors are those aspects related to the building science or associated with the time of day or building characteristics. Later, Fabi et al. (2012) presented a more general explanatory framework to investigate the relationship between a multitude of potential adaptive and non-adaptive actions and influencing factors with the purpose of providing an outline to explore what stimulates occupants' behavior in a building. The authors proposed five categories of drivers for describing occupants' interaction with windows: physical environmental, contextual, psychological, physiological, and social. More recently, Polinder et al. (2013) categorized the influential factors that can be useful to explain occupant behavior into internal and external driving forces. Internal driving forces arise from the interaction between biological and psychological aspects and are investigated in the domains of biology, social science, and economics. Internal driving forces comprehend biological, psychological, and social factors. External driving forces act on the individual to stimulate their reaction and comprise three factors: building and building equipment properties, physical environment, and time.

In contrast to these approaches and as mentioned in the introduction, a given action (e.g., closing a window) could be identified as either an adaptive action or a non-adaptive action depending on the corresponding trigger. Therefore, this section aims at providing a framework to categorize factors influencing occupants' presence and actions into adaptive triggers, non-adaptive triggers, and contextual factors. As shown in Fig. 2.1, contextual factors are seen here as moderators of triggers and behavior. Sub-groups are defined based on the five categories mentioned by Fabi et al. (2012). An overview of those triggers and contextual factors that have been studied in the past, or based on discussions between the authors that can influence occupancy and/or occupants' behavioral patterns is given in Fig. 2.2.

Thereby, adaptive triggers are physiological triggers or physical environmental triggers. Physiological triggers originate from a signal in the body that prompts an occupant to take an action, e.g., the onset of shivering. Physical environmental

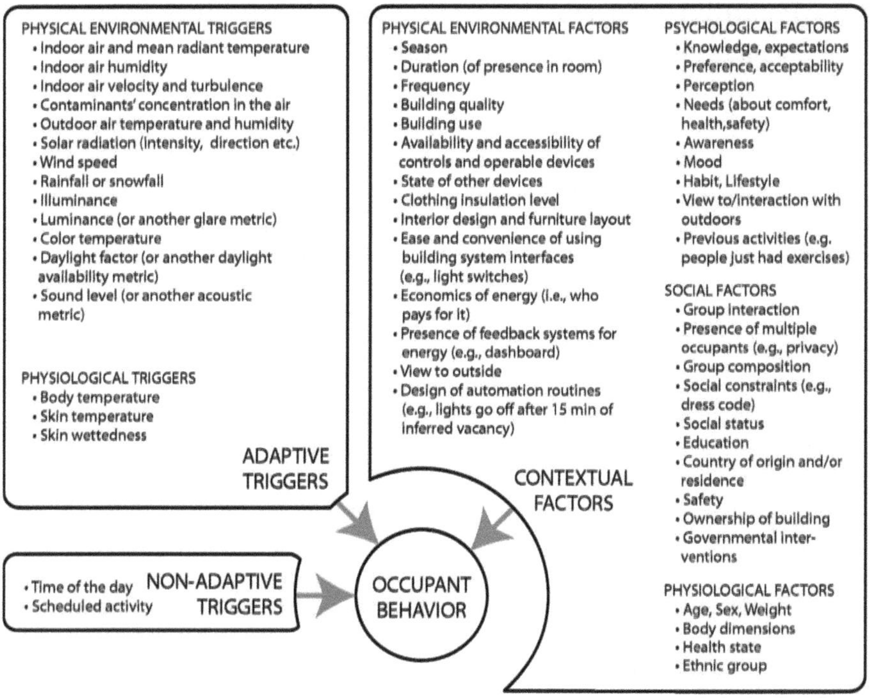

Fig. 2.2 Potential influencing factors driving occupants' behavior in a building

triggers are those physical properties that describe the indoor and outdoor environments and have been proven to stimulate occupant action in response to thermal, olfactory, visual, and aural stimulus (air and mean radiant temperatures, solar radiation, air pollutants, illuminance, sound pressure, etc.) with the aim of voluntarily modifying the surrounding built environment to restore or improve the comfort conditions. In contrast, non-adaptive triggers are factors that are independent of the physical environmental triggers (e.g., time of day, people leaving a room to attend a meeting elsewhere).

Contextual factors can be grouped into physical environmental factors, which remain unchanged over a period of time; psychological factors, which are related to individual and social factors; and physiological factors, which are not immediate triggers (O'Brien and Gunay 2014). Physical environmental contextual factors are fixed characteristics of the space and include things such as distance from the occupants' seated position to a control interface and ease of opening a window. Psychological factors "refer to thoughts, feelings and other cognitive characteristics that affect the attitude, behavior and functions of the human mind" (Sumpi and Amukugo 2016). They have been proven to explain occupants' interactions with building envelope, services, and equipment through expectation, awareness, preference, etc. Social factors refer to those items that affect a person's lifestyle and his/her attitude in relating to others, such as group interaction, education, social status, etc. Physiological factors are all those aspects "relating to the branch of biology that deals with the normal functions of living organisms and their parts" (Stevenson 2010) and which have been related to the occurrence of an occupant action inside a building, such as age, sex, state of health, etc. All of these factors have a significant influence on the indoor environmental quality, overall building energy performance, effectiveness of control strategies, and, eventually, on occupants' satisfaction and productivity.

A literature review was conducted in order to build a framework of triggers and contextual factors. The review, summarized in Table 2.2 through Table 2.4, provides the grounds for a list of factors (with examples) that influence occupancy and occupants' behavior in buildings. This list gathers together items from the final report of the IEA Annex 53, Total Energy Use in Buildings: Analysis & Evaluation Methods (Polinder et al. 2013) and additional contributions collected by the authors of this chapter. The items were grouped according to the framework described above, as represented in Fig. 2.2.

2.4 Literature Review of Relationship Between Action Types and Influencing Factors

In an ideal world, a monitoring campaign with respect to occupancy and occupants' actions would include all potential influencing factors described in the previous section. In reality, however, financial resources and therewith the number of observed factors are limited. Before starting a monitoring campaign it is thus important to balance the cost of potentially necessary devices (data points) and the results potentially obtained by their implementation. The decision to include or exclude a certain variable is best made based on results from previous studies.

An overview of the type of actions and their influencing factors as studied in the past is presented in Tables 2.2, 2.3 and 2.4. They present behavioral actions and identified influencing factors related to individual, personal, and spatial adjustments, respectively. These tables may serve as an important guide for selecting the type of actions to investigate, as well as deciding on the data points to include in data collection and analysis. The first two columns describe the action and context analyzed. The last two columns state whether an influencing factor was found to have a statistically significant influence on the action, a potentially significant influence (a significance level was not reported in the literature or a significance test was not conducted), or a non-statistically significant influence. Note that the statistical method used to determine the statistical significance of an influencing variable and/or the sample size may vary between studies depending on the field and publication, and so that significance levels are not necessarily comparable (refer to Chap. 3 for a more in-depth treatment of statistical significance). In addition, it is important to mention that the quality and depth of conveyed surveys vary. To give

Table 2.2 Behavioral actions and influencing factors identified related to individual adjustments

Name of action	Building type[a]	Factors found to have influence on behavior[b]	Factors found to have no influence on behavior
1. *Individual adjustments*			
Selection of clothing	Office	Indoor air temperature*** (Haldi and Robinson 2008) Outdoor air temperature*** (De Carli et al. 2007; Haldi and Robinson 2008; Schiavon and Lee 2013; Schweiker and Wagner 2015) Daily mean outdoor air temperature* (Haldi and Robinson 2011)	Sex (De Carli et al. 2007)
Adjustment of clothing	Residential	Time of day* (Gauthier 2016)	Indoor air temperature (Gauthier 2016) Outdoor air temperature (Gauthier 2016) Weekend/weekday schedule (Gauthier 2016)
Adjustment of clothing	Office	Thermal comfort[I] (Baker and Standeven 1994) Number of persons in room[I] (Schweiker and Wagner 2016) Personality traits[I] (Schweiker et al. 2016)	Indoor air temperature (Haldi and Robinson 2011) Outdoor air temperature (Haldi and Robinson 2011)
Adjustment of clothing	Retail	Outdoor air temperature[I] (Morgan and de Dear 2003) Sex[I] (Morgan and de Dear 2003)	
Adjustment of body posture	Any or none	While people can reduce heat loss by becoming more compact (e.g., crossing arms) to decrease effective body surface area—and this is a common adaptive action—no behavioral studies were found in the literature	
Activity level	Residential	Indoor air temperature* (Gauthier 2016) Probability of heating on/off* (Gauthier 2016)	Outdoor air temperature (Gauthier 2016) Weekend/weekday schedule (Gauthier 2016)

(continued)

Table 2.2 (continued)

Name of action	Building type[a]	Factors found to have influence on behavior[b]	Factors found to have no influence on behavior
Activity level	Office		Indoor air temperature (Haldi and Robinson 2011) Outdoor air temperature (Haldi and Robinson 2011)
Change in orientation or position to improve visual comfort	Any or none	While it is generally understood that occupants may change their orientation or position to improve visual comfort, no experimental/in situ results were found in the literature	
Using earplugs or earphones to avoid excessive aural discomfort	Any or none	While it is generally understood that occupants use earplugs or earphones to improve comfort, no experimental/in situ results were found in the literature	
Drinking hot/cold/beverages	Residential	Time of day* (Gauthier 2016)	Indoor air temperature (Gauthier 2016) Outdoor air temperature (Gauthier 2016) Weekend/weekday schedule (Gauthier 2016)
Drinking cold drinks	Office	Indoor air temperature*** (Haldi and Robinson 2008)* (Haldi and Robinson 2011) Outdoor air temperature*** (Haldi and Robinson 2008)* (Haldi and Robinson 2011)	
Drinking hot drinks	Office	Weighted running mean of outdoor air temperature* (Haldi and Robinson 2011)	Indoor air temperature (Haldi and Robinson 2011)

[a]Building types can be office, residential, hotel, educational, retail
[b]Significance levels are *** = $p < 0.001$, ** = $p < 0.01$, * = $p < 0.05$, [Ɨ] = significance level not stated

Table 2.3 Behavioral actions and influencing factors identified related to environmental adjustments

Name of action	Building type[a]	Factors found to have influence on behavior[b]	Factors found to have no influence on behavior
2. *Environmental adjustments*			
2.1. **Use of heating, ventilation and air-conditioning (AC) systems**			
Thermostat adjustments for heating	Residential	Indoor relative humidity* (Fabi et al. 2013) Outdoor air temperature*** (Fabi et al. 2013; van der Lans et al. 2013) Outdoor air humidity** (Hori 1995) Solar radiation* (Fabi et al. 2013) Wind speed* (Fabi et al. 2013) Time of day* (Fabi et al. 2013) Interaction frequency with heating controls*** (Hori 1995; Fabi et al. 2013) Window opening* (Hori 1995) Building insulation level*** (Müller et al. 2010) Ventilation type* (Keul et al. 2011) Type of metering (sub-metered vs. bulk-metered)[I] (Gunay et al. 2014) Expectations* (Keul et al. 2011) Ownership** (Keul et al. 2011) Sex[I] [8] Clothing** (Hori 1995; Fabi et al. 2013)	Time of day (Hori 1995)
Heating duration	Residential	Outdoor air temperature*** (Hori 1995) Outdoor air humidity*** (Hori 1995) Wind speed** (Hori 1995) Building insulation level*** (Müller et al. 2010) Heating system type*** (Hori 1995) Window opening* (Hori 1995) Level of control*** (Hori 1995) Understanding how controls function** (Hori 1995; Peeters et al. 2008; Keul et al. 2011) Clothing** (Hori 1995; Keul et al. 2011)	Ownership (Hori 1995) Government interventions (Müller et al. 2010)

(continued)

Table 2.3 (continued)

Name of action	Building type[a]	Factors found to have influence on behavior[b]	Factors found to have no influence on behavior
Switch on heating	Office	Outdoor air temperature (Nicol 2001) Country (Nicol 2001)	
Number of rooms heated	Residential	Interaction frequency with heating controls*** (Hori 1995) Level of control (Hori 1995) Type of metering (sub-metered vs. bulk-metered)[I] (Gunay et al. 2014)	
Type of rooms heated	Residential	Level of control*** (Hori 1995)	Sex (Hori 1995)
Percentage of times AC unit used	Residential	Indoor air temperature[I] (Jian et al. 2015) Indoor/outdoor temperature difference** (Stephens et al. 2011) Outdoor air temperature*** (Asawa et al. 2005; Schweiker and Shukuya 2009) Outdoor air humidity** (Asawa et al. 2005; Schweiker and Shukuya 2009) Wind direction** (Asawa et al. 2005) Wind speed* (Asawa et al. 2005) Season** (Asawa et al. 2005) Time of day** (Asawa et al. 2005; Schweiker and Shukuya 2009; Stephens et al. 2011) Cooling degree days* (Yun and Steemers 2011) Set point temperature of system[I] (Stephens et al. 2011) South orientated window*** (Schweiker and Shukuya 2009) Top floor*** (Schweiker and Shukuya 2009) Number of rooms with AC unit*** (Yun and Steemers 2011) Daily length of use[I] (Jian et al. 2015) Daily frequency of use[I] (Jian et al. 2015)	Indoor air temperature (Jian et al. 2015) AC unit used at home during childhood (Schweiker and Shukuya 2009) Climatic background (Schweiker and Shukuya 2009) Geographic background (Schweiker and Shukuya 2009) Household income (Yun and Steemers 2011) Occupancy patterns (Jian et al. 2015)

(continued)

Table 2.3 (continued)

Name of action	Building type[a]	Factors found to have influence on behavior[b]	Factors found to have no influence on behavior
		Preference for AC on*** (Hori 1995) Sex*** (Schweiker and Shukuya 2009) Origin moderate climate*** (Schweiker and Shukuya 2009) Geographic background* (Schweiker and Shukuya 2009) Health[I] (Kempton et al. 1992; Iwashita and Akasaka 1997)	
Switching on AC device	Residential	Indoor air temperature (Ren et al. 2014)[I] (Schweiker and Shukuya 2010a) Comfort range[I] (Bae and Chun 2009) Guests coming[I] (Schweiker and Shukuya 2010a)	
Switching off AC device	Residential	Leaving room[I] (Schweiker and Shukuya 2010a)	
Set point temperature adjustments for cooling	Residential	Outdoor air temperature*** (Schweiker and Shukuya 2010b) Geographical background*** (Schweiker and Shukuya 2010b) South orientation window*** (Schweiker and Shukuya 2010b) Preference for AC on*** (Schweiker and Shukuya 2010b) Sex* (Schweiker and Shukuya 2010b) Climatic background*** (Schweiker and Shukuya 2010b)	Floor (top, middle, ground) (Schweiker and Shukuya 2010b)
Existence of AC unit	Residential	Climate*** (Yun and Steemers 2011) Household income*** (Yun and Steemers 2011)	
No. of rooms with AC unit	Residential	Type of AC*** (Yun and Steemers 2011) Floor area*** (Yun and Steemers 2011)	Climate (Yun and Steemers 2011)

(continued)

Table 2.3 (continued)

Name of action	Building type[a]	Factors found to have influence on behavior[b]	Factors found to have no influence on behavior
2.2. **Window opening/closing**			
Window state	Residential	Indoor air temperature* (Rijal et al. 2007; Yun and Steemers 2008; Andersen et al. 2013)[Ɨ] (Dick and Thomas 1951; Rijal et al. 2007, 2008a, b; Schweiker et al. 2012a) Indoor air relative humidity* (Andersen et al. 2013) CO_2 concentration* (Andersen et al. 2013)[Ɨ] (Andersen et al. 2009) Outdoor air temperature*** (Herkel et al. 2008)* (Andersen et al. 2013)[Ɨ] (Dick and Thomas 1951; Brundrett 1978; Dubrul 1988; Fritsch et al. 1990; Rijal et al. 2007) Outdoor air temperature during day* (Yun and Steemers 2008) Outdoor air relative humidity* (Andersen et al. 2013) Solar radiation* (Andersen et al. 2013)[Ɨ] (Dubrul 1988; Andersen et al. 2011) Wind speed[Ɨ] (Dubrul 1988) Time of day* (Andersen et al. 2013)[Ɨ] (Dubrul 1988; Johnson and Long 2005) Season* (Andersen et al. 2013)[Ɨ] (Karava et al. 2007; Rijal et al. 2007; Herkel et al. 2008) Dwelling type[Ɨ] (Dubrul 1988; Johnson and Long 2005) Room orientation[Ɨ] (Dubrul 1988) Ventilation type[Ɨ] (Brundrett 1978; Dubrul 1988; van Dongen 2004) Heating system[Ɨ] (Dubrul 1988) Room type* (Andersen et al. 2013)[Ɨ] (Dubrul 1988; Andersen et al. 2009) Perceived illumination[Ɨ] (Andersen et al. 2009)	Indoor air temperature on arrival (Yun and Steemers 2008) Indoor air temperature during day (Yun and Steemers 2008)

(continued)

Table 2.3 (continued)

Name of action	Building type[a]	Factors found to have influence on behavior[b]	Factors found to have no influence on behavior
		Preference in terms of temperature[I] (Dubrul 1988) Smoking behavior[I] (Dubrul 1988) Presence at home[I] (Dubrul 1988) Age[I] (Dubrul 1988; Kvisgaard and Collet 1990) Sex[I] (Andersen et al. 2009)	
Window state	Office	Indoor air temperature*** (Haldi and Robinson 2008)* (Yun et al. 2008)[I] (Yun and Steemers 2008; Yun et al. 2009) Outdoor air temperature*** (Haldi and Robinson 2008) Time of day/arrival (Yun et al. 2009) Number of persons in a room[I] (Schweiker and Wagner 2016) Attitude towards actions (active/medium/passive window user) (Yun et al. 2009) Personality traits[I] (Schweiker et al. 2016) Perceived control over temperature* (Yun et al. 2008)	Indoor air temperature (Yun et al. 2008)
Window opening	Educational	Outdoor air temperature (Dutton and Shao 2010)	
Window opening	Office	Indoor temperature (Haldi and Robinson 2008) Outdoor air temperature (Haldi and Robinson 2008; Zhang and Barrett 2012a)	
Window closing	Office	Indoor air temperature (Haldi and Robinson 2009) Outdoor air temperature (Haldi and Robinson 2009) Arrival/intermediate/departure (Haldi and Robinson 2009)	
Degree of opening	Residential	Indoor air temperature* (Schweiker et al. 2012b) Outdoor air temperature* (Schweiker et al. 2012b)[I] (Dubrul 1988)	Indoor air temperature (Schweiker et al. 2012b) Outdoor air temperature (Schweiker et al. 2012b)

(continued)

Table 2.3 (continued)

Name of action	Building type[a]	Factors found to have influence on behavior[b]	Factors found to have no influence on behavior
2.3. **Solar shading device adjustments**			
Shading device position adjustment	Office	Outdoor daylight level[Ɨ] (Van Den Wymelenberg 2012) Direct sunlight from outdoors[Ɨ] (Maniccia et al. 1999), (Van Den Wymelenberg 2012) Indoor luminance contrast (glare)[Ɨ] (Van Den Wymelenberg 2012) Design of automation systems (e.g., lights go off after 15 min of inferred vacancy)[Ɨ] (Van Den Wymelenberg 2012) Orientation of the workstation[Ɨ] (Van Den Wymelenberg 2012) View to outside[Ɨ] (Van Den Wymelenberg 2012)	
Blind state	Residential	Solar radiation** (Andersen et al. 2009) Ownership* (Andersen et al. 2009) Perceived indoor air quality*** (Andersen et al. 2009) Age* (Andersen et al. 2009)	Outdoor air temperature (Andersen et al. 2009) Floor area (Andersen et al. 2009) Perceived illumination (Andersen et al. 2009) Thermal sensation (Andersen et al. 2009) Perceived noise (Andersen et al. 2009) Sex (Andersen et al. 2009)
Blind state	Office	Indoor air temperature*** (Haldi and Robinson 2008), (Inkarojrit 2008) Mean radiant temperature (Inkarojrit 2008) Outdoor air temperature*** (Haldi and Robinson 2008) Indoor solar intensity (Newsham 1994)	

(continued)

Table 2.3 (continued)

Name of action	Building type[a]	Factors found to have influence on behavior[b]	Factors found to have no influence on behavior
		Average luminance of the window or source luminance (Inkarojrit 2008) Background luminance (Inkarojrit 2008) Maximum luminance of the window (Inkarojrit 2008) Brightness sensitivity (Inkarojrit 2008) Solar penetration depth (Newsham 1994) Number of persons in room[Ɨ] (Schweiker and Wagner 2016) Personality traits[Ɨ] (Schweiker et al. 2016)	
Blind closing	Office	Indoor air temperature (Haldi and Robinson 2008) Indoor solar radiation (Mahdavi et al. 2008) Outdoor air temperature (Haldi and Robinson 2008) Solar radiation (Zhang and Barrett 2012b) Solar altitude (Zhang and Barrett 2012b) Glare (indicated by indoor illuminance or irradiance)*** (Haldi and Robinson 2010) Depth of solar penetration (Inoue et al. 1988; Reinhart and Voss 2003) Exterior solar irradiance (Reinhart and Voss 2003; Mahdavi et al. 2008) Availability of AC (Inkarojrit 2005) Façade orientation (Inoue et al. 1988; Mahdavi et al. 2008) Current shade position (Haldi and Robinson 2010) Number of occupants in space (Haldi and Robinson 2010)	Indoor air temperature (Lindsay and Littlefair 1992; Haldi and Robinson 2010) Outdoor air temperature (Haldi and Robinson 2010)
Blind opening	Office	Outdoor illuminance*** (Haldi and Robinson 2010) Solar radiation (Zhang and Barrett 2012b)	View to outside (Rubin et al. 1978)

(continued)

Table 2.3 (continued)

Name of action	Building type[a]	Factors found to have influence on behavior[b]	Factors found to have no influence on behavior
		Solar altitude (Zhang and Barrett 2012b) Morning upon arrival (Newsham 1994) Occupancy events (e.g., first daily arrival) (Haldi and Robinson 2010) Façade orientation (Inoue et al. 1988; Mahdavi et al. 2008) Current shade position*** (Haldi and Robinson 2010)	
Blind opening	Residential	Time of day (Bennet et al. 2014) Day of week (Bennet et al. 2014) Façade orientation (Bennet et al. 2014) Window size (Bennet et al. 2014)	Solar radiation (Bennet et al. 2014) Season (Bennet et al. 2014)
Blind closing	Residential	Time of day (Bennet et al. 2014) Day of week (Bennet et al. 2014) Façade orientation (Bennet et al. 2014) Window size (Bennet et al. 2014)	
2.4. **Light switching/dimming**			
Light switching/dimming	Residential	Outdoor air temperature** (Andersen et al. 2009) Solar radiation*** (Andersen et al. 2009) Perceived illumination** (Andersen et al. 2009) Thermal sensation* (Andersen et al. 2009) Weekend schedule[I] (Jian et al. 2015) Daily length of use[I] (Jian et al. 2015) Daily frequency of use[I] (Jian et al. 2015) Sex* (Andersen et al. 2009) Age* (Andersen et al. 2009)	Perceived noise (Andersen et al. 2009) Ownership (Andersen et al. 2009) Floor area (Andersen et al. 2009)

(continued)

Table 2.3 (continued)

Name of action	Building type[a]	Factors found to have influence on behavior[b]	Factors found to have no influence on behavior
	Office	Workplace illuminance[I] (Newsham et al. 2008) Outdoor daylight level[I] (Newsham et al. 2008) Interior design and furniture layout[I] (Maniccia et al. 1999) Ease and convenience of using building system interfaces (e.g., light switches)[I] (Maniccia et al. 1999)	
Light switch on	Office	Work plane illuminance[I] (Hunt 1979; Newsham 1994; Love 1998; Reinhart and Voss 2003; Lindelöf and Morel 2006) Time of day/events: morning arrival, after lunch, leaving[I] (Newsham 1994)	
Light switch off	Office	Departure[I] (Lindelöf and Morel 2006)	Workplace illuminance (Lindelöf and Morel 2006)
2.5. **Fan usage**			
Ceiling fan usage	Office	Efficiency of ceiling fan[I] (Schweiker et al. 2014) Number of persons in room[I] (Schweiker and Wagner 2016) Personality traits[I] (Schweiker et al. 2016)	
Desk fan usage	Office	Indoor air temperature*** (Haldi and Robinson 2008) Outdoor air temperature*** (Haldi and Robinson 2008)	
2.6. **Occupancy (schedule driven)**			
	Residential		–
		Time of day[I] (Johnson 1981; Al-Mumin et al. 2003; Yao and Steemers 2005) Time of week[I] (Johnson 1981) Season (and its effect on time spent outdoors)** (Marino et al. 2012) Number of persons in household[I] (Yao and Steemers 2005)	Elderly status[I] (Johnson 1981)

(continued)

Table 2.3 (continued)

Name of action	Building type[a]	Factors found to have influence on behavior[b]	Factors found to have no influence on behavior
		Building type (detached house, apartment, etc.)[I] (Farley 1978) Age[I] (Papakostas and Sotiropoulos 1997) Gender[I] (Papakostas and Sotiropoulos 1997) Marital status[I] (Johnson 1981) Culture[I] (Al-Mumin et al. 2003) Employment status[I] (Papakostas and Sotiropoulos 1997)	
	Office		
		Time of day[I] (Page et al. 2008; Duarte et al. 2013) Weekday vs. weekend[I] (Page et al. 2008; Davis and Nutter 2010; Duarte et al. 2013) Weekday (Monday, Tuesday, etc.)[I] (Davis and Nutter 2010; Duarte et al. 2013) Holidays[I] (Duarte et al. 2013) Season[I] (Duarte et al. 2013) Profession[I] (Feng et al. 2015) Office type (open plan vs. private)[I] (Duarte et al. 2013) Room type (e.g., break room, conference room, office)[I] (Duarte et al. 2013; Feng et al. 2015) Business/activity type[I] (Davis and Nutter 2010) Culture and occupant density[I] (Van Meel 2000) Culture and working hours (Iwasaki et al. 2006)	
2.7. **Use of electronic devices**			
Level of electricity consumption	Residential	Persons per dwelling*** (Jensen 2004; Gram-Hanssen 2005) Income** (Gram-Hanssen 2005) Area of the dwelling** (Gram-Hanssen 2005) Teenagers in the household** (Gram-Hanssen 2005) Age** (Gram-Hanssen 2005)	Sex (Gram-Hanssen 2005)

(continued)

Table 2.3 (continued)

Name of action	Building type[a]	Factors found to have influence on behavior[b]	Factors found to have no influence on behavior
Electricity consumption	Residential	Heating degree days* (Mora et al. 2015) Area of the dwelling*** (Mora et al. 2015) Income* (Mora et al. 2015) Number of household members* (Mora et al. 2015) Average age** (Mora et al. 2015) Age of household head** (Mora et al. 2015)	Type of house (Mora et al. 2015) Type of external wall (Mora et al. 2015) Year of construction (Mora et al. 2015) Structure (Mora et al. 2015) Type of windows (Mora et al. 2015) Energy saving lamps (Mora et al. 2015) Sex (Mora et al. 2015)
Number of appliances	Residential	Income (Gram-Hanssen 2005)	
Use of appliances	Residential	Household size[I] (Lutz et al. 1996) Education[I] (Vine et al. 1987)	
Use of appliances	Office	Duration of vacancy period following a departure[I] (Gunay et al. 2016) Computer type[I] (Webber et al. 2006; Moorefield et al. 2008) Company size[I] (Roberson et al. 2002) Occupancy rate[I] (Nordman et al. 1996; de Menezes 2013) Power management program settings[I] (Kawamoto et al. 2004; Lobato et al. 2011) Habits—non-behavioral factors[I] (Tetlow et al. 2015) Awareness, peer pressure, and competition[I] (Yun et al. 2013; Lasternas et al. 2014)	Indoor and environmental conditions (Tetlow et al. 2015)
Use of appliances	Educational	Awareness and peer pressure, competition, and economics[I] (Brewer et al. 2011)	

(continued)

Table 2.3 (continued)

Name of action	Building type[a]	Factors found to have influence on behavior[b]	Factors found to have no influence on behavior
2.8. **Domestic hot water usage**			
Shower duration	Residential	Weekday or weekend[I] (Aguilar et al. 2005) Season[I] (Kondo and Hokai 2012) Household size[I] (Ministerie of VROM 2009) Low-flow shower head[I] (Campbell et al. 2004; Ministerie of VROM 2009) Boiler[I] (Ministerie of VROM 2009) Income[I] (Ministerie of VROM 2009) Geographic background[I] (Foekema et al. 2007) Age[I] (Ministerie of VROM 2009) Sex[I] (Foekema et al. 2007)	
Frequency of bath/shower	Residential	Geographic background[I] (van Dongen 2004) Household composition[I] (Ministerie of VROM 2009) Age[I] (Ministerie of VROM 2009) Sex[I] (Foekema et al. 2007)	

[a]Building types can be office, residential, hotel, educational, retail
[b]Significance levels are *** = $p < 0.001$, ** = $p < 0.01$, * = $p < 0.05$, [I] = significance level not stated

an example, the authors of this chapter prefer a consistent and meticulous usage of terms; however, many studies neither report the exact approach used in survey data collection, nor specify whether they collected data of, for example, indoor air temperature or indoor operative temperature. In these cases of ambiguity the terms used in the original source are used.

In addition to the results of previous studies related to specific actions or influencing factors, Table 2.2, 2.3 and 2.4 show the variation in the number of studies dealing with one action or another. This organization can help in choosing the type of action to be investigated—for example, numerous studies have dealt with window opening and closing behaviors in both residential and office settings, whereas individual adaptive actions, such as drinking behaviors, change of activity levels, etc. have not been investigated as extensively and solely in the residential context. Therefore, based on all three tables, some of the major research gaps requiring additional study could be identified. That said, even for those cells with

Table 2.4 Behavioral actions and influencing factors identified related to spatial adjustments

Name of action	Building type[a]	Factors found to have influence on behavior[b]	Factors found to have no influence on behavior
3. *Spatial adjustments*			
3.1. **Movement from a room to another**			
	Any or none	While it is generally understood that occupants may relocate within a building to improve comfort, no experimental/in situ results were found in the literature	–
3.2. **Rotation or minor movements within a room**			
	Any or none	While it is generally understood that occupants may rotate, change their posture, and relocate within a room to improve comfort, no experimental/in situ results were found in the literature	–

[a]Building types can be office, residential, hotel, educational, commercial (shopping)
[b]Significance levels are *** = $p < 0.001$, ** = $p < 0.01$, * = $p < 0.05$, ^{I} = significance level not stated

numerous references, it might be meaningful to conduct additional studies that include other predictor variables, whether based on data from a distinctive cultural background, geographic context, or building characteristics. Comparative studies (e.g., Schweiker and Shukuya 2009; Schweiker et al. 2016; Schweiker and Wagner 2016) have demonstrated that many actions are contextually sensitive or differing due to personal characteristics, and that numerous studies in different contexts (e.g., climates, cultures, building types, mechanical systems) are required to properly understand a particular action.

Regarding the data points to include in data collection and analysis, it seems obvious to include those variables which have been found to be significant predictors for the respective action. At the same time, it should be noted that a variable found to be non-significant in a given context in one specific study (e.g., residential buildings in cold climates) could have a significant influence on the given action in the same context in a second study. Only by repetition can preliminary results deriving from one experiment or field study be either confirmed or challenged.

By all means, the authors recommend that the reader not make decisions based only on these tables, but also refer to the corresponding studies in order to learn more about their data collection process, analysis methods, general circumstances, and limitations. Such a step is important and necessary before deciding whether predictor variables should be included or not. In addition, the presented tables are not exclusive in the sense that important predictors might not be listed or studies reporting different results may have been overlooked. The authors recommend that researchers use this comprehensive literature review prior to embarking on major occupant monitoring campaigns, but that they also perform an additional review on their own as new research continually emerges.

Research exploration of influencing variables for seldom explored domains, such as spatial movements or changes in body posture, are particularly difficult. Tables 2.2, 2.3 and 2.4 offer some guidance for such exploration. It is important to analyze which predictors are found to be influencing a particular action and/or comparable actions the most. For example, indoor thermal conditions appear to influence most thermal comfort-related adaptive actions. A thorough analysis of Tables 2.2, 2.3 and 2.4, together with rational thinking of potential pathways and influences, would facilitate the research exploration of these and other unexplored actions

Importantly, the number of variables reported to have no influence on action is rather small compared to those reported to have an influence. Based on this chapter's literature review, it remains unknown whether in fact there are few variables that have no influence on specific actions, or whether this is a matter of known scientific bias due to a higher probability of significant results being reported. It is highly recommended that researchers report all variables measured and analyzed, regardless of their significance, in order to improve the knowledge of the research area and to avoid repetitive, fruitless work.

2.5 Conclusion

This chapter has summarized previous occupant behavior research with a focus on providing basic definitions and a comprehensive list of occupant actions and their influencing factors. The chapter should be interpreted in the context that the field of occupant behavior monitoring and modeling is in active development, where many domains have been studied in only a few buildings and contexts or by only a few researchers, if at all. Accordingly, researchers who are embarking on new monitoring campaigns should combine the information that was presented in this chapter and in the most recent literature with common sense and an appreciation for balancing research rigor with practicality (costs, time, etc.). Moreover, for monitoring campaigns that are destined to support modeling and simulation, researchers should be familiar with the modeling capabilities of common building simulation tools because simulation tools must actually calculate the model predictors for the model to be of use. For instance, even if individual occupant sensitivity to glare is a good predictor for window blind actions, metrics capable of quantifying this phenomenon are either too complex to be calculated (e.g., the daylight glare probability, DGP) or unavailable as an output in many building performance simulation (BPS) tools. Similarly, most BPS tools do not model odor; thus, this would not be a practical predictor for window opening action models. On the other hand, variables such as age can still be useful in order to select modeling parameters for specific design tasks—for example, the design of a retirement home.

The studies reviewed in this chapter strongly support the notion that occupant behavior is influenced by a large number of complex variables that extend well beyond physical phenomena to include social and psychological factors.

Researchers should take considerable care and rigor in both measuring and documenting these phenomena. Model users must be made aware of the contexts upon which observational studies and models were built, otherwise models may be applied in entirely different and inappropriate building contexts.

Among the occupant action domains explored in this chapter, some have benefitted from decades of research in diverse contexts, while others have been seldom researched. The actions that have been studied more extensively include: window opening in both commercial and residential buildings, occupant control of heating and cooling in homes, and window blind and lighting control in office buildings. Many domains require much more research, including: occupant posture, occupant repositioning in response to discomfort, and lighting use in residential buildings, to name a few.

One of the conclusions of this chapter is that it is crucial for future occupant monitoring campaigns to be published with sufficient detail to inform future researchers and save redundant effort. Contrary to common practice, as evidenced in Tables 2.2, 2.3 and 2.4, both the influencing factors that were found to be significant and those found to be insignificant should be documented. If several comprehensive studies indicate that a given variable is not a significant predictor in a particular domain, future studies can avoid diverting limited resources towards that variable. Moreover, the statistical significance of predictive variables should be determined and reported, as currently there is a large degree of variability in the literature in this regard. It is likewise critical to have consistent reporting of other technical terms. For instance, this chapter's literature review revealed that numerous papers did not specify whether "interior temperature" refers to indoor air temperature, operative temperature, mean radiant temperature, or even internal body temperature. Precision and consistency in the usage of terms is essential for the field of occupant research to move forward.

This chapter also introduced the most commonly studied occupant domains and provided a summary of existing literature in this area. The authors attempted to summarize the existing literature so that researchers can maintain focus in the potentially overwhelming sea of literature that has built over the last decades. The next chapters provide practical and fundamental guidance on research methods for collecting occupant data using a variety of unique approaches.

References

Aguilar C, White DJ, Ryan DL (2005) Domestic water heating and water heater energy consumption in Canada

Al-Mumin A, Khattab O, Sridhar G (2003) Occupants' behavior and activity patterns influencing the energy consumption in the Kuwaiti residences. Energy Build 35:549–559

American Optometric Association (2016) How your eyes work. http://www.aoa.org/patients-and-public/resources-for-teachers/how-your-eyes-work?sso=y. Accessed 20 Oct 2016

Andersen RK (2012) The influence of occupants' behaviour on energy consumption investigated in 290 identical dwellings and in 35 apartments. In: Proceedings of Healthy Buildings 2012, Brisbane, Australia
Andersen RV, Toftum J, Andersen KK, Olesen BW (2009) Survey of occupant behaviour and control of indoor environment in Danish dwellings. Energy Build 41:11–16. doi:10.1016/j.enbuild.2008.07.004
Andersen RV, Olesen BW, Toftum J (2011) Modelling window opening behaviour in Danish dwellings. In: Proceedings of Indoor Air 2011
Andersen RK, Fabi V, Toftum J et al (2013) Window opening behaviour modelled from measurements in Danish dwellings. Build Environ 69:101–113. doi:10.1016/j.buildenv.2013.07.005
Andersen S, Andersen RK, Olesen BW (2016) Influence of heat cost allocation on occupants' control of indoor environment in 56 apartments: studied with measurements, interviews and questionaires
Asawa T, Hoyano A, Takezawa H, Shimizu K (2005) Analysis of the behavioral characteristics of both window opening and air conditioning use at detached houses: Relationship between outdoor microclimate and residents' living open to outdoor. Part 2. J Environ Eng AIJ 87–94
Bae C, Chun C (2009) Research on seasonal indoor thermal environment and residents' control behavior of cooling and heating systems in Korea. Build Environ 44:2300–2307. doi:10.1016/j.buildenv.2009.04.003
Baker N, Standeven M (1994) Comfort criteria for passively cooled buildings a pascool task. Renew Energy 5:977–984
Bennet I, O'Brien W, Gunay HB (2014) Effect of window blind use in residential buildings: observation and simulation study
Bishop RC, Frey DJ (1985) Occupant effects on energy performance of monitored houses. In: Proceedings of the 10th National Passive Solar Conference, the solar, pp 395–400
Blight TS, Coley DA (2013) Sensitivity analysis of the effect of occupant behaviour on the energy consumption of passive house dwellings. Energy Build 66:183–192. doi:10.1016/j.enbuild.2013.06.030
Brewer RS, Lee GE, Johnson PM (2011) The Kukui Cup: a dorm energy competition focused on sustainable behavior change and energy literacy. In: 2011 44th Hawaii International Conference on System Sciences (HICSS), pp 1–10
Brundrett GW (1978) Window ventilation and human behaviour. In: Proceedings of the First International Indoor Climate Symposium in Copenhagen: Indoor Climate, Effects on Human Comfort, Performance and Health in Residential, Commercial and Light-Industry Buildings, p 31
Campbell HE, Johnson RM, Larson EH (2004) Prices, devices, people, or rules: the relative effectiveness of policy instruments in water conservation. Rev Policy Res 21:637–662. doi:10.1111/j.1541-1338.2004.00099.x
Carlucci S, Lobaccaro G, Li Y et al (2016) The effect of spatial and temporal randomness of stochastically generated occupancy schedules on the energy performance of a multiresidential building. Energy Build 127:279–300. doi:10.1016/j.enbuild.2016.05.023
Davis JA, Nutter DW (2010) Occupancy diversity factors for common university building types. Energy Build 42:1543–1551
De Carli M, Olesen BW, Zarrella A, Zecchin R (2007) People's clothing behaviour according to external weather and indoor environment. Build Environ 42:3965–3973. doi:10.1016/j.buildenv.2006.06.038
de Menezes ACK (2013) Improving predictions of operational energy performance through better estimates of small power consumption. Anna Carolina Kossmann de Menezes
Deuble MP, de Dear RJ (2012) Green occupants for green buildings: the missing link? Build Environ 56:21–27. doi:10.1016/j.buildenv.2012.02.029
Dick JB, Thomas DA (1951) Ventilation research in occupied houses. J Inst Heat Vent Eng 19:279–305
Duarte C, Van Den Wymelenberg K, Rieger C (2013) Revealing occupancy patterns in an office building through the use of occupancy sensor data. Energy Build 67:587–595

Dubrul C (1988) Inhabitant behaviour with respect to ventilation—a summary report of IEA Annex VIII
Dutton S, Shao L (2010) Window opening behaviour in a naturally ventilated school. In: Proceedings of 4th National Conference of IBPSA-USA, pp 260–268
Emery AF, Kippenhan CJ (2006) A long term study of residential home heating consumption and the effect of occupant behavior on homes in the Pacific Northwest constructed according to improved thermal standards. Energy 31:677–693. doi:10.1016/j.energy.2005.04.006
Fabi V, Andersen RV, Corgnati S, Olesen BW (2012) Occupants' window opening behaviour: a literature review of factors influencing occupant behaviour and models. Build Environ 58:188–198
Fabi V, Andersen RV, Corgnati SP (2013) Influence of occupant's heating set-point preferences on indoor environmental quality and heating demand in residential buildings. HVAC&R Res 19:635–645. doi:10.1080/10789669.2013.789372
Farley J (1978) High rise apartment or detached house: what are the implications for children's behavior? New Dir Environ Des Res (EDRA 9 Proc) Wash Environ Des Res Assoc 145–161
Feng X, Yan D, Hong T (2015) Simulation of occupancy in buildings. Energy Build 87:348–359
Foekema H, van Thiel L, Lettinga B (2007) Watergebruik thuis 2007 (Drinking water use at home 2007)
Fritsch R, Kohler A, Nygård-Ferguson M, Scartezzini J-L (1990) A stochastic model of user behaviour regarding ventilation. Build Environ 25:173–181
Gauthier S (2016) Investigating the probability of behavioural responses to cold thermal discomfort. Energy Build 124:70–78. doi:10.1016/j.enbuild.2016.04.036
Gram-Hanssen K (2005) Husholdningers elforbrug-hvem bruger hvor meget, til hvad og hvorfor?. Statens Byggeforskningsinstitut, SBi
Guerra Santin O, Itard L, Visscher H (2009) The effect of occupancy and building characteristics on energy use for space and water heating in Dutch residential stock. Energy Build 41:1223–1232. doi:10.1016/j.enbuild.2009.07.002
Gunay HB, O'Brien W, Beausoleil-Morrison I, Perna A (2014) On the behavioral effects of residential electricity submetering in a heating season. Build Environ 81:396–403. doi:10.1016/j.buildenv.2014.07.020
Gunay HB, O'Brien W, Beausoleil-Morrison I, Gilani S (2016) Modeling plug-in equipment load patterns in private office spaces. Energy Build 121:234–249. doi:10.1016/j.enbuild.2016.03.001
Haldi F, Robinson D (2008) On the behaviour and adaptation of office occupants. Build Environ 43:2163–2177. doi:10.1016/j.buildenv.2008.01.003
Haldi F, Robinson D (2009) Interactions with window openings by office occupants. Build Environ 44:2378–2395. doi:10.1016/j.buildenv.2009.03.025
Haldi F, Robinson D (2010) Adaptive actions on shading devices in response to local visual stimuli. J Build Perform Simul 3:135–153. doi:10.1080/19401490903580759
Haldi F, Robinson D (2011) Modelling occupants' personal characteristics for thermal comfort prediction. Int J Biometeorol 55:681–694. doi:10.1007/s00484-010-0383-4
Herkel S, Knapp U, Pfafferott J (2008) Towards a model of user behaviour regarding the manual control of windows in office buildings. Build Environ 43:588–600
Hoes P, Hensen JLM, Loomans MGLC et al (2009) User behavior in whole building simulation. Energy Build 41:295–302. doi:10.1016/j.enbuild.2008.09.008
Hori S (1995) Adaptation to heat. Jpn J Physiol 45:921–946
Hunt DRG (1979) The use of artificial lighting in relation to daylight levels and occupancy. Build Environ 14:21–33. doi:10.1016/0360-1323(79)90025-8
IJspeert JK, De Waard PWT, Van den Berg T, De Jong P (1990) The intraocular straylight function in 129 healthy volunteers; dependence on angle, age and pigmentation. Vision Res 30:699–707
Inkarojrit V (2005) Balancing comfort: occupants' control of window blinds in private offices. University of California, Berkeley

Inkarojrit V (2008) Monitoring and modelling of manually-controlled Venetian blinds in private offices: a pilot study. J Build Perform Simul 1:75–89. doi:10.1080/19401490802021012
Inoue T, Kawase T, Ibamoto T et al (1988) The development of an optimal control system for window shading devices based on investigations in office buildings. ASHRAE Trans 94:1034–1049
Iwasaki K, Takahashi M, Nakata A (2006) Health problems due to long working hours in Japan: working hours, workers' compensation (Karoshi), and preventive measures. Ind Health 44:537–540
Iwashita G, Akasaka H (1997) The effects of human behavior on natural ventilation rate and indoor air environment in summer—a field study in southern Japan. Energy Build 25:195–205. doi:10.1016/S0378-7788(96)00994-2
Jensen JO (2004) Livsstil, boform og ressourceforbrug. Aalborg Universitetsforlag
Jian Y, Li Y, Wei S et al (2015) A case study on household electricity uses and their variations due to occupant behavior in Chinese apartments in Beijing. J Asian Archit Build Eng 14:679–686. doi:10.3130/jaabe.14.679
Johnson BM (1981) DBR INTERNAL REPORT NO. 464
Johnson T, Long T (2005) Determining the frequency of open windows in residences: a pilot study in Durham, North Carolina during varying temperature conditions. J Expo Sci Environ Epidemiol 15:329–349
Karava P, Stathopoulos T, Athienitis AK (2007) Wind-induced natural ventilation analysis. Sol Energy 81:20–30
Katzschner L, Bosch U, Röttgen M (2006) Behaviour of people in open spaces in dependence of thermal comfort conditions. In: Weber W (ed) Proceedings of 23rd Conference on Passive and Low Energy Architecture
Kawamoto K, Shimoda Y, Mizuno M (2004) Energy saving potential of office equipment power management. Energy Build 36:915–923. doi:10.1016/j.enbuild.2004.02.004
Kempton W, Feuermann D, McGarity AE (1992) "I always turn it on super": user decisions about when and how to operate room air conditioners. Energy Build 18:177–191. doi:10.1016/0378-7788(92)90012-6
Keul A, Salzmann R, Lehmden A (2011) Komfort und Luftqualität im Niedrigenergie-Ziegelgebäude. Mauerwerk 15:176–178. doi:10.1002/dama.201100497
Kondo S, Hokai S (2012) A model for predicting daily hot water consumption. In: Proceedings of the 5th IBPC, pp 867–873
Kvisgaard B, Collet PF (1990) The user's influence on air change. In: Air Change Rate and Airtightness in Buildings. ASTM International
Lasternas B, Zhao J, Yun R et al (2014) Behavior oriented metrics for plug load energy savings in office environment. In: 2014 American Council for an Energy-Efficient Economy (ACEEE) Summer Study on Energy Efficiency in Buildings, pp 160–172
Lerman S (1980) Radiant energy and the eye. Macmillan
Lindelöf D, Morel N (2006) A field investigation of the intermediate light switching by users. Energy Build 38:790–801. doi:10.1016/j.enbuild.2006.03.003
Lindsay C, Littlefair PJ (1992) Occupant use of venetian blinds in offices
Lobato C, Pless S, Sheppy M, Torcellini P (2011) Reducing plug and process loads for a large scale, low energy office building: NREL's Research Support Facility; Preprint
Love JA (1998) Manual switching patterns in private offices. Light Res Technol 30:45–50. doi:10.1177/096032719803000107
Lutz JD, Liu X, McMahon JE et al (1996) Modeling patterns of hot water use in households, 25
Macintosh A, Steemers K (2005) Ventilation strategies for urban housing: lessons from a PoE case study. Build Res Inf 33:17–31. doi:10.1080/0961321042000322771
Mahdavi A, Mohammadi A, Kabir E, Lambeva L (2008) Occupants' operation of lighting and shading systems in office buildings. J Build Perform Simul 1:57–65. doi:10.1080/19401490801906502
Maier T, Krzaczek M, Tejchman J (2009) Comparison of physical performances of the ventilation systems in low-energy residential houses. Energy Build 41:337–353

Majcen D, Itard L, Visscher H (2013) Actual and theoretical gas consumption in Dutch dwellings: what causes the differences? Energy Policy 61:460–471. doi:10.1016/j.enpol.2013.06.018

Maniccia D, Rutledge B, Rea MS, Morrow W (1999) Occupant use of manual lighting controls in private offices. J Illum Eng Soc 28:42–56. doi:10.1080/00994480.1999.10748274

Marino AJ, Fletcher EN, Whitaker RC, Anderson SE (2012) Amount and environmental predictors of outdoor playtime at home and school: a cross-sectional analysis of a national sample of preschool-aged children attending Head Start. Health Place 18:1224–1230

Ministerie of VROM (2009) Energiegedrag in de woning

Moorefield L, Frazer B, Bendt P (2008) Office plug load field monitoring report

Moosmann C (2014) Visueller Komfort und Tageslicht am Bueroarbeitsplatz: Eine Felduntersuchung in neun Gebaeuden. KIT Scientific Publishing

Mora D, Carpino C, De Simone M (2015) Behavioral and physical factors influencing energy building performances in Mediterranean climate. Energy Procedia 78:603–608. doi:10.1016/j.egypro.2015.11.033

Morgan C, de Dear RJ (2003) Weather, clothing and thermal adaptation to indoor climate. Clim Res 24:267–284

Müller A, Biermayr P, Kranzl L et al (2010) Heizen 2050: Systeme zur Wärmebereitstellung und Raumklimatisierung im österreichischen Gebäudebestand: Technologische Anforderungen bis zum Jahr 2050

Murdoch JB (2003) Illuminating engineering: from Edison's lamp to the LED. Visions Communications

Newsham GR (1994) Manual control of window blinds and electric lighting: implications for comfort and energy consumption. Indoor Environ 6:135–144. doi:10.1177/1420326X9400300307

Newsham GR, Aries MBC, Mancini S, Faye G (2008) Individual control of electric lighting in a daylit space. Light Res Technol 40:25–41

Nicol F (2001) Characterising occupant behavior in buildings: towards a stochastic model of occupant use of windows, lights, blinds heaters and fans. In: Seventh International IBPSA Conference, 1073–1078

Nordman B, Piette MA, Kinney K (1996) Measured energy savings and performance of power-managed personal computers and monitors

O'Brien W, Gunay HB (2014) The contextual factors contributing to occupants' adaptive comfort behaviors in offices—a review and proposed modeling framework. Build Environ 77:77–87

Page J, Robinson D, Morel N, Scartezzini J-L (2008) A generalised stochastic model for the simulation of occupant presence. Energy Build 40:83–98

Papakostas KT, Sotiropoulos BA (1997) Occupational and energy behaviour patterns in Greek residences. Energy Build 26:207–213

Peeters L, der Veken J, Hens H et al (2008) Control of heating systems in residential buildings: current practice. Energy Build 40:1446–1455

Polinder H, Schweiker M, van der Aa A et al (2013) Occupant behavior and modeling

Reinhart CF, Voss K (2003) Monitoring manual control of electric lighting and blinds. Light Res Technol 35:243–258. doi:10.1191/1365782803li064oa

Ren X, Yan D, Wang C (2014) Air-conditioning usage conditional probability model for residential buildings. Build Environ 81:172–182. doi:10.1016/j.buildenv.2014.06.022

Rijal HB, Tuohy P, Humphreys MA et al (2007) Using results from field surveys to predict the effect of open windows on thermal comfort and energy use in buildings. Energy Build 39 (7):823–836

Rijal HB, Tuohy P, Nicol F et al (2008a) Development of an adaptive window-opening algorithm to predict the thermal comfort, energy use and overheating in buildings. J Build Perform Simul 1:17–30

Rijal HB, Tuohy PG, Nicol F et al (2008b) Development of adaptive algorithms for the operation of windows, fans and doors to predict thermal comfort and energy use in Pakistani buildings. ASHRAE Trans 114:555–573

Roberson JA, Homan GK, Mahajan A et al (2002) Energy use and power levels in new monitors and personal computers

Rubin AI, Collins BL, Tibbott RL (1978) Window blinds as a potential energy saver: a case study. US Department of Commerce, National Bureau of Standards

Schakib-Ekbatan K, Zakici FZ, Schweiker M, Wagner A (2015) Does the occupant behavior match the energy concept of the building? Analysis of a German naturally ventilated office building. Build Environ 84:142–150. doi:10.1016/j.buildenv.2014.10.018

Schiavon S, Lee KH (2013) Dynamic predictive clothing insulation models based on outdoor air and indoor operative temperatures. Build Environ 59:250–260. doi:10.1016/j.buildenv.2012.08.024

Schweiker M (2010) Occupant behaviour and the related reference levels for heating and cooling. Tokyo City University, Japan

Schweiker M, Shukuya M (2009) Comparison of theoretical and statistical models of air-conditioning-unit usage behaviour in a residential setting under Japanese climatic conditions. Build Environ 44:2137–2149

Schweiker M, Shukuya M (2010a) A critical review on thermal factors as predictors for occupant behaviour: towards a purpose-rank based model of reference levels. In: Windsor Conference 2010

Schweiker M, Shukuya M (2010b) Comparative effects of building envelope improvements and occupant behavioural changes on the exergy consumption for heating and cooling. Energy Policy 38:2976–2986

Schweiker M, Wagner A (2015) A framework for an adaptive thermal heat balance model (ATHB). Build Environ 94:252–262

Schweiker M, Wagner A (2016) The effect of occupancy on perceived control, neutral temperature, and behavioral patterns. Energy Build 117:246–259

Schweiker M, Brasche S, Bischof W et al (2012a) Development and validation of a methodology to challenge the adaptive comfort model. Build Environ 49:336–347

Schweiker M, Haldi F, Shukuya M, Robinson D (2012b) Verification of stochastic models of window opening behaviour for residential buildings. J Build Perform Simul 5:55–74

Schweiker M, Brasche S, Hawighorst M et al (2014) Presenting LOBSTER, an innovative climate chamber, and the analysis of the effect of a ceiling fan on the thermal sensation and performance under summer conditions in an office-like setting. Proceedings of 8th Windsor conference: counting the cost of comfort in a changing world Cumberland Lodge. Windsor, UK, pp 924–937

Schweiker M, Hawighorst M, Wagner A (2016) The influence of personality traits on occupant behavioural patterns. Energy Build 131:63–75. doi:10.1016/j.enbuild.2016.09.019

Socolow RH (1978) The twin rivers program on energy conservation in housing: Highlights and conclusions. Energy Build 1:207–213. doi:10.1016/0378-7788(78)90003-8

Stephens B, Siegel JA, Novoselac A (2011) Operational characteristics of residential and light-commercial air-conditioning systems in a hot and humid climate zone. Build Environ 46:1972–1983. doi:10.1016/j.buildenv.2011.04.005

Stevenson A (2010) Oxford dictionary of English, 3rd edn. Oxford University Press

Sumpi N, Amukugo HJ (2016) A conceptual framework as basis for the development of a psychosocial educational programme to facilitate the reintegration of incarcerated women who had dumped babies and/or committed infanticide. Int J Health 4:102–111

Takahashi I, Saito M, Matsuoka H et al (2000) Difference in thermal sensation and behavioral pattern of occupants between passive and active cooling strategies. In: Proceedings of PLEA 2000, Cambridge, UK 593–598

Tetlow RM, van Dronkelaar C, Beaman CP et al (2015) Identifying behavioural predictors of small power electricity consumption in office buildings. Build Environ 92:75–85. doi:10.1016/j.buildenv.2015.04.009

Toftum J (2010) Central automatic control or distributed occupant control for better indoor environment quality in the future. Build Environ 45:23–28

Van De Kraats J, Van Norren D (2007) Optical density of the aging human ocular media in the visible and the UV. JOSA A 24:1842–1857
Van Den Wymelenberg K (2012) Patterns of occupant interaction with window blinds: a literature review. Energy Build 51:165–176
van der Lans AA, Hoeks J, Brans B et al (2013) Cold acclimation recruits human brown fat and increase nonshivering thermogenesis. J Clin Invest 3395–3403
van Dongen JEF (2004) Occupant behaviour and attitudes with respect to ventilation of dwellings
Van Meel J (2000) The European office: office design and national context. 010 Publishers
Vine E, Diamond R, Szydlowski R (1987) Domestic hot water consumption in four low-income apartment buildings. Energy 12:459–467. doi:10.1016/0360-5442(87)90005-3
Webber CA, Roberson JA, McWhinney MC et al (2006) After-hours power status of office equipment in the USA. Energy 31:2823–2838. doi:10.1016/j.energy.2005.11.007
Winn B, Whitaker D, Elliott DB, Phillips NJ (1994) Factors affecting light-adapted pupil size in normal human subjects. Invest Ophthalmol Vis Sci 35:1132–1137
Yao R, Steemers K (2005) A method of formulating energy load profile for domestic buildings in the UK. Energy Build 37:663–671
Yun GY, Steemers K (2008) Time-dependent occupant behaviour models of window control in summer. Build Environ 43:1471–1482. doi:10.1016/j.buildenv.2007.08.001
Yun GY, Steemers K (2011) Behavioural, physical and socio-economic factors in household cooling energy consumption. Appl Energy 88:2191–2200. doi:10.1016/j.apenergy.2011.01.010
Yun GY, Steemers K, Baker N (2008) Natural ventilation in practice: linking facade design, thermal performance, occupant perception and control. Build Res Inf 36:608–624
Yun GY, Tuohy P, Steemers K (2009) Thermal performance of a naturally ventilated building using a combined algorithm of probabilistic occupant behaviour and deterministic heat and mass balance models. Energy Build 41:489–499. doi:10.1016/j.enbuild.2008.11.013
Yun R, Lasternas B, Aziz A et al (2013) Toward the design of a dashboard to promote environmentally sustainable behavior among office workers. In: Berkovsky S, Freyne J (eds) Persuasive Technology: 8th International Conference, PERSUASIVE 2013, Sydney, NSW, Australia, April 3–5, 2013. Proceedings. Springer, Berlin, pp 246–252
Zhang Y, Barrett P (2012a) Factors influencing the occupants' window opening behaviour in a naturally ventilated office building. Build Environ 50:125–134. doi:10.1016/j.buildenv.2011.10.018
Zhang Y, Barrett P (2012b) Factors influencing occupants' blind-control behaviour in a naturally ventilated office building. Build Environ 54:137–147. doi:10.1016/j.buildenv.2012.02.016

Chapter 3
Designing Research

David Shipworth and Gesche M. Huebner

Abstract The aim of this chapter is to set out a process that researchers can follow to design a robust quantitative research study of occupant behavior in buildings. Central to this approach is an emphasis on intellectual clarity around what is being measured and why. To help achieve this clarity, researchers are encouraged to literally draw these relationships out in the form of a concept map capturing the theoretical model of the cause and effect between occupant motivations and energy use. Having captured diagrammatically how the system is thought to work, the next step is to formulate research questions or hypotheses capturing the relationship between variables in the theoretical model, and to start to augment the diagram with the measurands (things that can actually be measured) that are good proxies for each concept. Once these are identified, the diagram can be further augmented with one or more methods of measuring each measurand. The chapter argues that it is necessary to carefully define concepts and their presumed relationships, and to clearly state research questions and identify what the researcher intends to measure before starting data collection. The chapter also explains the ideas of reliability, validity, and uncertainty, and why knowledge about them is essential for any researcher.

3.1 Introduction

The aim of this chapter is to set out a process that researchers can follow to design a robust quantitative research study on occupant behavior in buildings. The material is introductory, and is intended to provide an overarching framework for thinking about the research design process. It is not sufficient in itself, but refers to other

D. Shipworth (✉) · G.M. Huebner
University College London, Energy Institute, 14 Upper Woburn Place,
London WC1H 0NN, UK
e-mail: d.shipworth@ucl.ac.uk

G.M. Huebner
e-mail: g.huebner@ucl.ac.uk

A. Wagner et al. (eds.), *Exploring Occupant Behavior in Buildings*,
DOI 10.1007/978-3-319-61464-9_3

chapters in this book and to other more detailed sources of information on specific elements. Whilst this chapter is necessarily highly abridged and incomplete in many areas, it should steer the reader away from some of the main errors and misunderstandings to which the field as a whole is prone.

It is important to note that this chapter takes a broadly quantitative social and physical realist approach to researching occupant influences on energy demand in buildings. This arises from this book's origin—to improve the representation of building occupants within building energy simulation models, which are themselves quantitative and realist in their representation of the world. The aim is therefore to establish relationships (ideally causative ones) between the external environment, the building and its internal environment, occupant behavior, and building energy consumption.

Taking a realist approach means that there are occupants' actions that directly affect energy demand in buildings, and that these actions can be explained in part through the use of concepts that are independent of the researcher and the individual occupants themselves. Concepts are central to research design and will be discussed throughout this chapter. A concept can be thought of as an abstract idea that captures the central elements of what it refers to. Examples of concepts include temperature, comfort, glare, environmental attitudes, financial costs, etc. A realist approach takes the view that while these concepts can never be measured perfectly (i.e., without any error) they can be measured and used to—imperfectly and incompletely—predict occupant behavior. The chapter therefore does not take a solely social constructivist approach of saying that occupant influences on energy use in buildings are purely a construct of human social processes with no meaning or existence outside of the individual occupants engaged in them. It thus places this work more in the context of such academic disciplines as physics, psychology, quantitative social science, and behavioral economics, and less in such academic disciplines as qualitative sociology, anthropology, and ethnography.

Central to the approach taken here is an emphasis on intellectual clarity around what is being measured and why, and literally drawing this out in the form of a theoretical model of the cause and effect relationships between occupant motivations and energy use in the form of a concept map. Having captured diagrammatically how the system is thought to work, the next step is to formulate research questions or hypotheses capturing the relationship between variables in the theoretical model, and to start to augment the diagram with the measurands (things that can actually be measured) that are good proxies for each concept. Once these are identified, the diagram can be further augmented with one or more methods of measuring each measurand. In research adopting a realist approach, be it qualitative or quantitative, it is necessary to carefully define concepts and their presumed relationships, and to clearly state research questions and what the researcher intends to measure before starting data collection. Some forms of analysis are only applicable to quantitative research approaches, such as quantifying reliability; however, an awareness of the ideas of reliability, validity, and uncertainty is essential for any researcher.

This chapter is a synthesis of descriptions of the research and model building approaches used in both the physical and social sciences. Casti (1992) defined a mathematical model of a system as "... the specification of observables describing such a system and a characterization of the manner in which these observables are linked" (p. 2). This is a useful starting point for a discussion on the process of designing a program of research to understand how occupant behavior influences energy demand in buildings. This definition has two key elements that relate to designing research: firstly, specifying what the measurement will be, i.e. the observables (how they are measured is the realm of research methods); and secondly, determining if the variables are causally related (this is one of the functions of research design). While Casti's definition is a useful starting point, it is insufficient. Another essential element is theory. As noted by Ruttkamp (2002), "The only way in which we can have scientific contact with the world...is through actions involving selection, abstraction, and generalisation, which are always executed within some theoretical framework or disciplinary matrix...." (p. 17). This third element, theory, allows making sense of the observables, and the relationships (links) between them, within an explanatory (i.e., theoretical) framework that permits transferring these insights between instances. These three elements, methods, research design, and theory, need to be brought together in order to design any program of research.

3.2 Why Do the Research (Research Aims and Questions)

Determining a good research aim or research question is an essential and often neglected first step in the research process. Bouma (2000) distinguishes between when research *aims* are appropriate, and when research *questions* and/or hypotheses are. Where research is exploratory or descriptive, then a research aim is appropriate. When research is more explanatory or seeking to establish causation, then research questions and hypotheses are appropriate. In the context of occupant behavior in buildings, both forms of research are common, although descriptive research predominates. Unlike a research question, which specifies relationships between two or more concepts, a research aim describes a more general area of enquiry, and leaves open greater scope for exploratory data analysis through looking for patterns between different elements of the data collected. It needs to be remembered, however, that such descriptive or exploratory work can only be used to describe correlations between the concepts measured. If establishment of causation is desired, subsequent, more experimentally based work needs to be conducted.

Examples of research questions in the context of occupant behavior:

> "Do occupants open windows more frequently as CO_2 levels rise?" or "Do occupants tilt blinds when sun shines directly on their computer monitor?"

As Bouma (2000) notes, a good research question postulates relationships between two concepts and facilitates the process of designing a research study to answer that question. Two separate aspects need to be addressed. The first aspect relates to the things that are measured (for example, CO_2 levels and window opening). These concepts need to be operationalized in ways that allow measuring them—for example, in the case of CO_2 and window opening, using appropriate sensors or through observations. The second aspect relates to the nature of the relationship between the concepts. The capacity to determine whether the relationship is correlational or causal is determined by the choice of research design. This is discussed in detail below.

A well-framed research question makes the construction of hypotheses far easier. Hypotheses are appropriate in cases where a quantified measure of confidence in the answer is desired, and take the form of a statement (a declarative sentence) of what the researcher expects to happen.

Possible research questions and resulting hypotheses: A research question such as, "Do occupants open windows more frequently as CO_2 levels rise?" may give rise to a range of hypotheses such as, "As CO_2 levels rise (hypothesized cause) occupants will open windows more frequently (hypothesized effect)"; "As CO_2 levels rise occupants will open windows for longer periods of time"; and/or "As CO_2 levels rise occupants will open windows wider".

It is typical for one research question to give rise to many hypotheses, as hypotheses need to be sufficiently specific to be measurable without ambiguity. This usually takes the form of a measure of statistical confidence between the data gathered and the theoretical model of occupant behavior and energy outcomes being explored. In such hypotheses testing, it is usually a measure of the lack of fit between the measured data and the inverse of the hypotheses—the null hypothesis—that is used. The null hypothesis is the embodiment of scientific skepticism; it assumes that there is no relationship between the things being measured, here CO_2 levels and window opening, and only accepts that there is one if there is enough evidence to reject this conservative assumption. Many introductory textbooks have been written about statistical hypotheses testing and with this has come a degree of standardization of key parameters, such as the levels of confidence needed (frequently a p-value of 0.05, i.e., 95% confidence, is cited). Statistical confidence is a measure of how confident one is that the study can correctly determine if an intervention failed to work. To be 95% confident means that there is only a one in twenty chance of a false positive finding, i.e., saying the intervention worked when in fact it did not. This is also called a "type I error". Statistical power is the converse of this. It is a measure of how confident one is that the study can correctly determine if an intervention worked. To have 80% statistical power means that there is only a one in five chance of a false negative finding, i.e., saying the intervention did not work when in fact it did. This is also called a "type II" error. In the energy in buildings area both of these are important. Neither incurring the costs of energy savings measures that are ineffective (type I error), nor discarding interventions that work (type II error) would be a good outcome.

For research on the influences of occupant behavior on energy demand in buildings it is important to realize that such high levels of statistical confidence may or may not be appropriate, and the reader is referred to the recent pronouncements by the American Statistical Association (Wasserstein and Lazar 2016) for a more rounded discussion of this topic.

> Example of the differing effect of statistical confidence: A 1:20 (5%) chance of having incorrectly counted occupants leaving a building in the event of a fire is inappropriately low (in a building of 100 occupants this could leave 5 trapped inside)—whereas requiring only a 1:20 (5%) chance of incorrectly identifying the number of people in a room for the purposes of estimating fresh air volumes is inappropriately high (only an approximate estimate of occupancy is needed to adjust fresh air volumes appropriately). In each case, the appropriate levels of statistical confidence and statistical power need to be assessed against the risks of making each type of error and the costs involved in reducing them.

3.3 Identifying the Concepts to Measure and How They Link Together (Theory)

One of the most important elements in the process of identifying concepts to measure and how they link together is the drawing out of a theoretical model. This step should be undertaken both when doing more descriptive work based on exploratory data analysis and when seeking to understand cause and effect using research questions and hypotheses. It should take the form of a diagram of concepts and links showing how they are related. There are many software packages in which such a theoretical model can be drawn, one of the most useful is Cmap, a free, dedicated concept mapping software package (Novak and Cañas 2006). The advantage of using concept mapping software is that it allows the labeling of concepts, as well as links between concepts, thus creating a map of how different factors interact. This theoretical model can either be one that reflects a *Theory* [i.e., an established theory that is to be tested in a specific context like the Theory of Planned Behavior (Ajzen 1991)] or a *theory* (i.e., the researcher's own mental model of how occupant behavior influences energy use in buildings).

In constructing this map, it is important that as many causal steps as possible be included. For example, the model might link occupant thermal comfort to home energy demand. This could be as simple as: occupant (cold) thermal discomfort ——(leads to)——> occupants turning up the thermostat——(leads to)——> greater energy use. This is intuitively reasonable, but it makes a lot of assumptions: how occupants will respond to cold thermal discomfort (through adjusting the thermostat); the home (that changing temperatures at the thermostat changes the temperature where the occupant is); the thermostat (that it is connected and working); and the boiler and heating system (that it can deliver the heat output necessary to raise the temperature where the occupant is). It is important that as many of these assumptions and causal links as possible be expanded in the theoretical model to allow the researchers to decide what to measure along the causal chain and to understand if they do not find a relationship between their primary variables of interest (say, thermal comfort and home energy use) that they are aware that the breakdown in the causal chain can be anywhere along it, and it is not just that occupants do not act as expected.

Ideally, the model would go from occupant motivations through to energy use. This roots the model in psychological, social, or physiological drivers, and explains how these are translated through occupant behavior and interaction with (or in reaction to) elements of the building to changes in energy and power use. Such rooting of the causal model in occupant motivations helps in identifying potential points of intervention with occupants to change how they respond to (or interact with) the building, while the modeling of the building's response to this interaction allows testing of whether the assumptions about the building controls and physics are as imagined.

An example of such a theoretical model represented in the Cmap software is provided in Fig. 3.1. This is based on the Theory of Planned Behavior, one of the

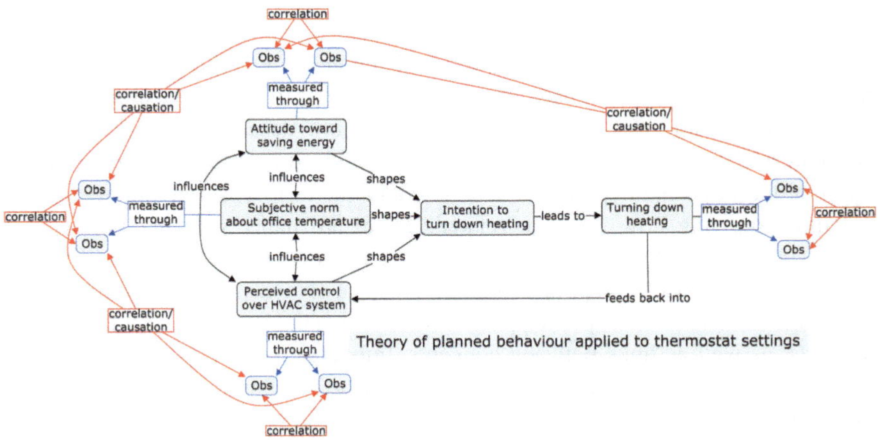

Fig. 3.1 Graphical representation of the theory of planned behavior. *Black boxes* and links represent the established theory. *Blue boxes* and links represent measurable properties. *Red boxes* and links represent analytical relationships for testing validity

most well-known and tested theories in psychology to understand the antecedents of behavior. It postulates that the attitude toward a behavior, subjective norms, and perceived behavioral control shape an individual's behavioral intentions and, ultimately, their behavior.

3.3.1 Concepts

In both the physical and social sciences, concepts play the important role of being the thing that researchers are frequently trying to measure. In the area of occupant influences on energy use in buildings, concepts range from social norms of behavior, through thermal/aural/visual comfort, to temperature/sound pressure levels/illuminance levels, to building management systems and heating, ventilation, and air conditioning (HVAC) systems, and to energy, power and carbon emissions. It may seem alien to link together things as seemingly disparate and diffuse as social norms with things as apparently concrete as temperature and energy—but this is only because the latter have been reified (i.e., made concrete through an agreed process of measurement) and their methods of definition and measurement so widely accepted that it has been forgotten that they were once as ill-defined and vague.

Put simply, concepts are those things researchers are usually *trying* to measure. They are not, however, usually the things that are actually measured because it is usually only possible to measure proxies to concepts. This is why concepts and variables are not the same thing. This will be discussed further in Sect. 3.6. The theoretical model should have concepts as its nodes, with such concepts connected by a series of links indicating the relationships between the concepts.

3.3.2 One to One Relationships (Links)

In the theoretical model, links describe how it is thought that concepts influence each other. While not necessary, it is often useful to attach signs to these links indicating whether the relationship is thought to be positive (+), negative (−), or unknown/variable (?). Doing this makes constructing research questions and hypotheses simple, as they are then just verbal descriptions of the relationships between concepts in the model. In the example discussed above regarding window opening behavior, the theoretical model may have a link between CO_2 levels and window opening. This could be translated into the research question, "Do occupants open windows more frequently as CO_2 levels rise?" If the theoretical model labeled the relationship with a +, this would give rise to the hypotheses: "As CO_2

levels rise (hypothesized cause) occupants will open windows more frequently (hypothesized effect)" and "As CO_2 levels rise occupants will open windows wider". Thus, the sign on the link in the theoretical model indicates the expected relationship between the concepts.

It is important to note that there can be many more links in a model than nodes (concepts), as each node can have links to many others. That said, it is important not to end up linking each node to every other node, as that conveys little information—it merely says, "everything is connected to everything else". Believing that each node should be connected to every other node can arise from two issues. Firstly, this can arise through including temporal relations (i.e., feedback loops) in the model. One variable can influence another in the short term, but the second variable can then influence the first, either directly, or indirectly, at a different timescale. It is often important to define the time scale of interest and exclude feedback processes that occur over longer or shorter periods. The second reason "everything is connected to everything else" models arise is because the concepts used are not defined precisely enough–this is a question of scope, and it may be that for the purposes of a study intermediary steps are out of scope.

3.3.3 One to Many Relationships (Hierarchies)

Another major form of relationship to be aware of when constructing the theoretical model is hierarchies. This is where concepts have a natural nested structure. When studying occupants in buildings such hierarchies are rife—people within offices, offices within premises, premises within buildings, buildings within companies, etc. Identification of these relationships is important, as it will influence the unit of analysis (the entity on which data are collected) and the definition of the target population (the group from which the sample is drawn), and will inform the sort of analysis run on the data. If, for instance, the assumption is that the actions of occupants are strongly shaped by the building they are in, then it would not make sense to draw a sample of 1000 people from only four buildings and think they constitute a representative sample of the population. One thousand people is usually sufficient to be a statistically representative sample of a large population, but only where those people are independent and sampled at random from the whole population. In this case the sample consists just of four occupant-building combinations. If assuming that buildings do not influence occupant actions then such a sample of 1000 is fine, but if assuming that buildings strongly shape occupant actions then the population should be of buildings, and a sample size of four buildings is very inadequate. This illustrates why having a representation of hierarchy in the theoretical model matters for the research design.

3.4 Units of Analysis, Populations, and Scope

Having established the theoretical model, the next step is to delineate the scope of its applicability. This will require a clear statement of population of interest, i.e., the population of units of analysis the theoretical model is supposed to represent. This is likely to be as constrained by the resources available for the study as by what the researcher would theoretically like to represent. The geographical scope of applicability, along with the temporal scope need to be defined—both will tell those using the study where and when the results are no longer applicable. Finally, the required degree of precision needs to be decided, which will determine the sampling strategy and the sample size required. Each of these concepts is discussed in turn in the sections that follow.

3.4.1 Units of Analysis

The unit of analysis is the thing that data are collected about. In the context of occupant behavioral impacts on energy demand in buildings this can be quite a range of units: from companies, through campuses, to buildings, premises, floors, individual offices, down to individual occupants. For domestic buildings, the unit of analysis may be homes, rooms, or individual occupants. The challenge in this area is that there are strong hierarchical relationships between these levels, and so the behavior of the same individuals in different buildings may vary more than the behavior of different individuals in the same building. Where this is the case, then it is probably more appropriate to think of the building as being the unit of analysis.

3.4.2 Population of Interest and Scope

Where the building is the unit of analysis, then those characteristics of the building that shape occupants' influence on energy demand help define the population of buildings that the findings of the study apply to. For example, if occupant behavior in naturally ventilated buildings with high thermal mass is studied, the population may be these buildings, and thus the sampling strategy needs to sample from a population of such buildings to generate generalizable results. The limits of *where* (geographical scope) and *when* (temporal scope) the findings would apply need to be defined. The geographical scope could be determined by external conditions ranging from climate regions to the extent of external pollution (a factor influencing window opening behaviors). Hence, the findings may be restricted just to naturally ventilated buildings in temperate climates with low levels of external pollution. There may also be temporal limitations, either seasonal (results only applying in spring, summer and autumn) or in terms of a specific longevity (results only applying for the next decade due to expected changes in technology or society).

3.4.3 Descriptive or Inferential Statistics

The overwhelming majority of statistical work in the buildings field is classed as descriptive statistics. Descriptive statistics report on the statistical characteristics of the data gathered. If the study is of 100 buildings, then descriptive statistics describe those 100 buildings. Examples are reports on frequencies (e.g., counting how many double-glazed windows are present) or correlations between variables (e.g., windows opened for longer when ambient temperatures increase). The common element being that the findings only relate to the units of analysis studied, and nothing can be said about whether the findings apply more generally.

Inferential statistics, on the other hand, seek to make statements about things that have not been studied directly. Inferential statistics, also called inductive statistics, describe the statistical characteristics of similar unobserved buildings, such as the population from which the units of analysis (e.g., occupants/buildings) were drawn. To be able to say something about a population by studying a sample it is necessary to know how well the sample represents the population. This is the field of sampling and sample size calculations. Whilst not specific to experimental research, sampling and sample size calculations are crucial in any experimental research design.

To recap, inferential statistics is the method to make inferences from the collected data to more general conditions. It is what is commonly described when using statistical measures such as confidence intervals and p-values for research findings.

Confidence intervals are a function of the sampling error (also known as "standard error") and depend on the size of the sample—the bigger the sample the smaller the sampling error. Confidence intervals express the range of values within which the parameter of interest (e.g., the mean) of the population from which the sample is drawn can be said to fall, based on that same parameter in the sample (e.g., the sample mean).

Modern statistical software has many advantages, but one of its disadvantages is that it will provide answers to questions without first testing whether the assumptions on which those answers are based have been met.

For example, before calculating and reporting confidence intervals for findings can be meaningful, specific assumptions must hold. These include that the standard deviation of the population is known (not just that of the sample); that each member of the sample was randomly and independently selected from the population; and that the sample is (or can be transformed to be) approximately normal. Where these assumptions do not hold, calculation of confidence intervals is still possible, but requires changing the default settings in most statistical software. For example, if the standard deviation of the population cannot be determined from other published statistics and if the sample size is small, then the Student's t-distribution can be used instead of the z-distribution. This widens the confidence intervals and helps account for the uncertainty arising from using the sample standard deviation rather than that of the population. In the context of built environment studies these assumptions are frequently violated and non-standard approaches are needed.

It is important to remember that confidence intervals only represent one particular aspect—and frequently a fairly minor aspect—of the uncertainty that is inherent in the research process. Issues such as instrument accuracy and precision (discussed above) are not captured in the calculation of confidence intervals. It is an unfortunate reflection on contemporary academia that to quantify is to reify; the capacity to quantify one element of uncertainty (sampling error) is somehow thought to make it more real than other forms of uncertainty which, while less easy to quantify, are no less real and frequently far more important.

Determination of sample sizes for inferential statistics in a building occupancy study is challenging because of the hierarchical structure of the problem as discussed above. In order to understand this, a brief recap on some of the fundamental concepts of statistics is required. Inferential statistics, of which sample size calculations are a part, is about making the statements about a population based on a measured subsample of that population. All calculations of sample sizes are predicated on the assumption that there is a well-defined population, and that an unbiased sample from that population can be selected through a random selection process in which each member of the population has an equal probability of being selected into the sample. In practice, this is virtually impossible to do, and so judgment is called for in assessing the extent to which the way with which units of analysis were selected into the sample may bias the outcome.

The aim of any research should be to match the underlying assumptions of the statistical methods used; thus, the researcher should seek to clearly define their population of interest, and, wherever possible, to draw members from that population with equal probability. It is common to see comparisons of the descriptive statistics of a sample (i.e., reporting on house type, household size, income, other demographics) compared to those of a nationally representative survey, like a census, with authors reporting that because the sample looks like the census (usually through visual comparison of histograms) that the sample is representative. While this provides some reassurance, it is not strictly speaking correct—particularly in energy in buildings work. Usually such demographic factors explain only a limited share of the observed variance between households' energy consumption, and so some measure of demographic similarity does not necessarily translate into similar patterns of energy consumption. It is also worth noting that reporting values such as confidence intervals is also not meaningful or necessary when all members of the population are surveyed (i.e., in a census).

The choice between descriptive and inferential statistics is an important one that will fundamentally shape the research and the conclusions that can be drawn. While most researchers would like their findings to apply more generally, the work involved in doing so is considerable and so the decision to do so should not be taken lightly.

3.4.4 Required Precision

Of particular importance in this context is defining the precision with which the outcome variables need to be known for the findings to be relevant to the substantive problem being addressed. Precision, often called "reliability" in the social sciences, is a measure of how much spread there would be in the data if exactly the same thing were to be measured with the same instrument many times. It is different from accuracy, which is a measure of how well these measurements correspond with the true value. Most instruments have some level of imprecision (say, ±1 °C on a thermistor), which puts a fundamental limit on how precisely a measurement can be specified.

Precision is important because most interventions in buildings will be subject to some form of cost benefit analysis, with the intervention implemented if it can be shown that the benefits outweigh the costs. In this context, it is important to know in advance the likely costs, thus providing a prior estimate of the size of the benefits (energy savings, indoor air quality improvements, etc.) required for the intervention to be deemed worthwhile.

> Implications of precision: If the intervention is only expected to change, say, internal temperature by 1 °C, and temperature can only be measured to ±1 °C, then, without large numbers of measurements, it is unlikely to detect an effect of the intervention with that level of imprecision—a different instrument would need to be used (e.g., one that measures temperature to ±0.1 °C).

Similarly, it is important to determine the statistical confidence required of the findings. This will vary with context. If the objective is to publish in refereed journals, then 95% statistical confidence is frequently expected. If the objective is to decide between two alternate courses of action incurring similar costs, then statistical confidence greater than 50% (i.e., on the balance of probabilities) may be all that is required, depending on the balance of risks associated with false positive (type I) and false negative (type II) errors for each option.

To recap, false positive (type I) errors occur where the intervention being trialed did not actually work, but the study concluded that it did. The risk here is of implementing an intervention that does not work, thus wasting time and money. The more worried one is about this, the higher the level of statistical confidence needed. False negative (type II) errors occur where the intervention being trialed actually did work, but the study concluded that it did not. The risk here is of throwing out a good idea and missing out on potential improvements to the building. The more worried one is about this, the higher the level of statistical power needed.

In general, the higher the precision with which the results need to be known, the more expensive the trial will be. High costs could arise from the need to measure things more precisely or the need to reduce the uncertainty in generalizing to the population of interest, which will require a larger sample of the chosen units of analysis.

3.5 Sampling and Sample Size

3.5.1 *Sample Frames*

As discussed above, each study should specify the population to which the findings are thought to apply. Once this is specified, then if generalization from a sample to a population (inferential statistics) is to be used, a sample frame is needed from which to draw a sample. Factors identified in the theoretical model as influencing the outcome variable(s) of interest will need to be addressed (exemplified or nullified) in the construction of a sample frame. The sample frame is (ideally) a list of all units of analysis in the population. In some cases, depending on the unit of analysis, this may be difficult to obtain. Where such a list (sample frame) is available, then the sample is drawn from this list using the sampling strategy. Where such a list is not available, less statistically correct methods will need to be used such as quota sampling—for example, choosing a certain number of buildings in each of a range of categories that the theoretical model says will be important.

3.5.2 *Sampling Strategies*

There is a wide range of sampling strategies. These broadly divide into probability-based methods, which are needed for generalizing from the sample to the population, and non-probability sampling methods, which are often used for pragmatic and costs reasons.

Of the probability-based methods, the "gold standard" is pure random sampling. This is the ideal case, as every member of the population (as represented in the sample frame) has an equal probability of being included in the sample. It would amount to drawing the sample purely randomly from the sample frame, all chosen units consenting to being monitored and then monitoring them all with no missing data. It needs to be stressed that all inferential statistics are based on the assumption that the sample is drawn at random from the population and any deviation from this is a compromise of this most basic assumption on which inferential statistics is based.

Because pure random sampling is often both very difficult and very expensive, a range of alternative methods have been developed that are still statistically generalizable. A full description of such methods is beyond the scope of this chapter; examples include systematic sampling (sampling every *n*th member of the sampling frame, but starting at a random point between 1 and *n*, so each member has an equal probability of being sampled); stratified random (where a random sample is drawn from different strata of interest, e.g., low-, medium-, and high-rise buildings, or urban, sub-urban, and rural buildings, but with the proportions of the population reflected in the strata of the sample); and cluster sampling (where groups of co-located members of the population are selected, e.g., ten buildings in each of five cities).

The best of non-probabilistic sampling methods is quota sampling, where a set of important criteria drawn from the theoretical model are identified and units of analysis selected on a first come first served basis until a quota is reached in each cell of the sample frame. For example, in a study of occupants and their adaptive responses to thermal comfort, Gauthier and Shipworth (2015) used a sample frame of age, weight and gender, and recruited people (the unit of analysis) to populate that frame.

The second most robust is purposive sampling, in which population members are recruited based on certain characteristics considered useful to the study. This may vary from deliberate selection of extreme cases to get a sense of the breadth of possible responses; to heterogeneous sampling, i.e., taking a spread of participants to cover the whole range of possible responses; to homogenous sampling in which some forms of variance are deliberately excluded through selection of a sample; to critical case, or typical case sampling. Other, less robust forms of sampling include snowball (where participants recommend others they know to participate); self-selection (the widely used practice of allowing people to volunteer, or opt-into a trial); and convenience (where trial participants are based on whoever is to hand—hence the proliferation of studies of people and buildings on university campuses!). Each of these methods carries significant "health warnings" to the robustness of the trial, with all three methods having the potential to introduce significant biases into the results.

3.5.3 *Spatial Sampling*

Spatial sampling varies from the geographic dispersal of research subjects with the population ranging from local to global, through to the spatial density of deployment of sensors collecting environmental variables in an occupied space. In both cases, the required density of sampling depends on the rate of change of the variable of interest in space and on the sensitivity of the other variables in the theoretical model being used for the research design to changes in those variables. In many instances, existing standards or established models will provide guidance on such sensitivities. For example, thermal comfort, as represented in the predicted mean vote (PMV) model, is far more sensitive to changes in ambient temperature than it is to changes in relative humidity. Thus, even if both ambient temperature and relative humidity were to change at equal rates in the space, it would not be necessary to sample relative humidity as frequently. Spatial sampling is conceptually similar to any other form of sampling (population or temporal), where the factors driving the size of the sample are the effect size the researcher is trying to measure (what magnitude of change is considered worthwhile detecting) and the variance in the space (how much different locations vary from each other). If measuring a variable that varies a great deal, or if the theoretical model is thought to be very sensitive to that variable, or if trying to detect a small change in the outcome variable of interest of the model, then a larger sample is needed.

Sample size calculators can be used to determine spatial sample sizes, but this is seldom done for a range of reasons. Firstly, spatial data are usually highly spatially auto-correlated, i.e., the value of a variable in two adjacent points in space is likely to be pretty similar. Secondly, usually there is good prior knowledge of how a variable is likely to change in space both inside and outside buildings—particularly environmental variables such as temperature and light levels. Thirdly, the units of analysis (people, buildings, etc.) are seldom randomly distributed within the geographic scope of the study. All of this, coupled with the expense and impracticality of monitoring a large number of physical locations, makes purposive sampling both more acceptable and more pragmatic. For most studies, the aim is to measure variables experienced by the unit of analysis; hence, environmental parameters are best measured where the units of analysis (e.g., people or buildings) are located. Doing this reduces the uncertainty that arises from having to estimate these values from data collected in another time and place. This is the basis of the so-called "right here right now" approach to gathering thermal comfort data. While sampling at the unit of analysis is ideal, there are often times when it is not practical, and instead sampling is done at fixed points in the environment. This could be by using secondary weather station data, or by monitoring values inside buildings at fixed heights and locations away from people.

> Qualitative rules in determining a sensor strategy: firstly, it is important to estimate the accuracy and precision with which each variable needs to be known (see "required precision" above). Accuracy differs from precision in that it refers to any systemic bias in the readings. In physical monitoring an example would be a poorly calibrated sensor which is always reading above or below the "true" value. In psychology, it may arise from a psychological trait such as centrality bias (where people tend to avoid picking the end values of scales). A sensor located away from the unit of analysis may well record values that are consistently different from those at the unit of analysis. It is important to think through how large a difference is tolerable before the

findings are no longer fit for purpose. Secondly, as discussed above, it is important to consider how much imprecision is acceptable. The greater the imprecision in the measurements, the less likely it is to find statistically significant results. Imprecision clouds data with noise, making the signal harder to detect. If trying to find a small signal (for instance, a weak influence of occupant behavior on energy use in buildings), then as much precision as possible is needed in the measurements. The final issue to consider is under-specification of the measurand. This is addressed in Sect. 3.6.2.

3.5.4 *Temporal Sampling*

The principles of temporal sampling are similar to those of spatial sampling, except applied in the time dimension. Again, rate of change is the key determinant, along with the sensitivity of the variables of interest to that change, and the response-time of the system. It may not be necessary to frequently sample a variable that changes rapidly, where it is acting on a system that changes slowly or where the outcome variables of interest in the theoretical model are comparatively insensitive to that variable. Conversely, if the variable changes rapidly, and the system and its outcome variable of interest is responsive to that change, then sampling at high frequency may be required.

As with spatial variability, temporal variability can be highly auto-correlated, i.e., values of a variable sampled closely in time can be very similar. For this reason, temporal sampling rates will primarily be driven by the rate at which the variable is thought to change. An additional element to add to the concept map of the theoretical model is an a priori estimate of the rate of change of each variable to be measured. This can be based on previous studies or preliminary fieldwork/pilot studies. The second factor that determines the sampling rate is the characteristic timescale of change of the system. Nicol et al. (2012) argue that there is no point in taking comfort votes from people at intervals of less than half an hour because for practical purposes their comfort state does not change sufficiently between such intervals to warrant it. While this may or may not be true, if the objective of the study and the theoretical model are consistent with this, then there would be little point measuring data at higher temporal frequencies unless assessing this claim was part of the objectives of the study.

While such rules of thumb can be used to determine regular temporal sampling rates for most studies, there are instances where different temporal sampling strategies are appropriate. This becomes particularly apparent in wireless sensor networks where minimizing energy use by sensors can be critical. Here, more sophisticated sampling rates can be used such as variance-based sampling. Such approaches vary the rate of sampling in proportion to the rate of change of the variable of interest. When the variable is static or changing slowly, then sampling can be quite infrequent. When the variable is changing rapidly, then the sensor can increase the rate of collecting and transmitting data to capture the additional information when it is useful. These sampling strategies are currently under development in computer science—those considering using them would need to liaise with their sensor developers to implement such strategies (see also Chap. 4). It is also necessary to determine the thresholds at which the sampling rates should change; this is frequently expressed as a change in the variable relative to the historical observed range of variance for each variable.

Other bases for determining sampling rates include matching or replicating other studies in the field to ensure comparability, sampling as frequently as battery/memory/financial constraints will allow (a conservative strategy given it is always possible to down-sample to lower frequencies, if desired), and adaptive designs in which sampling is initially high, but is reduced after preliminary data analysis if the rate is in excess of requirements.

3.5.5 Sample Size Calculations

One of the most frequently asked—and unfortunately most difficult to answer—questions in research design is, "How large should my sample be?". This is important, because if no relationship (descriptive design) or causation (experimental design) is found, it could be for a range of reasons. Firstly, there could simply be no effect; secondly, it could be because variables were not measured precisely enough; and thirdly, it could be because the study was underpowered. An underpowered study is one in which too few participants have been tested to detect an effect with the desired level of statistical confidence and power.

Hence, in order to determine an adequate sample size, sample size estimations are essential. In the following sections, methods for calculating sample sizes will be discussed, as will concepts such as confidence intervals, p-values, types of statistical errors, and statistical power. Calculation of sample sizes is a significant research area in its own right, and one addressed extensively in the quantitative social sciences and psychology fields. For the purposes of this chapter, the focus is on two of the main areas for which sample sizes are calculated: internal validity and inferential statistics.

Internal validity refers to the extent to which the findings from the study can be correctly attributed to the interventions being experimentally tested. While it is not exclusively the case, people gathering data through surveys are frequently more concerned about questions of inferential statistics, while people conducting experiments are frequently more concerned about questions of internal validity. As these are all complicated topics in their own right, the reader is referred to standard texts in the field such as (Groves et al. 2004).

3.5.6 External Validity

External validity is the assessment of the extent to which the findings from the sample can be considered to apply to similar, but not identical, units of analysis. These units of analysis can be considered as forming a range of similar but distinct populations that the findings can be said to hold for. These should not be confused with the general population (say all people or buildings in the country). These populations of similar units of analysis are defined by how closely the units of analysis are to those in the study. For example, a study of the thermal comfort of sixth grade school children may involve a sample of 200 students in ten schools. External validity arguments could be made that the same findings would hold for other students (e.g., the grades above or below) in those schools, or that they may even hold for students in other (similar) schools. Such arguments ultimately rest on qualitative arguments and citations of other studies' findings to support such claims. Citing confidence intervals and p-values for other (related) populations is inappropriate, as the argument for the external validity of these findings to these other groups is ultimately not a statistical one.

In this context, the above discussion about hierarchies, inferential statistics, external validity, units of analysis, and sample frames needs to be borne in mind. The sample frame needs to represent the population of the units of analysis, whether occupants in a building or buildings in sector of the building stock. Once having found or developed such a sample frame, a random subsample can be drawn of the size needed to achieve a certain level of statistical confidence (see note below on calculation of sample sizes). It is important to remember here that "random" is a well-defined term, and a suitable random number generator should be used to draw the sample. Then, the members of the chosen sample should be approached and recruited into the study. If not enough units of analysis (e.g., people or buildings) are willing to participate, it is not acceptable to simply draw more potential participants from the sample frame, as this simply serves to drive up the nonresponse rate (as discussed below). Whilst it is tempting to conduct "opt-in" trials, where volunteers are sought to participate in the project, this immediately violates the underlying assumption that each member of the population has an equal probability of being part of the trial, for by definition those who choose to participate are different from those who do not. The correct approach is to attempt to recruit all of those drawn at random from the sample frame, and then carefully note the percentage of those who accept to those who do not. This percentage, known as the response rate, needs to be as high as possible in order to minimize nonresponse bias. If only one in 10 people asked agrees to participate, then by definition 9/10 people have chosen not to—thus again violating the underlying assumption that the sample represents the population. There is a considerable literature in the quantitative social sciences about how to maximize participation rates in surveys and experiments. Amongst the best works in this area are those by Dillman, for example, "The Tailored Design Method" (2000). Whilst these methods are primarily designed for use in social surveys, they are also in many cases equally applicable to the recruitment of participants into field studies in buildings.

There are many situations where the above approach of using sample frames, random samples, and avoidance of self-selected samples is either unworkable or (arguably) unnecessary. Where the study is of something that self-selection is unlikely to influence, then it can be argued that any sample of sufficient size, random or non-random, can be generalized from. Where use of non-random samples is unavoidable, then the researcher is left balancing different forms of uncertainty. Using or increasing trial participant numbers through use of self-selection, snowballing, or other non-random methods increases precision by increasing sample sizes; however, it does not increase accuracy. Addressing this means either acknowledging that the findings only pertain to, say, building occupants who volunteered to participate, or arguing that the causal mechanisms at play are independent of the act of volunteering to participate. For example, where non-randomly sampled participants are then random allocated to experimental groups, conclusions can be robustly drawn about the outcome of the experiment on the participants—but these findings can only be inferred to apply to people likely to volunteer for such experiments.

Whilst there are instances in which the aim is to generalize from a sample to a population of people within an individual building, frequently the goal is also to try and generalize across a particular class of building within the building stock. As discussed above, this is enormously challenging, particularly in the non-domestic buildings area. The best global example of such a non-domestic building survey is the long-running Commercial Buildings Energy Consumption Survey (CBECS) in the USA. Although constructing such a survey may seem like an impossible task, it is important to note that if the theoretical model states that buildings shape users' responses to them, then it is very important to include a representative sample of such buildings in the study. Failure to do so means that no sensible statistical claims about the generalizability of the findings can be made. Effectively, the study is a conglomeration of case studies rather than a survey. It is for this reason that most of the reported confidence intervals from studies in this field do not make statistical sense, as they do not define the population to which they are claiming statistical generalizability—and if they do, they do not have a sufficiently large and representative sample drawn from that population to support such claims. It is important in this context not to disregard studies that fall short of the statistical requirements for generalizability. For logistical reasons, few studies in the buildings field achieve such requirements and, as mentioned above, sampling uncertainties are only a small proportion of the uncertainty in reported findings irrespective of sample sizes.

This covers some, but not all, of the range of issues identified in Fig. 3.2 on threats to the validity of inferential statistical findings. A detailed description of the measures undertaken to address each of these threats is beyond the scope of these guidelines and is covered in standard undergraduate texts on social survey design, for example (Sarantakos 2012).

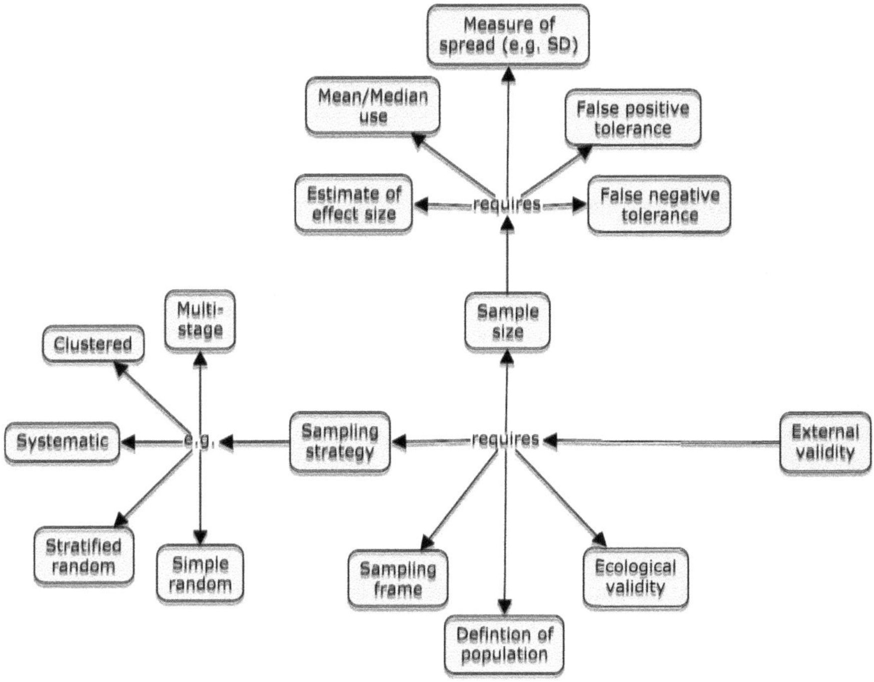

Fig. 3.2 Threats to the validity of inferential statistical findings

3.5.7 An Illustrative Example of Sample Size Calculations

An example of how to calculate sample sizes for a trial is provided that is loosely based on the British **energy**wise project. Very simplified calculation methods are presented here for the purposes of exposition.

Project Summary: Energywise assesses how much electricity fuel poor customers in Great Britain will save if provided with a smart meter and some energy-saving appliances. The project uses a randomized control design. The unit of analysis is the home (house + household).

Sample size calculation for establishing inferential statistical validity.

Aim: To ensure that the findings observed in the sample will hold in the wider population with a given degree of statistical confidence.

Step 1: Determining the population size

This was set at 260,000 based on the estimate of the number of customers on the Priority Services Register (a proxy for fuel-poverty) in the UK Power Networks' distribution zones. While an underestimate of the number of fuel poor in Great Britain, for populations over 20,000 estimated sample sizes change little.

Step 2: Calculating the sample size

For sample size statistics for inferential statistical validity the following equation was used (PSU 2014):

$$n = \frac{\left[\frac{P[1-P]}{\frac{A^2}{Z^2}+\frac{P[1-P]}{N}}\right]}{R}$$

Where:	Inputs:
n = sample size required	
N = population size	N = 260,000 (see above)
P = variance in population	P = 0.5. Assuming 50% of participants save more than 6% and 50% less
A = precision	A = 5%
Z = confidence level	Z = 1.6449 for 90%
R = Estimated Response rate	Adjusted after calculation
This produces a value of n of 271 survey participants required in the trial	

3.5.8 *Internal Validity*

Internal validity is a key concern in experimental research designs, such as randomized control trials. A key mechanism for ensuring internal validity is the provision of intervention and control groups that are initially statistically identical, differing only in the application of the intervention to the intervention group. One key test of internal validity is the test of the likelihood that observed differences between the intervention and control groups are statistically significant, and at what level of statistical confidence.

The capacity to statistically distinguish between the intervention and control groups is only one of the issues to be considered with respect to internal validity. Sarantakos (2012) provides a good overview of the range of issues known as "threats to internal validity" that must also be considered when designing such trials, as well as descriptions of the measures that have to be undertaken to address such issues.

To illustrate the process of calculating sample sizes for internal validity an example is again provided based on the British **energy**wise project.

Step 1: Determining the level of statistical confidence and power needed for internal validity on the trial

The consortium members were asked: "Are you more worried about:

(a) mistakenly accepting an intervention that doesn't work because the evidence wasn't strong enough? or

(b) mistakenly rejecting an intervention that does work because the evidence wasn't strong enough?"

The first of these relates to false positive (type I) errors, and the second to false negative (type II) errors.

To properly assess these, a risks-based approach to costs and benefits is needed, i.e., the probability of the error needs to be multiplied by the magnitude of the consequences expressed in human or monetary terms. This is an area where the judgment of the researcher is called for.

In the **energy**wise project the following approach was adopted

> Tell me, in percentage terms, how sure you want to be that an intervention actually delivers the energy savings we measure?

(A) On the balance of probabilities (i.e., 50–65% confident)
(B) Pretty confident (i.e., 65–80% confident)
(C) Beyond reasonable doubt (80–95% confident)
(D) Almost certain (>95% confident)

> Tell me, in percentage terms how sure you want to be that we don't mistakenly reject an intervention that actually does work?

(A) On the balance of probabilities (i.e., 50–65% confident)
(B) Pretty confident (i.e., 65–80% confident)
(C) Beyond reasonable doubt (80–95% confident)
(D) Almost certain (>95% confident)

The consensus amongst the project partners, on both the risk of false positives and false negatives, was that the group wanted to be "pretty confident" which was translated into a statistical confidence of 0.25 and a level of statistical power of 0.75.

It is difficult to overstate the importance of conducting this often overlooked step in sample size calculations. In many cases in energy use in buildings, occupant behavioral energy savings are only one element of the operational decision to install a given technology. They are frequently a "nice to have" benefit of, for instance, upgrading a building control system, or making a decision that incurs comparatively little additional cost. In this context, requiring 95% confidence of a trial is operationally inappropriate because the risks of failure are small (although, for academic publication purposes, it may be necessary).

Step 2: Determining the effect size

For the **energy**wise project, data on effect size was taken from the Energy Demand Research Project: Final Analysis report published by the UK energy regulator Ofgem (Raw and Ross 2011). This study, known as the EDRP, was the most up-to-date study on the effect size of smart meters available in Britain at the time. The following quote shows how uncertain the potential savings may be: "In the case of electricity consumption… a full range of 0–11% (energy savings) for some periods and customer groups" (p. 4).

In light of this, and because of the nature and extent of the intervention in the **energy**wise trial, an energy saving of about 6% from the intervention group was used.

Step 3: Determining the mean and standard deviation of electricity consumption

The inputs of the mean and standard deviation were taken from government statistics, specifically the "Review of typical domestic consumption values" consultation document (Villalobos 2013).

The standard distribution of domestic electricity users in the UK was used (UK Profile class 1 electricity consumption). This provided the following values:

- Arithmetic mean: approximately 3200 kWh/annum

No figure for standard deviation was provided, and so an estimate was made based on the inter-quartile range as follows:

- Average inter-quartile: (1200 + 1600)/2 = 1400 kWh/annum
- The ratio of interquartile range to standard deviation range is 34%/25% = 1.36
- Estimate of standard deviation is therefore ~1.36*1400 = 1900 kWh.

This, however, was for an average home, and needed to be adjusted for fuel-poor homes which were the subject of the study, as data on the mean and standard deviation is not available for this subpopulation. An adjustment was made based on the following logic: fuel poor customers are a subpopulation of all UK Electricity Profile Class 1 customers. They will, however, have a lower mean and a narrower standard deviation, as they are a more homogeneous group living in smaller homes. It was thus estimated that **energy**wise trial participants would have a mean electricity consumption of 3000 kWh and a standard deviation of around 1500 kWh. Note that these adjustments were merely educated guesses, as no further information was available on which to base these corrections.

Step 4: Sample size calculation for establishing internal validity

These sample size calculations were done using the G*Power 3.1.7 sample size calculation software as reported in Faul et al. (2007, 2009).

The analysis presented here uses the simplest test possible: a one-tailed t-test comparison of the difference between two independent means (two groups) using the input parameters above. Figure 3.3 shows how sample size scales with the degree of statistical power desired. The value of 0.75 used in the calculation above corresponds to the estimated sample size of 506 on the graph.

Note that this is a very rough initial estimation of the sample size needed to distinguish a 6% effect size between two equally sized groups with a statistical confidence of 0.25 and a statistical power of 0.75.

The sample size calculation for internal validity generates an estimated intervention group and control group size of 253 each, i.e., 506 in total. The sample size calculation for external validity generates an estimate of 271 in total. The test for internal validity is the larger of the two, and is therefore the factor determining

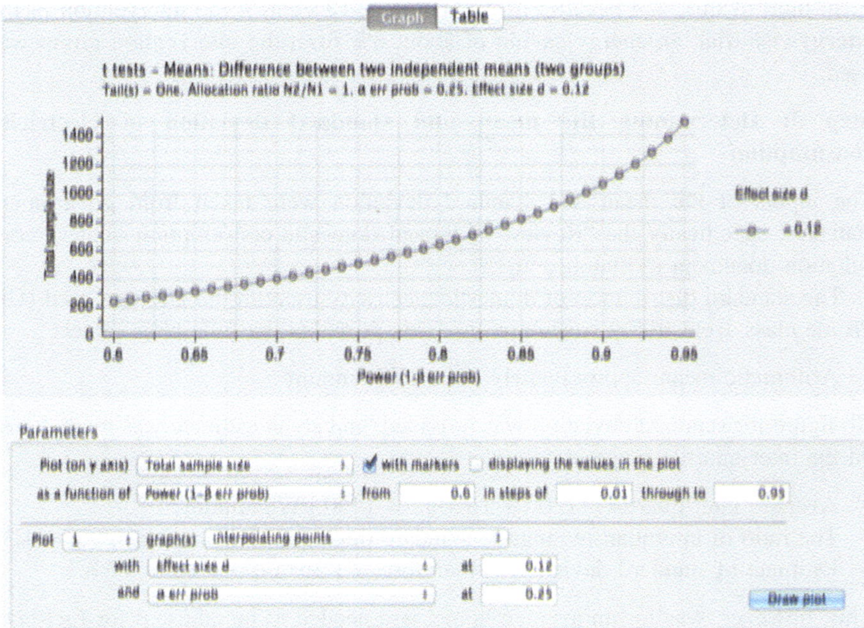

Fig. 3.3 Effect of varying statistical power on the estimated sample size

sample size. In addition to this, an allowance for the estimated number of participants leaving the trial ("dropouts") needs to be made and added to the sample size.

3.5.9 Dropouts and Response Rates

There are two adjustments that need to be made to the sample size calculation in order to determine the number of participants that need to be recruited. These are the expected dropout rate and the expected response rate.

The sample size calculation is based on the number of participants needed to conduct the analysis of the data at the end of the study. However, dropouts are likely, and hence the initial sample needs to be increased by the expected number of dropouts. In shorter term experimental work, such as a week-long survey of occupant behavior in an office building, comparatively few people may drop out of the study. In contrast, however, if conducting a year or multi-year study of occupant behavior in homes, 30–50% of participants may either move house, or choose to leave the study. The number of dropouts will be a function of the respondent burden (i.e., the inconvenience that participants have to put up with) and the duration of the study. The higher the respondent burden and the longer the study, the greater the likelihood of dropouts. When estimating the number of dropouts, the most useful method is to look to similar studies and make adjustments based on what expectation of the respondent burden and duration from the dropout rates reported in those. The calculated sample size should be increased by the expected dropout rate. In the example used above, the sample size of 506 would be increased by 30% to reach the number of people to account for later dropouts (in this case to 723).

The estimate of the likely response rate, (i.e., the ratio of who was invited to participate to the number that accepted that invitation) will vary depending on the method used to recruit participants. It is worth noting that expectations around what is an acceptable response rate vary from field to field. In the quantitative social sciences, particularly at the level of national statistics, statisticians will frequently start to become concerned when response rates drop below around 70%. In contrast to this, it is not uncommon in building occupancy studies for response rates either to be unknown, or to be substantially below 10%. The critical issue here is that any reduction below 100% represents a certain degree of self-selection of the sample.

3.6 How to Measure Concepts (Methods)

Having looked at research questions, established the theoretical model, and determined the boundaries of the applicability of the study, the next step is to determine how to measure the concepts in the theoretical model. This is the realm of research methods. Other chapters in this book talk in detail about different specific research methods and these should be referred to as appropriate. This section is going to focus on issues of clearly defining what is being measured and ways of trying to quantify some of the uncertainty in the measurements.

3.6.1 *Concepts and Constructs*

In research on occupants in buildings, relevant concepts include temperature, comfort, glare, productivity, and adaptive response. These are used to construct a theoretical model of how occupants respond to their physical environments.

It is useful to draw a distinction between concepts and constructs. Markus (2008) distinguishes between *concepts*, which he defines as the reification of all actual or potential instances of a set of experiences in the real world, and *constructs*, which are the instances of these in a specific population. Within a population, concepts and constructs are the same thing; however, the distinction becomes particularly important in international comparative work where concepts transfer between populations and constructs may not.

The benefit of such a distinction is that the area of occupant behavior in buildings is a highly international one in which researchers may frequently attempt to measure the same concepts, acknowledging that how those concepts are constructed and operationalized will necessarily need to take into account differences in climate and culture.

3.6.2 *Operationalizing Constructs into Measurands*

Operationalizing constructs is the process of determining how best to measure them. Sometimes they can be measured directly with a single instrument, for example, air temperature. Frequently, however, it is necessary to combine outputs from a range of instruments to measure the construct of interest. When multiple instruments are needed to measure a construct the term latent variables or hidden variables are frequently used to describe them.

Trochim (2006) captures this in one of his diagrams, reproduced in Fig. 3.4.

In Fig. 3.4, the theoretical model is represented in the top half of the diagram, while the translation of this into a concrete program of research is represented underneath. The aim of operationalization is to translate the theoretical model into measurable things as validly and as reliably as possible. Over time and multiple research programs, elements in the observation box will inform and change the theory box. In any individual research study, the observations of the research program reflect the theoretical model being evaluated.

Implicit in the construction of the theoretical model as advocated above is the need to clearly specify the study's outcome variable of interest. This must be done before the onset of the experiment to avoid "fishing" for significant effects after the experiments. How the outcome variable will be measured needs to be defined in

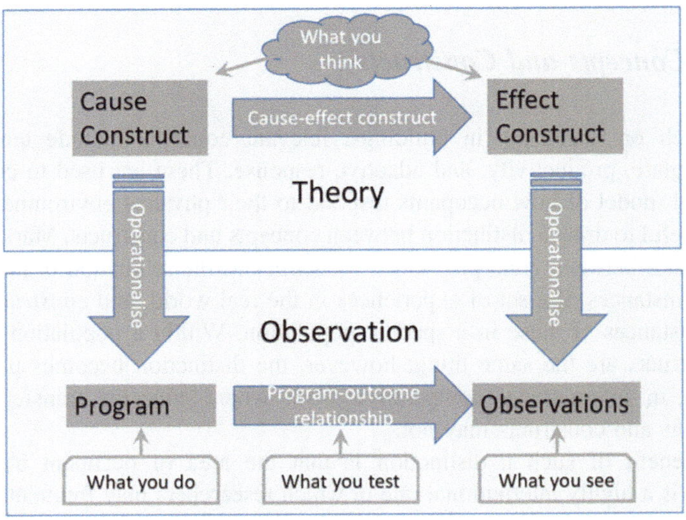

Fig. 3.4 Theory -> observation relationship (adapted from Trochim 2006)

detail. This will need to be done in the context of the research aim or question the study is designed to answer.

One of the key uncertainties that arises in the operationalization of constructs is what is called "under specification of the measurand".

The term "measurand" is used in the field of metrology (the science of measurement) and is defined by the International Bureau of Points and Measures (Joint Committee for Guides in Metrology (JCGM) 2008a) as "quantity intended to be measured"—in this case, the construct.

Under specification of the measurand is the failure to specify *exactly* what it is that should be measured. For example, external temperature with respect to a building is often not specified exactly. External temperature can vary considerably around the envelope, and thus any measurement of the concept of external temperature is subject to considerable error as each researcher will operationalize the concept differently. If specified more precisely—say, external ambient temperature measured within a Stevenson screen at 1.5 m above ground level one meter away from the building envelope at each compass point with the arithmetic mean value taken—there would be far less (but still some) leeway to measure differently. Under specification of the measurand is not a problem of measuring; it is a problem of operationalizing concepts—and leads to uncertainty in comparing the results of different studies and in replicating studies.

3.6.3 Latent Variables

Latent variables—also known as "hidden variables" or "hypothetical constructs"—are variables that cannot be measured directly. Some authors distinguish between the terms, using the term "hidden variable" as something that physically exists and could therefore in principle be measured directly, but for cost or other reasons may not be, and "hypothetical variables" as those that do not physically exist, but are useful explanatory tools, for example, attitudes or inflation.

Latent variables are common in all fields of research including in building occupancy studies. They vary from things like the volume of a room (which is constructed from a series of individual linear measurements and knowledge of geometry), to operative temperature (which requires measuring both air and radiant temperature), to psychological variables such as environmental attitudes or perceived control, which are usually measured through a set of questions.

The construct of interest is, for practical or other reasons, not directly observable and must be measured by combining the outputs from multiple individual instruments. In psychology there is a considerable methodological literature about how scales (i.e., sets of questions) should be developed and a considerable body of statistical science behind their evaluation.

3.6.4 Instruments

Various aspects of instrument selection, development, and placement will be discussed in detail in other chapters, particularly Chap. 4 through Chap. 8. One general point that should be made is that, conceptually, social research methods (participant observation, social surveys, interviews, focus groups) are also instruments in that they are designed to measure specific things that are subject to the same forms of uncertainties (imprecision, inaccuracy, etc.) as their physical counterparts. Thinking of physical, physiological, psychological, and social instruments in the same way is useful in supporting cross-disciplinary collaboration and establishment of a common vocabulary of measurement in this highly interdisciplinary and socio-technical area of study.

3.6.5 Quantifying Uncertainty

The International Bureau of Points and Measures emphasizes the fact that any quantitative measure consists of three components. The first part consists of some multiplication of the number of base units (for example, a home might use 2000 kWh of electricity per annum). The second part stipulates an error margin around that value (e.g., ±100 kWh per annum). The third part stipulates the probability that the "true" value lies within that error margin (e.g., 0.9). Any quantitative assessment that fails to clearly identify each of these three elements for each measurement is incomplete and makes the result difficult to interpret. This ideal is one that is frequently hard to achieve in practice, but the ideas that it entails are important for researchers to understand. In particular, the third component is a reminder that instruments never perfectly capture the true value that is intended to be measured (i.e., the measurand). Accepting that all measurements are approximate and never perfect has two consequences. Firstly, that it is necessary to estimate the degree of precision required in order for findings to be useful. This is a function of the purpose of the study and can be established before any considerations of methods is undertaken. Secondly, that it is necessary to decide whether the measurements taken and models used allow making statements that fall within this required degree of precision. Without the quantification of the uncertainties surrounding the study answers it cannot be judged whether the measurements and models are suitable for any given purpose.

Uncertainty quantification is a complicated and specialist field that is beyond the scope of this book. An excellent introductory reference on instrument error and error propagation in the physical sciences is Taylor (1997) and an authoritative guideline on error propagation using Monte Carlo analysis is provided by the Joint Committee for Guides in Metrology (JCGM) (2008b). Interestingly, in the area of instrument validity and reliability, the social sciences have developed better frameworks for assessment, e.g., the Multi-Trait Multi-Method (MTMM) approach (Campbell and Fiske 1959).

3.7 How to Measure Relationships (Research Design)

Once having identified concepts, turned them into constructs, and operationalized them into things that can be measured, the challenge remains of determining the nature of the relationship between the concepts in the research question. There are essentially three types of relationships that could exist between the concepts measured: there could be a causal relationship, the concepts could be correlated, or they could be entirely independent. It is the role of research design to determine the nature of the relationship between the concepts. Research design is the process of devising a process that directly satisfies a brief, in this case, the research question or research aim.

Broadly speaking there are two forms of research design: descriptive (or correlational) research designs, and experimental (or causative) research designs.

It is important to note that both descriptive and experimental research designs use the same research methods. For example, a sensing campaign supported by occupant surveys can support analysis that is either descriptive of the relationships between the variables or shows causal relationships between variables. In order to establish causation, all other possible explanatory factors (all confounding variables) need to be eliminated implying that nothing else could have caused this observed relationship. This is conventionally and best done using experimental designs.[1] Such designs look to isolate the effect of one variable on another by holding all others constant in a controlled environment. This is a powerful and valuable approach, but not without limitations. The primary critique of such methods is their potential lack of ecological validity, i.e., that the findings from such studies do not reflect "real world" conditions and so what is observed in the lab or experimental field trial may not be observed in uncontrolled conditions. The more naturalistic the environment is in which the occupant experiences the experiment,

[1]There are some methods of analysis that some analysts argue can establish causation outside of an experimental context. Lead amongst these is Judea Pearl and his application of statistical graphical modeling methods such as Bayesian networks (Pearl 2000). This is both a highly advanced field of statistical analysis, and a hotly contested topic that is beyond the scope of this book.

the greater the ecological validity (see also Chap. 7). However, a more naturalistic setting makes control of confounding variables more difficult.

While it may seem intuitive that two variables are causally related, it is all too often the case that they are linked through a third variable which causes them to vary simultaneously.

> A good example of confounding variables is the relationship between CO_2 levels and thermal comfort in a room. As occupants come into a room CO_2 levels will rise alongside temperature. If the relationship between CO_2 and thermal comfort is measured, it would show that they are highly correlated. There are also valid metabolic arguments as to why CO_2 may change metabolic rate and cognitive function and consequently impact on thermal sensation. In standard field monitoring conditions, it is very difficult to disentangle the rise in CO_2 with the associated rise in temperature, and thus to determine whether it is the CO_2, the temperature rising, or both that is impacting on people's thermal sensation. Therefore, standard monitoring field studies are not a good research design to try and answer this particular research question. Here the experimental precision of laboratory conditions is preferable, allowing independent variation of CO_2 levels from temperature levels in order to isolate the effect of one variable from the other.

A concept map illustrating some of the key concepts in both descriptive (correlational) and experimental (causative) research designs is provided in Fig. 3.5.

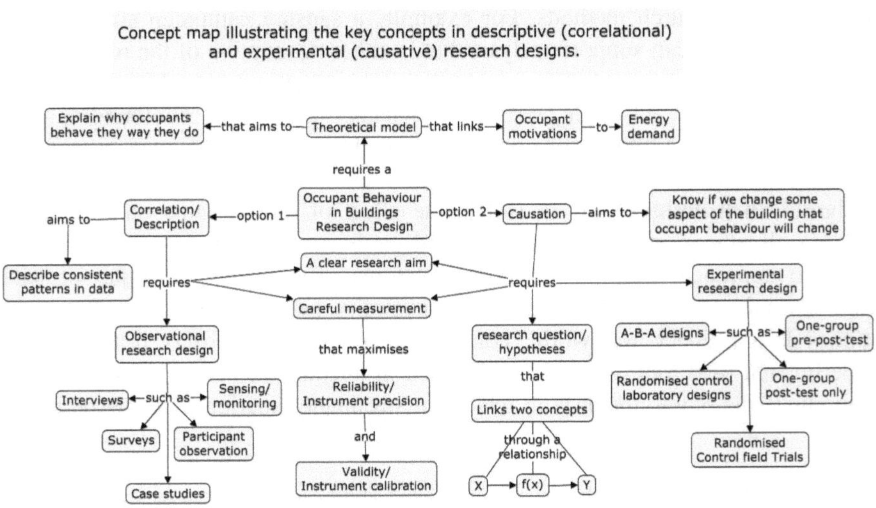

Fig. 3.5 Concept map illustrating some of the key concepts in both descriptive (correlational) and experimental (causative) research designs

3.7.1 *Descriptive (Correlational) Designs*

Descriptive (correlational) research designs are the mainstay of studies into the impact of the occupant behavior on building energy demand. This would classically take the form of gathering data through installed sensors, virtual sensors, or data gathered for other purposes (frequently termed "administrative" data), potentially augmented with some occupant surveys delivered either on paper or electronically through smart-phones or computers. The data would then be analyzed for correlations between the variables. Such a study design allows us to understand relationships in the data, but not to say that a change in one observable causes a change in the other. There are times when this seems counterintuitive. This is usually where the theoretical model or mental model feels like the only possible explanation for an observation. For example, it is tempting to interpret window opening behavior as always being related to regulating the thermal or indoor air quality environment within the building, particularly where this is the purpose of the study. However, alternative reasons can also explain why occupants may be opening windows—for example, out of habit, or in a residential setting to talk with people outside, or to listen to the birds in the garden. The sun coming out would correlate both to a rise in internal temperature and to increased bird activity in the garden. It is very easy when interpreting data from an energy perspective to mislabel a correlation (here between internal temperature and window opening behavior) as causative. Drawing conclusions of causation in instances where other potential mechanisms have not been controlled for can easily lead to wrong conclusions. There are a wide range of descriptive research designs which are covered in detail in many textbooks [e.g., Bryman (2008); Saunders et al. (2015)]. Three of the main types of design are covered briefly here.

3.7.2 *Case Studies*

One of the most widely used designs in the research on occupants in buildings is the case study. As the name suggests, a case study focuses on one individual instance (say, an individual building, campus, or community) and applies multiple methods to understand the workings of that particular case. Case studies offer no capacity for generalization because they are a sample of one. An excellent reference on the use of case study research is Yin (2013).

Some argue that if the case is in some senses archetypal, then lessons learned can be translated to similar cases. This is intuitively reasonable, but scientifically indefensible, as studying one case can say nothing about whether other similar cases work in the same way. It is tempting to assume that if other cases share similar characteristics and those characteristics are found to be explanatory of the behavior of the individual case, then the results must surely apply more broadly. This assumption only holds, however, under a certain theoretical or mental model of those factors which are important across the set of similar cases—an assumption that seldom holds true in practice.

Case studies are enormously powerful for identifying factors the commonality of which can then be explored using more sample-based research designs. One approach which seeks to span the gap between individual case studies, and a population-based sample, is the Qualitative Comparative Analysis method developed by Ragin (1987). This approach has now developed into a suite of methods which seek to systematically draw out commonalities between a small set of case studies. The approach is widely used in international comparative analysis and frequently is based on numbers of case studies ranging from 10 to 50.

3.7.3 Cross-Sectional Design

A cross-sectional design is one that gathers data at a particular point in time from a range of units of analysis (occupants, buildings, etc.). A one-off social survey is a classic example of the approach. When correctly designed, such approaches can support generalizations from the sample to the population. The design and construction of social surveys is covered in Chap. 8.

Cross-sectional designs can either be conducted once or at multiple points in time, thus creating a repeat cross-sectional research design. Repeat cross-sectional design does not measure the same people at each point in time, but rather generates a new representative sample from the population each time the survey is conducted. This distinguishes them from longitudinal surveys in which the same people are measured repeatedly through time. Repeat cross-sectional designs have the advantage that it is easier to draw a cleaner representative sample at each time point, than it is to try and maintain a panel of the same participants through time. If the aim of the research is to understand changes at the population level, then repeat cross-sectional designs are most appropriate.

Classic introductory texts on social survey designs include: Sarantakos (2012), which covers nearly all aspects of social research to a good undergraduate level of understanding, and Foddy (1993), which is excellent for details on survey and interview questions.

3.7.4 Longitudinal Surveys

Longitudinal surveys are ones that measure the same units of analysis through multiple points in time. There are many different kinds of longitudinal surveys, including panel surveys, where "panel" is the name given to the sample of units of analysis (people, buildings, etc.) being drawn to represent the population and then followed through time with repeated surveys; and cohort studies, where a group of units of analysis sharing a common characteristic (say, a sample of buildings of a certain type built in the same year) are followed through their lifetime. Again, further details on such designs are provided in Chap. 8.

3.7.5 Causative (Experimental) Designs

An experiment is a procedure to test a hypothesis. The main difference between experimental research and other types of research is the aim of establishing causality, i.e., insight into cause-and-effect relationships, by testing what happens to an outcome variable if a specific factor is manipulated. The outcome variable is usually called the "dependent variable". The independent variable, also called the "treatment" or the "intervention", is under direct control of the researcher and is used for creating experimental conditions.

An example to illustrate the experimental approach and its variables: It might be interesting to know whether a pop-up window displayed on the screen at the end of the working day with a prompt to turn off the computer before leaving leads to a higher number of turned off computers (intervention group) than when providing no such prompt (control group).

The dependent variable would be the number of computers turned off at the end of the day, monitored over specific time period (e.g., two weeks) and averaged over that period. The pop-up window constitutes the independent variable.

Extraneous variables are factors not of interest to the researcher, but that need to be controlled for as they can also impact on the dependent variable and their effect can be confounded with the effect of the independent variable (hence they are also called "confounding variables"). The age and type of computer might be confounding variables in that people with old computers that take a long time to boot-up, or that do not permit being shut down with programs still open, make it less likely that someone will shut a computer down. Random assignment of participants to groups is one method of eliminating the effect of extraneous variables. If the sample is large enough, one would expect the same distribution of the extraneous variable in the group receiving the intervention and the one not, e.g., the same number of older computers, in both groups. However, in relatively small groups, randomization might not work. Where this is the case, another method is to control for the effect of those variables in the analysis of the data. By including them as variables in the statistical analysis the effect of the intervention can be tested while holding the extraneous variable constant. This allows the effect of the extraneous variable to be analyzed and accounted for. Extraneous variables are more of a problem when they are not obvious and when randomization cannot be relied on to ensure an equal distribution across all groups, e.g., because the sample is too small.

Random assignment is a critical feature of experimental work; it ensures that the groups are the same in important characteristics and that differences in the outcome measures are attributable to the intervention and not differences between the groups per se. The **energy**wise trial example given above is an example of one of the most common (and best forms of) randomized experimental design: a randomized design comparing control and intervention group in a post-test.

Pre-tests can be used in experimental studies. In the example given, the number of computers turned off before the intervention might be counted to establish a baseline. Since this could easily be done after all employees have left in the evening there would not be any concern that, in doing so, employees' attention would be drawn to the need to switch off computers and hence influence the trial's outcome. This is, however, a concern in other settings where, by including a pre-test, a topic is made salient to trial participants (e.g., making them more aware of energy use), and thus the pre-test could impact on the post-test. Pre-tests are also associated with higher costs, time, and effort, and hence are not necessarily advisable. However, they can be useful in other respects: in the example, a pre-test might reveal that all computers are switched off anyway, and hence, that there is no point in running the study!

Two other forms of experiment exist. The first is the quasi-experiment. It has the same elements as a true experiment, but lacks the crucial aspect of randomization, i.e., participants are not randomly assigned to conditions. Instead, assignment to conditions is via self-selection. This poses a serious problem because the assumption that groups are equal no longer holds, and hence there might be confounding variables. While the extent to which groups differ on certain easily measured variables (age, gender, income, etc.) can be assessed, it is quite plausible that there remain confounding variables which are hard to assess because they are difficult to measure. Ultimately, it is not known what made participants decide to choose one intervention over another, or to be in the control group. Despite this significant disadvantage, quasi-experiments are common in applied settings because they avoid a lot of the logistics of establishing a true experiment, or allow analysis of things that would be impossible or unethical to conduct experiments on. For an excellent example of a quasi-experimental design see the recent thermal comfort study by Luo et al. (2016).

The third main type of experiment is the natural experiment, where a naturally occurring condition is contrasted with a comparison condition. Here the cause cannot be manipulated, i.e., the independent variable is not set by the researcher. For example, an earthquake might destroy several high-rise buildings in one city, and so a study might test if inhabitants of that city are less likely to buy flats in high-rise buildings over the next two years than inhabitants of a city of a similar size (and ideally, similar in other characteristics such as wealth, presence of industry, etc.) that was not affected by an earthquake. The big advantage of natural experiments is that they allow the study of the effect of phenomena that otherwise could not be studied; however, groups are not necessarily equal (or even similar), were not randomly assigned, and there might be a wide range of confounding variables.

3.8 Pre-analysis Plans

One of the key points that Wasserstein and Lazar (2016) notes (on behalf of the American Statistical Association) in their article on good practice in the use of tests of statistical significance is that proper inference requires full reporting and transparency. Wasserstein emphasizes that

> Conducting multiple analyses of the data and reporting only those with certain p-values (typically those passing a significance threshold) renders the reported p-values essentially uninterpretable. Cherry-picking promising findings, also known by such terms as data dredging, significance chasing, significance questing, selective inference and "p-hacking," leads to a spurious excess of statistically significant results in the published literature and should be vigorously avoided. …Whenever a researcher chooses what to present based on statistical results, valid interpretation of those results is severely compromised if the reader is not informed of the choice and its basis (p. 10).

This is mirrored in an article by Simmons et al. (2011) in which they argue that researchers have a lot of degrees of freedom to make decisions during the data collection and analysis that distort the research process and artificially inflate the probability that they will find positive results. To combat this, Taubman et al. (2010) and many others have argued for development and publication of a data analysis plan prior to conducting the research, also called a "pre-analysis plan", or PAB. They note, "by planning and disclosing the hypotheses to be tested and specifications to be used in advance of seeing the data, the plan should avoid (or at least minimize) issues of data mining and specification searching" (p. 3).

An analysis plan will usually include the following sections:

- Overview of the study (including: aim; research/experimental design; outcome measure; sample).
- Ethical considerations (including in experimental research ethical aspects arising from things like withholding intervention from one group, negative effects of an intervention, and privacy aspects).
- Statement of hypotheses to be tested (including: expected average effects; causal chain of process and mechanisms; heterogeneous effects on sub-groups).
- Estimating equations to be used (including: stating the spatial and temporal sampling frequency to be used; estimating average treatment effects; estimating treatment effects using interaction terms; what predicts the outcome variable of interest).
- Testing for balance if experimental design is used (including: randomization/balance checks).

- Procedures for addressing missing or low quality data, covariate imbalance and questions with Limited Variation (including: item non-response; covariate imbalance; questions with limited variation).
- Variable construction (including how each variable is to be constructed from the raw data).

Such analysis plans should be prepared in advance of the study and, in the ideal case (and as required by some journals), published online to ensure full accountability of analysis and so editors can check that no additional analysis has been conducted to "massage" the data to achieve desired outcomes.

The other issue which analysis plans serve to improve is statistical conclusion validity. Statistical conclusion validity refers to the extent to which statistics are used properly and appropriate conclusions drawn from analysis. It relies on other forms of validity that extend to the choice of analysis methods, with a particular emphasis on whether the underlying assumptions of these analysis methods (frequently normality of distributions) hold in the case of the analysis conducted [see (Sackett et al. 2007)].

As with many aspects of best practice in research design, production and publication of such analysis plans is often not done in building occupancy research. This risks leading to high levels of cherry picking of favorable findings by running multiple analyses and publishing only those "of interest" (i.e., frequently those with positive relationships between variables). Such skewing of the research process makes both interpretation and replication of findings difficult or impossible and undermines the quality of work in the field. It should be stressed that performing exploratory data analysis by conducting tests not outlined in the original analysis plan is an entirely acceptable form of scientific practice—however, it is one that should only be used to generate hypotheses for testing in future well-designed studies in which such forms of analysis are written into original analysis plan.

3.9 Conclusion

Research design is an essential, but often misunderstood and overlooked component of the research process. This chapter lays out a systematic approach to research design centered on the construction of a concept map diagrammatically representing the theoretical cause-effect model that the research is seeking to test. Making this explicit through concept mapping requires representing the concepts being explored as nodes, and the relationships between those concepts as links. Once the theoretical model is mapped out, then research questions are easily articulated as the relationships between the concepts in the model. Hypotheses can be drawn from the research questions that the research can be designed to test. This approach also provides a framework for the writing of pre-analysis plans, which help researchers clearly articulate their proposed methods of analysis prior to collecting their data, thus helping to guard against malpractice, such as searching for statistically significant relationships between variables that were not the original intent of the study.

Occupant behavior in buildings research must be fit for purpose. To be fit for purpose, the purpose must be known and the findings of the research must fall within acceptable margins of error for that purpose. Therefore, to be useful, research must not only produce findings, but also quantify the uncertainty in those findings to show they lie within the acceptable margins of error for that purpose. To achieve this requires both quantifying uncertainty, but more importantly designing-out enough uncertainty to fall within required error margins. The procedures outlined in this chapter address both these elements. Accepting that things cannot be measured perfectly, mapping the theoretical model, choosing an appropriate research design, and selecting and applying appropriate methods all help in reducing uncertainty.

The procedures outlined in this chapter constitute best practice in research design and may seem intimidating to many new and established researchers in this field. Indeed, many of these methods represent the cutting edge of best practice in research in the more pure-science fields such as the social sciences, psychology, physics, and metrology. Studying the actions and influences of occupants on energy use in buildings is a theoretically and scientifically challenging task as scientifically demanding as any in the pure sciences. It is all too easy for the influences of occupants to become lost in a sea of confounding influences on energy demand, ranging from the impact of the weather, through the performance of the building fabric, to the behavior energy producing and consuming technologies and their control systems and the complex temporal interdependencies of all of these. To disentangle these influences and isolate the influence of occupants requires theoretical clarity and rigorously designed and conducted research in order to establish the foundations and significant findings of the field.

References

Ajzen I (1991) The theory of planned behavior. Organ Behav Hum Decis Process 50:179–211
Bouma G (2000) The research process, 4th edn. Oxford University Press, Melbourne, Australia
Bryman A (2008) Social research methods, 3rd edn. Oxford University Press, New York
Campbell D, Fiske D (1959) Convergent and discriminant validity by the multi-trait multi-method matrix. Psychol Bull 6:81–105
Casti J (1992) Reality rules: picturing the world in mathematics: the fundamentals, the Frontier. Wiley, New York
Dillman D (2000) Mail and internet surveys: the tailored design method. Wiley, New York
Faul F, Erdfelder E, Lang A-G, Buchner A (2007) G*Power 3: a flexible statistical power analysis program for the social, behavioral, and biomedical sciences. Behav Res Methods 39:175–191
Faul F, Erdfelder E, Buchner A, Lang A-G (2009) Statistical power analyses using G*Power 3.1: tests for correlation and regression analyses. Behav Res Methods 41:1149–1160

Foddy W (1993) Constructing questions for interviews and questionnaires: theory and practice in social research. Cambridge University Press, Cambridge

Gauthier S, Shipworth D (2015) Behavioural responses to cold thermal discomfort. Building Research and Information 43(3):355–370

Groves R, Fowler F, Couper M, Lepkowski J, Singer E, Tourangeau R (2004) Survey methodology. Wiley, Hoboken, New Jersey

Joint Committee for Guides in Metrology (JCGM) (2008a) 200:2008 International vocabulary of metrology—basic and general concepts and associated terms (VIM)

Joint Committee for Guides in Metrology (JCGM) (2008b) Evaluation of measurement data—supplement 1 to the guide to the expression of uncertainty in measurement—Propagation of distributions using a Monte Carlo method. BIPM: 90

Luo M, de Dear R, Ji W, Lin B, Ouyang Q, Zhu Y (2016) The dynamics of thermal comfort expectations. Build Environ 95:322–329

Markus K (2008) Constructs, concepts and the worlds of possibility: connecting the measurement, manipulation, and meaning of variables. Measur: Interdisc Res Perspect 6(1–2):54–77

Nicol F, Humphreys M, Roaf S (2012) Adaptive thermal comfort: principles and practice. Routledge, London

Novak J, Cañas A (2006) The theory underlying concept maps and how to construct and use them, Institute for Human and Machine Cognition (IHMC)

Pearl J (2000) Causality: models, reasoning, and inference. Cambridge University Press, Cambridge, England

PSU (2014) An equation for determining final sample size from program evaluation tipsheet #60—how to determine a sample size. URL: http://extension.psu.edu/evaluation/pdf/TS60.pdf. Accessed 2014-10

Ragin C (1987) The comparative method: moving beyond qualitative and quantitative strategies. University of California Press, Berkley

Raw G, Ross D (2011) Energy demand research project: final analysis. Ofgem, London. http://www.ofgem.gov.uk/Pages/MoreInformation.aspx?docid=21&refer=Sustainability/EDRP

Ruttkamp E (2002) A model-theoretic realist interpretation of science. Kluwer, Dordrecht

Sackett P, Lievens F, Berry C, Landers R (2007) A cautionary note on the effects of range restriction on predictor intercorrelations. J Appl Psychol 92(2):538–544

Sarantakos S (2012) Social research, 4th edn. Palgrave, Basingstoke

Saunders M, Lewis P, Thornhill A (2015) Research methods for business students. 7th Edn. Trans-Atlantic Publications

Simmons J, Nelson L, Simonsohn U (2011) False-positive psychology: undisclosed flexibility in data collection and analysis allows presenting anything as significant. Psychological Science 22

Taubman S, Allen H, Wright B (2010) The short-run impact of extending public health insurance to low income adults: evidence from the first year of The Oregon Medicaid Experiment—Analysis Plan

Taylor J (1997) An Introduction to error analysis: the study of uncertainties in physical measurements. University Science Books, Sausalito

Trochim W (2006) The research methods knowledge base, 2nd edn. Internet WWW page, at URL: http://www.socialresearchmethods.net/kb/. (version current as of October 20, 2006)

Villalobos (2013) Review of typical domestic consumption values. Consultation document. Ofgem Ref: 113/13. 3 July 2013

Wasserstein R, Lazar N (2016) The ASA's statement on p-values: context, process, and purpose. The American Statistician: 00-00

Yin R (2013) Case study research: design and methods, 5th edn. SAGE Publications, Thousand Oaks

Chapter 4
Sensing and Data Acquisition

Bing Dong, Mikkel Baun Kjærgaard, Marilena De Simone, H. Burak Gunay, William O'Brien, Dafni Mora, Jakub Dziedzic and Jie Zhao

Abstract Occupant sensing and data acquisition are essential elements for occupant behavior research. A wide range of different types of sensors has been implemented to collect rich information on occupants and their interactions with the built environment, such as presence, actions, power consumption, etc. This information establishes a foundation to study the physiological, psychological, and social aspects of occupant behavior. This chapter summarizes existing occupancy and occupant behavior sensing and data acquisition technologies in terms of field applications, and develops nine performance metrics for their evaluation. The reviewed technologies focus on both occupants' presence and interactions with the built environment, and are grouped into six major categories: image-based,

B. Dong (✉)
Built Environment Science and Technology (BEST) Lab, Department of Mechanical Engineering, The University of Texas at San Antonio, One UTSA Circle, San Antonio 78249, USA
e-mail: bing.dong@utsa.edu

M.B. Kjærgaard
Center for Energy Informatics, Mærsk McKinney Møller Institute, University of Southern Denmark, Campusvej 55, 5230 Odense M, Denmark
e-mail: mbkj@mmmi.sdu.dk

M. De Simone
Department of Mechanical, Energy and Management Engineering, University of Calabria, Via Pietro Bucci 46/C, 87036 Rende CS, Italy
e-mail: marilena.desimone@unical.it

H.B. Gunay · W. O'Brien
Department of Civil and Environmental Engineering, Carleton University, 3432 Mackenzie Building, 1125 Colonel By Drive, ON K1S 5B6 Ottawa, Canada
e-mail: burak.gunay@carleton.ca

W. O'Brien
e-mail: liam.obrien@carleton.ca

D. Mora
Hydraulic and Hydrotechnical Research Center (CIHH), Technological University of Panama, Avenida Domingo Diaz, Panama City, Panama
e-mail: dafni.mora@utp.ac.pa

A. Wagner et al. (eds.), *Exploring Occupant Behavior in Buildings*,
DOI 10.1007/978-3-319-61464-9_4

threshold and mechanical, motion sensing, radio-based, human-in-the-loop, and consumption sensing. This chapter provides an overview and discussion of different current state-of-the-art and future sensing technologies for researchers.

4.1 Introduction

With the rapid development of computer and electrical engineering, sensors and data acquisition devices in buildings have become increasingly common for collecting building performance data, including energy and power, thermal comfort, visual comfort, and indoor air quality. However, there is a lack of sensor technologies specifically aimed at occupant behavior research, such as counting people or monitoring window blinds positions. Occupant sensing, a unique sensing device and data acquisition system, remains a relatively uncommon element in building automation systems (BASs). Accordingly, many researchers have their own in-house occupant sensors to achieve a specific research goal.

For occupant research, it is crucial to collect information on how occupants interact with the built environment and building systems. This information ultimately helps building operators to better understand occupant behavior in buildings and make decisions that improve a building's performance in terms of energy and occupant comfort. Meanwhile, commonly used energy simulation tools often assume synthetic occupancy, lighting, and plug load schedules due to a lack of field data, which could lead to errors as great as 600% (Haldi and Robinson 2010). Such occupancy profiles are commonly based on surveys and manually observed data, which take a long time to gather and do not accurately reflect actual occupancy status. A study showed a 46% difference between standard diversity factors and actual occupant profiles in an office building (Duarte et al. 2013). Hence, researchers have started to utilize advanced sensing and data acquisition technologies for built environment research to better capture occupants' behavior and their interactions with building systems.

This chapter provides an overview of state-of-the-art occupant sensing and data acquisition technologies. The chapter first introduces nine occupant sensing performance evaluation metrics for choosing the most appropriate occupant sensor(s). Secondly, it introduces state-of-the-art occupant sensing technologies with regards

J. Dziedzic
Department of Energy and Process Engineering, Norwegian University of Science and Technology, 7491 Trondheim, Norway
e-mail: jakub.w.dziedzic@ntnu.no

J. Zhao
Delos Living LLC, 22 Little West 12th Street, NY 10014 New York, USA
e-mail: jie.zhao@delos.com

to sensor hardware, sensing principles, and test bed case studies. Thirdly, it covers current occupant data acquisition systems, including types of data transmission, data storage, robustness, and security. Fourthly, it describes other occupant behavior research-related sensing systems for indoor and outdoor environment monitoring. Lastly, the chapter discusses occupancy sensing challenges.

4.2 Sensing System Performance Metrics

Upon a comprehensive survey of the literature, methodologies of occupant sensing and data collection can be organized in six categories: image-based, threshold and mechanical, motion sensing, radio-based environmental, human-in-the-loop, and consumption sensing. An overview of the surveyed sensing technologies and their performance metrics is given in Table 4.1. Further details of these technologies are provided in Sect. 4.3. The nine metrics used for evaluating the technologies are: cost, deployment area, collection style, power type, sensing range, accuracy, data storage, data sensed, and deployment level. They are described below.

- Cost: The cost of an occupant sensing technology can be specified in quantitative values—for example, "$" represents a low-cost sensor, while "$$$$" represents an expensive sensor. Cost includes costs associated with acquiring the hardware, installing and integrating the hardware, and operating the technology. Papers and reports often only record hardware costs, but the others must also be estimated and recorded.
- Power type: The way a sensing technology is powered should be noted, including self-powered (battery) and external. For battery power it is important to consider the operation time under normal conditions and whether or not a rechargeable battery is used. For devices that obtain their power through an external source, it is important to consider the voltage and if a device uses a special connection, such as getting power through a phone line or over Ethernet.
- Data storage: Data storage options provided by a sensing technology can be either internal (an on-board storage device is used) or network (data from the sensor(s) are stored on a server or a distributed environment). If internal storage is used, the size and time to process data is important to consider due to limited storage space. If possible, the sample size of collected data should also be reported. If network storage is used, details of that architecture should be reported.
- Deployment type: Any description of an occupant detection technology should include a description of how it is deployed. This should feature three key components: the type of building it is deployed in (e.g., office, residential home, educational building), the type of room(s) it is deployed in (e.g., office, conference room, dining area), and the specific deployment location (e.g., in a doorway, on a person, on an appliance). Reporting on all three of these components are critical to evaluating a study and the device itself.

- Sensing range: Where applicable, an occupant-sensing technology should be tested to find maximum and minimum ranges, if applicable, as well as the area or view angle that it can cover. These values should be reported as specific quantities and, if possible, the effect they have on accuracy should be reported.
- Data sensed: There are five categories of data that can be sensed: presence state, counting (i.e., number of people), people tracking, building system state, and occupant actions. Most technologies only measure one of these categories at a time. The types of data sensed are critical when reporting on an occupant sensing technology and need to be reported precisely and clearly. The first and simplest category of sensing is binary occupant presence, i.e., whether or not presence is sensed in a space. A second category is value-based occupancy counting, which senses occupants' movement as in or out of a space and reports the count. A third category is individualized tracking data, where the location of every occupant is sensed. The final two categories are occupant actions and building system state as a result of occupant actions, which together collect data about the behavior and building interactions of occupants within the deployment area. This can include window, shading, and door operation, activity level, appliance usage, and other interactions.
- Collection style: The data collection method of an occupant detection technology should be reported as either periodic (sampling within a fixed period) or event-based (sampling only when triggered, such as lighting on/off).
- Accuracy and failure: Where possible, the absolute accuracy should be compared to a manual observation (or other relevant methods) and reported. Chapter 9 describes aspects of ground truth in detail. In addition, any report on a new technology or deployment area should provide a discussion of situations of sensor failure—e.g., failures due to environmental conditions, system failures (loss of power, exceeding storage capacity, etc.), or failures due to inaccuracies —and how these failures reflect on conclusions made or the ability of the device to be used in related future applications.
- Demonstrated control applications: An occupant action and presence detection method can be evaluated based on its demonstrated applications. This can be measured according to two criteria: (1) the number of papers reporting on it, and (2) whether it is commercially available. Based on a literature review, occupancy technologies are mainly applied to three areas: lighting, HVAC, and security applications. Market availability is included in Table 4.1.

4.3 Occupant Behavior and Presence Sensing

4.3.1 State-of-the-Art of Occupant Sensing Technologies

This section provides a brief literature and technology review of the categories of technologies listed in Table 4.1.

Table 4.1 Overview of sensing technologies and their performance metrics

		Cost	Power type		Data storage		Deployment type		Sensing range	
	Specific sensing technologies		Battery	Wired	Internal	Network	Industry/Comm./ Public	Residential	Distance from sensor	Angle from sensor
Image-based	Video	$$$	Y	Y	Y	Y	Y	N	Infinite	90–180°
	IR camera	$$$$	Y	Y	Y	Y	Y	N	Infinite	90–180°
Threshold and mechanical	IR beam	$	N	Y	N	Y	Y	N	20 m	N/A
	Piezoelectric Mat	$$	N	Y	N	Y	Y	N	N/A	N/A
	Reed switch	$	N	Y	N	Y	Y	Y	N/A	N/A
	Door badges	$$$	N	Y	N	Y	Y	N	N/A	N/A
Motion sensing	PIR	$$	Y	Y	Y	Y	Y	Y	10 m	110°
	Ultrasonic doppler	$$	Y	Y	Y	Y	Y	Y	20 m	360°
	Microwave doppler	$$	Y	Y	Y	Y	Y	Y	20 m	360°
	Ultrasonic ranging	$$	Y	Y	Y	Y	Y	Y	4 m	90°
Radio-based	RFID	$$$	Y	N	N	Y	Y	Y	3–200 m+	N/A
	UWB	$$$	Y	N	N	Y	Y	N	3–200 m+	N/A
	GPS	$$$	Y	N	N	Y	Y	N	Infinite	N/A
	WiFi/Bluetooth	$$$	Y	N	N	Y	Y	Y	32 m	N/A
Environmental	Air properties	$$	Y	Y	Y	Y	Y	Y	Per space	N/A
	Acoustic	$$	Y	Y	Y	Y	Y	Y	Per space	360°
Human-in-the-loop	Observation	$$$$	N/A	N/A	N/A	N/A	Y	Y	N/A	N/A
	Occupant data	$$	N/A	N/A	N/A	N/A	Y	Y	N/A	N/A
	Building data	$$	Y	Y	Y	Y	Y	Y	N/A	N/A
Consumption sensing	Energy	$$	Y	Y	Y	Y	Y	Y	N/A	N/A
	Water	$$	Y	Y	Y	Y	Y	Y	N/A	N/A

(continued)

Table 4.1 continued

		Data sensed					Collection style	Accuracy	Demonstrated control applications			
	Specific sensing technologies	Presence	Count	People-tracking	Actions	State			Product on the market	Lighting	HVAC	Security
Image-based	Video	Y	Y	Y	Y	Y	Periodic/events	High	Y	N	N	Y
	IR camera	Y	Y	Y	Y	Y	Periodic/events	High	Y	N	N	Y
Threshold and mechanical	IR beam	N	N	N	N	N	Events	Low	Y	Y	N	Y
	Piezoelectric Mat	Y	N	N	N	N	Events	Low	Y	N	N	N
	Reed switch	N	N	N	Y	Y	Events	Low	Y	Y	Y	Y
	Door badges	Y	Y	Y	Y	Y	Events	Medium	Y	Y	Y	Y
Motion sensing	PIR	Y	Y	N	N	N	Events	Medium	Y	Y	Y	Y
	Ultrasonic doppler	Y	N	N	N	N	Events	Medium	Y	Y	N	Y
	Microwave doppler	Y	N	N	N	N	Events	Medium	Y	Y	N	Y
	Ultrasonic ranging	Y	Y	N	N	N	Events	Medium	N	N	N	N
Radio-based	RFID	Y	Y	Y	N	N	Periodic	Medium	Y	N	N	N
	UWB	Y	Y	Y	N	N	Periodic	Medium	Y	N	N	N
	GPS	Y	Y	Y	N	N	Periodic	Medium	Y	N	N	N
	WiFi/Bluetooth	Y	Y	Y	N	N	Periodic	Medium	Y	N	N	N
Environmental	Air properties	Y	Y	N	N	N	Periodic	Low	Y	N	Y	Y
	Acoustic	Y	Y	N	Y	Y	Periodic	Medium	Y	Y	Y	Y
Human-in-the-loop	Observation	Y	Y	Y	Y	Y	Periodic/events	High	N/A	N	N	Y
	Occupant data	Y	Y	Y	N	N	Events	Low	Y	N	Y	Y
	Building data	Y	N	N	Y	Y	Events	Medium	Y	Y	Y	Y
Consumption sensing	Energy	Y	Y	N	Y	Y	Periodic	Medium	Y	N	Y	Y
	Water	Y	Y	N	Y	Y	Periodic	Medium	Y	N	Y	Y

Note "$" represents a low-cost sensor (<$10) and "$$$$" represent an expensive sensor (>$1000), N/A Not applicable, Y used, N not used

Image-based Sensing

Currently, the primary focus of image-based occupant detection technologies is to track people as they move through spaces, commonly known as "presence" (Kamthe et al. 2009; Erickson et al. 2014; Gade et al. 2012, 2013; Kumar et al. 2014). They are used to provide ground truth information for studies with other sensors (Hutchins et al. 2007; Erickson et al. 2009; Meyn et al. 2009; Lam et al. 2009; Dong and Lam 2011; Dong et al. 2015; Li and Dong 2017) and to track occupants—for example, to study occupant interactions with windows (Inkarojirit 2005; Konis 2012), window blinds, and shades (Reinhart 2001; Kapsis et al. 2013), or occupant evacuation (Proulx and Reid 2006).

In theory, image-based occupant detection technologies detect electromagnetic information and convey it in the form of a matrix, where the information in the matrix is relative to the coverage and resolution of the sensing technology. Technologies in this category include: infrared (IR) cameras, visible light cameras, and luminance cameras. Typically IR cameras use thermopile array sensors. Visible light cameras detect human body movement by measuring depth through a combination of multi-infrared and image-based sensors. The depth sensor projects a cloud of dots that enables the sensor to gather information about the background by analyzing the projected dot diameters, and then approximating distance from the measurement device by an infrared vision camera. After appearance of an object in the cloud spectrum, the object immediately disturbs the cloud, changing the dots' diameter and enabling the device to measure a body shape. Distinguishing information of an object with information with its background enables the sensor to monitor dynamics of body movement (Seer et al. 2014). Luminance cameras, meanwhile, take a photo and measure the luminance, linking the pixel of an image to the luminance value. In both the literature and industry, visible light and luminance cameras are more common than IR cameras for occupant detection due to their relatively cheaper price.

The most advanced versions of image-based technology use detection algorithms running within the packaged visible light camera hardware to detect the direction and number of people travelling through a space (Wang and Fesenmaier 2013). Simpler approaches use visible light cameras to detect motion to indicate occupant presence (Ding et al. 2011). Figure 4.1 shows a few examples of image-based camera deployment, where (a) is a micro camera through RaspberryPI at the University of Calabria (luminance camera); (b) is a commercially available camera network (visible light camera) at the University of Texas at San Antonio (UTSA); and (c) is a stereo vision camera network (visible light camera) at South Denmark University.

Besides the application of detecting occupancy, visible light cameras built for time-lapse photography have been used to monitor a large number of blinds, as this can be done without being invasive to the occupants and data can be collected relatively inexpensively. Indeed, a major advantage of using photography is that a lot of information can be captured with a single camera. The time-lapse

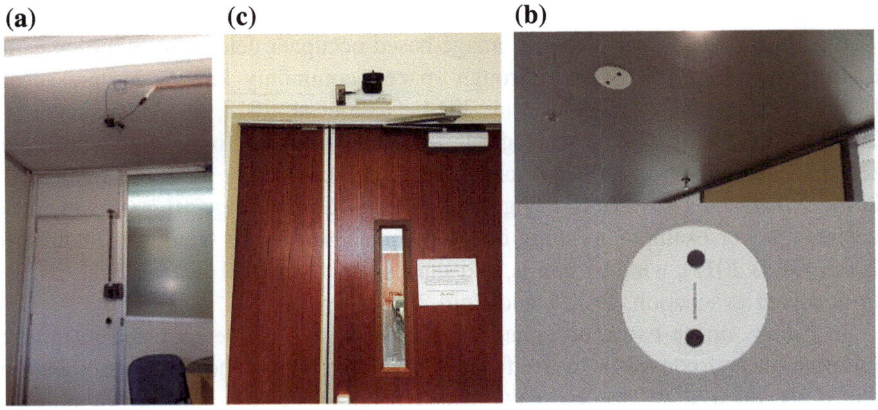

Fig. 4.1 Examples of various camera network deployment for occupant behavior studies. **a** Micro camera through RaspberryPI at University of Calabria (Italy) (Picture by Dafni Mora). **b** Stereo vision camera network at South Denmark University (Picture by Mikkel Baun Kjærgaard). **c** Commercially available camera network at UTSA (Picture by Bing Dong)

photographic method also poses several major challenges, however: photographs require significant time and cost to take and interpret, they do not yield information on indoor conditions, and they can be limited in their ability to detect e.g., Venetian blind slat angles. Manual interpretation severely limits the practicality of monitoring shades beyond days or weeks (Rea 1984), which is inadequate for developing robust occupant behavior models. Although several researchers have developed computer vision code to interpret window blind positions, this approach is imperfect, as challenges such as reflections and obstructions might still be present (Kapsis et al. 2013; Meerbeek et al. 2014).

Overall, the primary critiques of camera-based approaches are their relatively high cost, low coverage per sensor, complexity of the systems that must be attached to run the sensors, complexity of the algorithms required for advanced tracking, and failures in situations of weak or strong lighting conditions (for visible light and luminance cameras) or temperature variability (for IR cameras). While cameras are often considered as one of the most accurate forms of occupant data collection, this is only true when they are examined manually—a slow and tedious process even with tools that speed up the process. Several studies employing optical sensors (Hutchins et al. 2007; Erickson et al. 2009; Meyn et al. 2009; Brackney et al. 2012) also reported methodologies for addressing the problem of over- and under-counting inherent to cameras. For example, Meyn et al. (2009) used networked cameras that were set up to record on an automatic basis; even still, miscounts were frequent in weak lighting conditions and when occupants either stood or loitered in the camera view. Erickson et al. (2009), on the other hand, used a primitive camera capable of recording only a 64×64 pixel grayscale image, and then used a set of very lightweight image processing algorithms to detect occupant movement. Despite using different approaches, Meyn et al. (2009) and Erickson et al. (2009) both reported an over-count rate of 25% with respect to the ground truth.

Lastly, privacy concerns related to image-based sensors are significant and have led to the cancellation of at least one study (Jenkins 2007). Chapter 11 discusses privacy and research ethics at length.

Threshold and Mechanical Sensing

Threshold and mechanical sensors detect or change the state of a building component with which occupants frequently interact, such as a window (Caucheteux et al. 2013) or a door (Agarwal et al. 2010). Examples in this category include: reed contacts, which detect whether a door or a window has been opened or closed; door badges, which an occupant must swipe to access a room; piezo-electric mats, which produce an electric signal when an occupant stands or walks on them; and IR beams, which produce a signal when the beam is blocked at the entrance.

Reed contacts are low-cost and low-power sensors that are easy to mount on doors or windows. They can be used to detect whether or not a door or window is open or closed, but they cannot measure how much the door or window is open. Moreover, they may not be able to differentiate between closed and ajar positions—a subtle difference that may have profound implications.

Door badges are commonly used for access control in buildings. If the access control system is able to log the identity of the occupants that gain access and the logs can be exported, then access badges can be used for occupancy counting (Hay and Rice 2009). However, door badges are an expensive solution if implemented solely for occupancy counting and accurate only if a single person passes through an entry per card swipe.

Piezoelectric mats enable the sensing of occupants passing instrumented areas (Ranjan et al. 2013). The technology is low cost, but only accurate if people walk or stand on the mat long enough for it to observe them.

Finally, IR beams enable the counting of people passing instrumented entries where they will be counted along the specific line of the beam. The main problem with beams is under- or over-counting if multiple persons pass at the same time, or wrongly counting inanimate items blocking the beam. In the literature, these sensors are rarely used as the sole means of detecting occupants; rather, they are more often paired with other sensor types (Agarwal et al. 2011).

Figure 4.2 shows examples of threshold and mechanical sensors including: (a) typical window, door, and air conditioning switching sensor deployment using reed contacts, (b) a closer look at a door reed contact sensor, and (c) a closer look at an air conditioning reed contact sensor.

Motion Sensing

Motion sensors detect the presence or absence of an occupant through occupants' movements. The primary sensor types for this are the passive infrared (PIR) sensor, ultrasonic Doppler, microwave Doppler, and ultrasonic ranging (Agarwal et al. 2010, 2011; Hnat et al. 2012; Yavari et al. 2013). Figure 4.3 shows an example of an ultrasonic range sensor developed by UTSA that is used to count

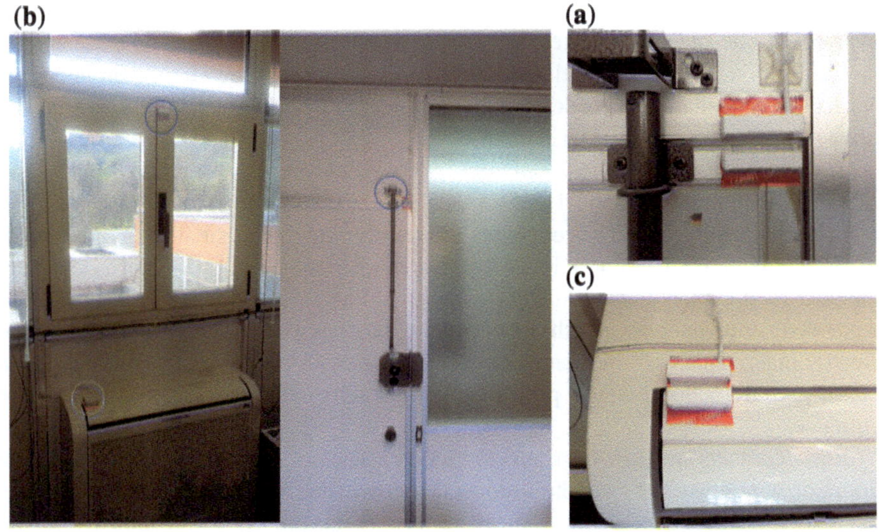

Fig. 4.2 Examples of threshold sensors (Pictures by Marilena De Simone). **a** Door contact sensor. **b** Window, door and air conditioning switching sensors deployment at University of Calabria, Italy. **c** Air conditioning switching sensor

Fig. 4.3 Ultrasonic range sensor developed by UTSA, USA

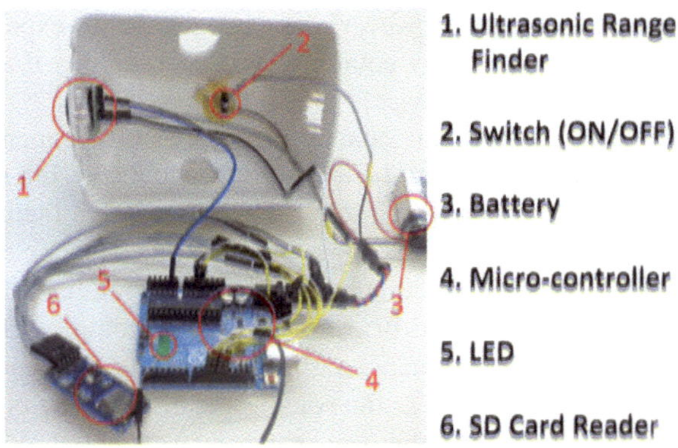

occupants in an office building. Privacy concerns for these technologies are lower than for cameras, but still exist if mounted to cover spaces with one or few known occupants.

PIR is by far the most commonly used sensor technology in this category. Most of the literature uses this sensor to conduct research such as: testing it as part of a network, using it for lighting control, using it to inform, validate, and verify occupant presence models, and using it as part of a test bed for network topologies (Agarwal et al. 2010, 2011; Dong and Lam 2011; Yavari et al. 2013; Dong et al. 2015).

PIR sensors are medium cost, but only accurate if mounted with good coverage of the areas of occupancy. These sensors often under-count because they require line of sight and go inactive when occupancy activity is low. Currently, advanced work with PIR sensors is looking at tracking individuals as they move through a space (Narayana et al. 2015).

The ultrasonic and microwave Doppler sensors measure frequency, i.e., the speed at which an object is moving towards or away from the sensor. Doppler sensors are technically developed and have greater sensitivity than PIR sensors, yet are not commonly applied for building automation. They also tend to over-count due to extreme sensitivity to smaller movements.

Ultrasonic range sensors, meanwhile, measure the distance to objects and have been used to measure motion through doorway passing events (Hnat et al. 2012). However, ultrasonic range sensors have very high sampling rate and generate a lot of data compared to other sensors, which makes data analysis a challenge. These sensors are medium cost and accuracy is only good if the mounting environment is free from ultrasonic noise and if no non-human objects are moved through the doorway.

Radio Signal Sensing

Occupant detection systems based on the measurement of radio signals have been demonstrated to provide occupancy information on user location, presence, count, identity, and movement (Li et al. 2012; Martani et al. 2012). Radio signals cover the range of electromagnetic wave frequencies from 10 kHz to 300 GHz (Misra and Enge 2011). Radio signals are transmitted from a transmitting node to a receiving node. The transmitted radio signal consists of a short series of pulses or a modulated radio signal.

Radio-signal sensing can provide three types of measurements that can be used for occupancy and occupant behavior detection:

- Proximity: Signal reception at a receiving node denotes proximity of the transmitting node to the receiving node;
- Distance: Signal properties or modulated content enable estimation of the physical distance from the transmitting node to the receiving node; and
- Distortion: Signal distortion properties at the receiving node denote that the presence of occupants has impacted signal properties.

This sensing system can be realized in different configurations, depending on if the occupant is carrying the receiving or transmitting node—or none in the case of signal distortion. For signal distortion, the transmitter and receiver might be co-located to realize a standard radar setup.

Different types of radio-based technologies have been standardized and commercialized and can be used for occupancy detection. Relevant radio technologies for occupancy detection include: Radio frequency identification (RFID), WiFi/Bluetooth, Ultra-wideband (UWB), and Global Positioning System (GPS). The main characteristics of the abovementioned four technologies are described below.

First, RFID is a technology for the automated identification of objects and people by using tags that carry a unique code or identification. A stationary reader then identifies the tags (Chiesa et al. 2002), which can be passive or active. Passive RFID tags operate without a battery within a limited range (approximately 1–2 m) and the cost of readers is relatively high. Active RFID tags are small transceivers, which can actively transmit their ID (or other additional data) in reply to an interrogation. Active tags have a much longer range (tens of meters), which makes them suitable for larger environments (Deak et al. 2012). The advantages of RFID technology are that it is non-contact and non-line-of-sight. It is also very cost-effective when scaling to many tracked objects (Koyuncu and Yang 2010). The technology can be used to collect both proximity measurements and distance measurements for more fine-grained positioning of occupants.

A second technology is WiFi, which is standard for short-range wireless communication. WiFi enables the positioning of WiFi devices by mapping fixed terrestrial private and public WiFi access points (APs) (Kjærgaard 2007). The advantage of WiFi-based positioning is that the infrastructure in the form of APs is already in place and complements the measurement error of GPS in the center of cities or indoors (LaMarca and de Lara 2008). WiFi APs can also position nearby devices by listening for WiFi signals from devices (Ruiz-Ruiz et al. 2014). A final option is to enable WiFi APs to act in a radar setup to detect occupant presence by measuring signal distortion (Sabek et al. 2015). Positioning based on WiFi has an accuracy of 2–10 m depending on the positioning method chosen (LaMarca and de Lara 2008).

Like WiFi, Bluetooth is standard for wireless short-range communication. The Bluetooth low energy (BLE) profile is an emerging technology designed as a low-power solution for control and monitoring applications (Gomez et al. 2012). BLE is used in both proximity beacons, which enable BLE devices to discover their location, and in wearable devices that enable stationary BLE sensors to discover the presence of BLE devices. In addition, proximity BLE devices can provide distance information that can be used to position each device. The device location can be determined using the method known as location fingerprinting, which boosts BLE position accuracy significantly (Faragher and Harle 2014).

A third radio technology is UWB, a wireless communication technology that makes the transmission of extremely short pulses possible, thus enabling very accurate distance and signal distortion measurements. The Federal Communications Commission (FCC) in the USA defines UWB as any signal that occupies more than 500 MHz of bandwidth in the 3.1–10.6 GHz band (Aiello and Rogerson 2003). For positioning, occupants wear transmitters while an infrastructure of UWB receivers collects distance measurements. For signal distortion, either individual or pairs of receivers measure signal distortions to detect occupants. The accuracy of the user positions is about 10–50 cm (Khoury and Kamat 2009).

Finally, GPS or other Global Navigation Satellite Systems (GNSS) are attractive options for positioning devices in outdoor environments, but are not suitable for indoor applications because they need a clear line-of-sight to orbital satellites to track position. To position occupants, the occupant needs to carry a receiver, either as a stand-alone unit or embedded in a smart device that receives signals from the satellites. Depending on the technology chosen, the accuracy in good sky conditions ranges from 1 cm to 10 m (Misra and Enge 2011).

In the end, it is important to consider that radio signals that are transmitted through air are affected by humidity, presence of other signals, and many other environmental factors that can have a high impact on results' accuracy. Several of the technologies above provide occupant location or tracks; such information may require further processing to detect the number of occupants in a particular area or the total flow of people through an area.

Mixed Sensing

Occupants interact with their indoor environment in various ways, where each interaction can be described as a stochastic process. Occupants emit heat and "pollutants" (e.g., CO_2 and odor) and generate sound, as well as open/close windows and turn lights on and off. These interactions and their effect on the indoor environment normally cannot be measured using a single sensing technology; often, a mixed sensing approach is adopted instead, where various types of sensors are used together (sensor fusion). There have been studies that combined different sensors such as multi-infrared, image-based, and acoustic sensors, to allow the monitoring of picture depth (Seer et al. 2014). For example, a device called Microsoft Kinect® projects a cloud of dots that gather information about the background by analyzing the projected diameters of the dots, and then approximating the distance from the measurement device by infrared vision camera. When paired with image-based sensors this device can precisely determine occupancy in an observed area. Figure 4.4 shows an example of deployment of Kinect sensors for a residential test bed.

As another example, Fig. 4.5 shows an information technology-enabled sustainability test bed (ITEST) developed by Dong and Lam (2011). It includes occupant sensing, data acquisition, data storage and management, and data processing. ITEST has PIR and an array of sensors, including total volatile organic compound (TVOC) concentration, cameras, CO_2, temperature, illuminance, relative humidity, and acoustic, that are used together to detect and predict occupant presence and numbers in an office building (Dong and Lam 2011).

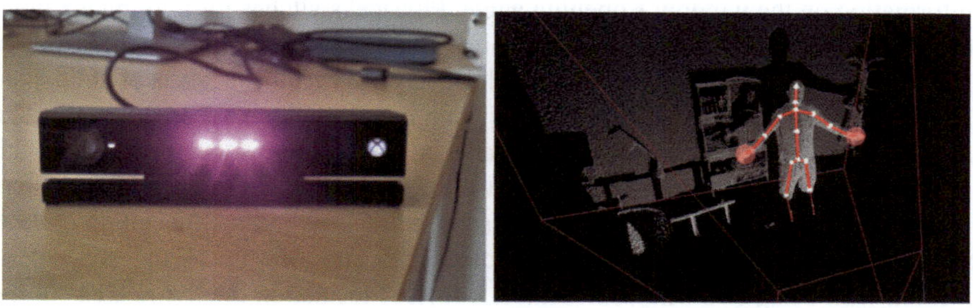

Fig. 4.4 Microsoft Kinect® with a sample raw data (Microsoft 2016) (picture by Jakub Dziedzic)

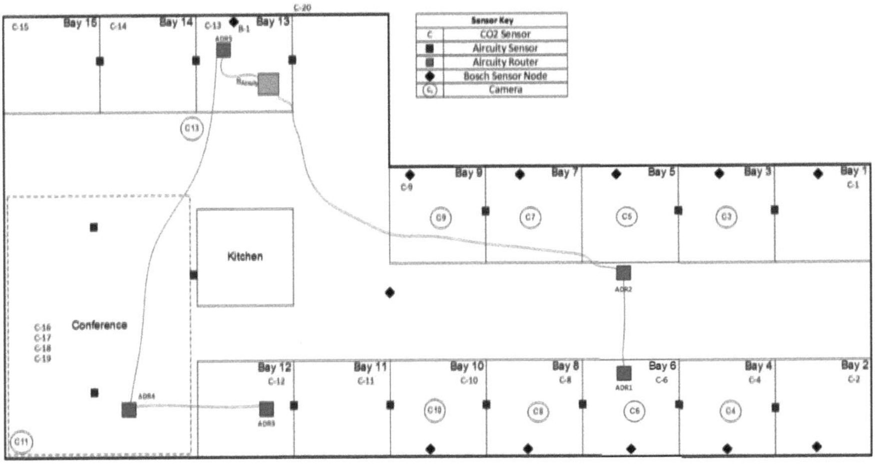

Fig. 4.5 An example of large-scale sensor network for occupant behavior detection at intelligent workplaces, Carnegie Mellon University (Dong and Lam 2011) (Drawing by Bing Dong)

Gilani and O'Brien (2016) deployed a wide array of sensing technologies, in this case to study occupant use patterns of shading and lighting systems with different user interfaces and control approaches. They deployed indoor and outdoor photo sensors measuring horizontal illuminance, motion detectors monitoring occupancy state, and indoor temperature and humidity sensors. In tandem with this mixed environmental sensing, occupants' interactions with their lights, blinds, and thermostats were monitored.

Similarly, the ZEB Living Laboratory at the Norwegian University of Science and Technology (NTNU) adopts a combination of different categories of methods for occupant behavior sensing, as shown in Fig. 4.6. Mechanical sensors (reed contacts, physical switches for the lighting system), motion sensors (PIR and mixed sensors such as Microsoft Kinect®), and electric energy meters (at the room/appliance level) are employed in the facility for user behavior detection. Reed contacts are installed on every window and on the main door to track their status—although such a system only allows open/closed status to be detected, with no information on the window's opening angle. Physical switches for lighting are input devices for the main controller of the facility; the lighting level for each light source is then managed by the main controller through power trim by means of solid-state relays. Since the status of each physical switch is continuously recorded by the data acquisition system, this information can also be used to assess user interaction with artificial lighting, and to obtain energy use for lighting down to each light source.

Further, PIR sensors and illuminance sensors are installed on the ceiling of each room in the ZEB Living Laboratory, pointing towards the floor, and are experimented with using two Microsoft Kinect® devices located in the living areas of the house. Continuous electric energy use for each of the appliances and for each room is monitored by means of electric energy meters connected to the data acquisition

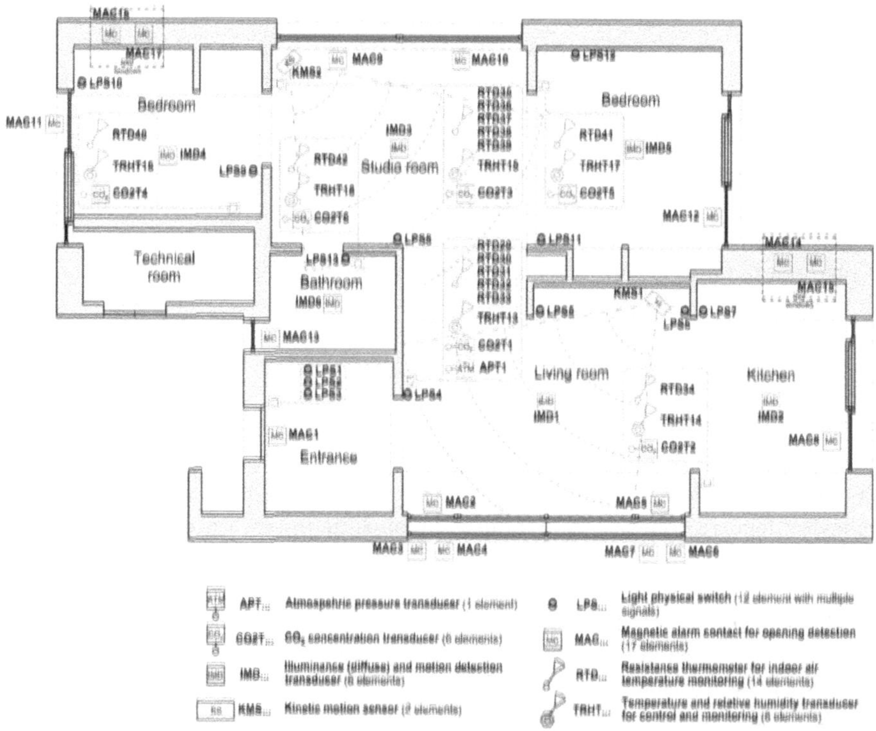

Fig. 4.6 Living lab—a multipurpose experimental facility the ZEB Living Lab, built by the Research Center for Zero Emission Buildings (www.zeb.no) (Drawing by Francesco Goia)

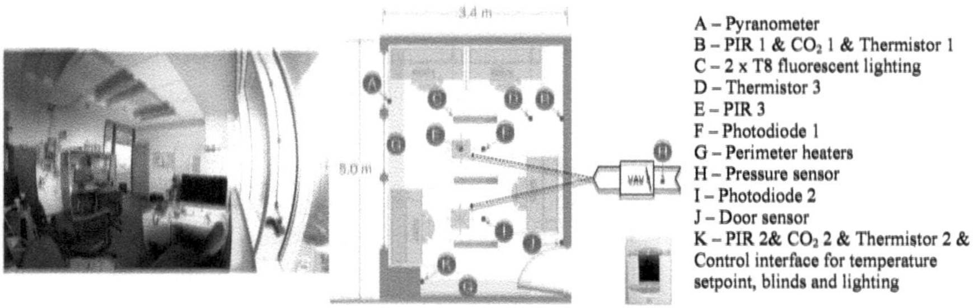

Fig. 4.7 Test bed at Carleton University (Graph by H. Burak Gunay)

system. Finally, the facility includes sensors for indoor environmental monitoring, such as indoor air temperature and temperature stratification, relative humidity, CO_2 concentration, and illuminance (diffuse). More information on the ZEB Living Laboratory can be found in Chap. 7.

The test bed facility from Carleton University (shown in Fig. 4.7) is designed with a stand-alone controls and automation network. The sensed data include commercial-grade temperature, relative humidity, CO_2, illuminance, solar radiation from a pyranometer, presence (on/off) from PIR sensors, and door on/off status from contact sensors. In addition, occupants' interactions with the motorized shades, electric lighting, and thermostat were recorded (Gunay et al. 2016).

The two main limitations of using a mixed sensing approach in test beds to collect occupant behavior and presence data are self-experimentation and the Hawthorne effect (discussed in detail in Chap. 6). For example, in the abovementioned example at Carleton University, the principal investigator's graduate students used the test bed for occupant behavior studies; therefore, the occupant data from the test beds should be cautiously interpreted.

4.3.2 Human-in-the-Loop

The human-in-the-loop method is defined by cases where occupancy and/or behavior data are collected with humans involved in the measurement. Methods in this category include: manual observation, Internet-based occupant data, and device interactions.

Manual observation covers the logging of data performed by a person directly sensing the information being relayed—for instance, counting the people walking through a hallway in person, or watching a video recorded in a building and annotating the video with occupancy information. Manual observation is often used as ground truth when evaluating the accuracy of other occupancy sensors. The method is costly due to the labor required, but potentially high accuracy if it is possible to precisely define the task so as to ensure consistency in interpretation and recording. While this method lacks some of the hard science possessed by the other categories of methods, it is, for instance, the only way to directly measure occupants' clothing level and assess personal preferences, and capture contextual factors such as physical environmental factors and psychological factors. Refer to Chap. 2 for details on contextual factors.

Internet-based occupant data covers the use of various types of data provided by occupants and collected by e.g., social networking applications, calendar data, or surveys. Although there are some privacy concerns associated with this approach (e.g., collecting sensitive information), many organizations already gather such data, which brings down the cost of occupancy sensing. Methods have been proposed combining social networking and calendar data to estimate cubicle occupancy (Ghai et al. 2012).

Device interactions cover data about occupant actions registered through their interactions with control interfaces. Common interfaces include thermostats, light switches, and controls for motorized blinds. Wall thermostats and other modern control interfaces often contain programmable buttons with which occupants' control decisions can be executed. These control decisions can include increasing/decreasing temperature setpoints, turning on/off lighting, and adjusting motorized blinds' position. The statistical analyses of data concurrently gathered from occupants' control actions render the potential for developing occupant behavior and presence models. These models have proven to be useful in building controls (Goyal et al. 2013) and design-related applications (e.g., O'Brien and Gunay 2015; Gilani et al. 2016).

A more common method of using sensors for monitoring blinds is to log occupant control of motorized blinds. This has the major advantage that the infrastructure is likely already in place, and so cost is minimal and installation during occupancy is not required. However, a major disadvantage of this method is that occupants use motorized window blinds much more than manual ones [approximately three times more according to Sutter et al. (2006)]. Thus, these results cannot be extrapolated to develop manual blind control models. A practical issue in large controls networks in commercial buildings is the database scan rate, which can be as slow as two scans per second. This can result in actions being missed—for example, an occupant may push the light switch button many times assuming that the controller missed the first signals. In addition to provoking occupant frustration, this may also affect occupants' activeness and thus cause the sensor to register false actions.

4.3.3 Consumption Sensing

Consumption sensing covers methods of measuring water and energy consumption in buildings. The accuracy of such methods depends on the level of metering spanning in granularity from one meter per building to one meter per receptacle/fixture. A better granularity of metering can also be obtained via algorithmic methods (i.e., non-intrusive load monitoring methods) that split total consumption into its individual components. The cost of such methods directly relates to the cost of installing relevant metering. Existing studies have shown that power consumption of electric appliances in offices and homes has a very high correlation with the occupancy status of the space (Zhao et al. 2014). Figure 4.8 shows an example of the system architecture of using individual power consumption meters and wearable devices to learn plug-in equipment occupant behavior. The system was deployed in the Center for Sustainable Landscape in Pittsburgh, USA, with 15

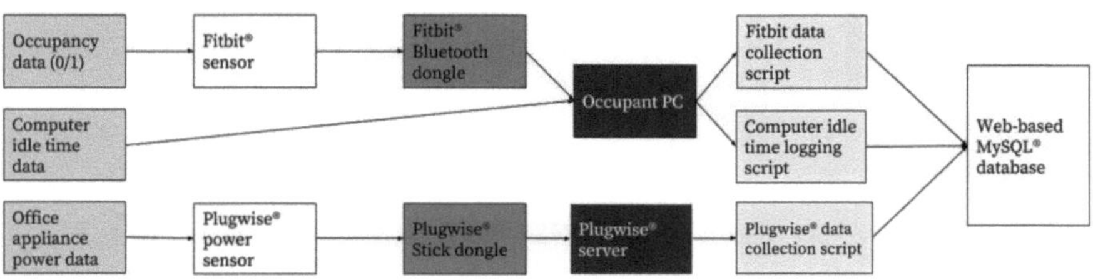

Fig. 4.8 Plug load meter data and ground truth data collection system architecture for learning occupant behavior (Zhao et al. 2014)

voluntary subjects. Wearable devices and keyboard/mouse monitoring software programs were provided to the subjects to collect ground truth data. Wireless individual plug load power meters were installed to collect subjects' computer, computer monitor, desk lamp, and other office equipment power consumption in real time. Then, the data were trained by using data mining algorithms to predict individual occupant behavior and group occupancy schedules. The study showed that the percentage of correctly classified individual office appliance usage behavior instances was above 90%. The correlation coefficient of the predicted group schedules versus the ground truth schedules was also above 0.90.

Other studies have considered the monitoring of water consumption (Ranjan et al. 2014), which requires smart water meters. The deployment of smart water meters is still far behind electricity meters.

4.4 Occupant Data Acquisition

As covered in the preceding sections, a wide range of sensing technologies is available for collecting occupant data. This section looks more in depth at methods of data acquisition, as sensors might be deployed in the area of interest for a particular study, or be part of the existing building automation and control network. Figure 4.9 illustrates four different technical configurations for occupant data acquisition: manual collection, wireless network, gateway/building automation system, and internet-enabled.

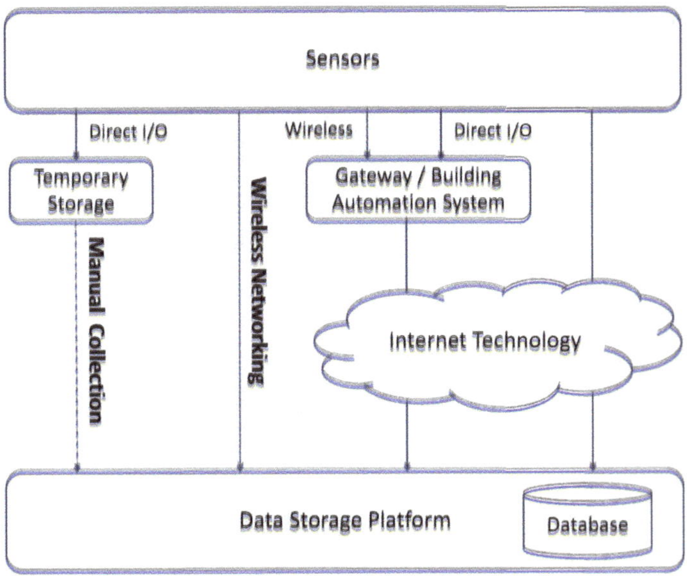

Fig. 4.9 Overview of the four different technical setups for acquisition of occupancy data (Figure by Mikkel Baun Kjærgaard)

Data acquisition cannot be discussed without consideration of data storage, however. Occupant data can be stored using different data storage platforms—for example, the occupant data from a building automation system (BAS) can be permanently stored in a commercial data archiver. The same data could also be stored in other ways, e.g., as individual files or in a database. When storing data, a number of parameters that affect the quality of the collected data must be considered. These parameters are:

- Latency: the time between measurement sampling and availability on the data storage platform for further processing;
- Granularity: the frequency of occupant data available on the storage platform;
- Robustness: the probability that occupant data is delivered to the storage platform; and
- Security: the probability that occupant data is manipulated or intercepted by a third party. Data security also has major ethical implications, as discussed in Chap. 11.

The sections below discuss the four different approaches to collecting and storing occupant data and evaluate them according to the parameters listed above.

4.4.1 Manual Data Storage

With this approach, data are collected locally on a temporary storage medium (e.g., flash storage). The collection from the sensors to the temporary storage medium might be implemented with a sensor node consisting of a smartphone or a small size computer board. The sensors can then be connected to the sensor node based on either local input/output (I/O) or local networking. The data collected on temporary storage are then manually collected and copied to a more permanent data storage platform. This setup is relevant to consider when occupancy data are collected in a remote location where it can be difficult or expensive to establish a permanent network connection from sensors to a data storage platform.

Latency can be an issue with this method, as data are not available on the storage platform until they are manually collected. However, the granularity of data will not be restricted—though the capacity of the temporary storage might limit it. In terms of robustness, the probability that all collected data arrive at the data platform is lowered by several potential reasons for data loss. First, the use of a temporary storage medium might result in data loss if it is damaged, lost, or stolen during the collection period. Second, the manual procedure for collecting data from the temporary storage might result in data loss if the person collecting the data does not follow the procedures correctly. Third, if the temporary storage is not large enough, data losses may occur, as logging systems often provide the choice to either write over the first record or stop recording upon reaching maximum capacity. In terms of security, the temporary storage medium is open for attacks by third parties that could potentially delete, copy, or manipulate data. These security problems can be minimized by the use of access control and data encryption on the storage medium. For example, in a case study by D'Oca et al. (2014) conducted on the effects of thermostat and window use behavior in houses, the researchers periodically visited the houses under investigation to gather and back up the data stored in the internal memory of the occupancy sensors.

4.4.2 Wireless Network

In this setup, data is collected from sensor nodes to the data storage platform using wireless networking. A range of wireless networking technologies is available (i.e., ZigBee, Bluetooth, WiFi, SigFox, EnOcean), each with different properties in terms of robustness, latency, throughput, range, and energy consumption. If the wireless networking technology is not capable of creating a direct link between a sensor node and a data storage platform, options also exists for multi-hop networking. Multi-hop transmission enables data to reach the storage platform passing through several sensor nodes.

The latency of wireless network-based setups is generally low, but increases in cases where special power savings schemes are used that involve sensor nodes going to sleep mode, or in the case of multi-hop networking where intermediate nodes might delay transmission. The maximum granularity of data depends on the throughput of the wireless networking technology or on energy considerations in battery-powered sensor nodes. As for robustness, the probability that all collected data arrive at the data platform is lowered by several potential causes for data loss. For instance, wireless networking can experience interruptions in connectivity due to interference issues or high traffic volumes, which result in data loss if data cannot be buffered at the sending sensor node. Finally, security is an issue, as wireless networking is open to attacks by third parties that could result in interruptions in connectivity or copying of data. This issue can be addressed by proper encryption of data, though encryption can impact sensor battery life, if applicable.

As an example case, consider Dong and Lam (2011), who studied occupant behavior and indoor conditions in a living lab set-up. Wireless sensors used were self-configured and had a flexible installation. The easy reconfiguration is especially important in a living environment due to spatiotemporally varying sensing needs. Wireless sensor nodes with sensors for temperature, relative humidity, light level, PIR-based motion activity, and absolute sound level were installed. The wireless sensors delivered data to a data storage platform wirelessly through a base station.

4.4.3 Gateway or Building Automation System

Gateways and Building Automation Systems (BASs) collect data from the sensors using direct I/O, local networking, or wireless networking. In modern buildings, occupant actions with light switches, thermostats, and motorized blinds can be registered by a control network. In addition, many commercial buildings are equipped with a range of sensors for monitoring the indoor climate. Some of the common building sensor types pertaining to occupant modeling include: motion detectors, CO_2 sensors, relative humidity sensors, photo sensors, thermistors, and current sensors. A modern BAS in commercial buildings provides access to real-time sensory data—representing a low-cost and non-invasive way to collect data. Sensing data can be stored and accessed from the BAS in different ways depending on the particular system: (1) from a local controller database, (2) a system-wide data archiver, or (3) a data archiver with cloud storage. In residential buildings, a gateway can be used to collect data from several different sources since a full BAS is not typically available. The gateway collects data and sends them to the data storage platform via Internet technology. The storage platform might be hosted on a server or on a cloud platform.

The latency of gateway and BAS setups can be low if measurements arrive quickly at the gateway and if the Internet connection from sensor node to storage platform has a low latency. The granularity of data is not limited by these setups if the Internet connection supports a high throughput. In terms of robustness, the probability that all collected data arrive at the data platform is generally high. The main issue is if the Internet gets disconnected either at the gateway or at the storage platform—buffering or temporary storage at the gateway or in the BAS can help avoid data loss in such situations. Although it is possible to budget the limited data storage capacity by adjusting the sampling frequency, there is uncertainty on how fast the data will overflow. Finally, security may be an issue, as Internet communication is open to attacks by third parties, which could result in interruptions in connectivity or copying of data. Like in wireless network setups, this issue can be addressed by proper encryption of data.

As an example case, Gunay et al. (2016) considered the three data storage solutions mentioned above. Inside the controller's database, the occupant data were stored as trend log objects. The major limitation of this approach is that the storage capacity of controllers tends to be small (e.g., less than 500 kB). Depending on the sampling frequency and the number of trend logs in the controller, one would need to back up the data from these local controllers within weeks before the newer data starts overriding the older values. Furthermore, some occupant data need to be stored on a per event basis (e.g., light switches, motion detections). Although it is possible to budget the limited data storage capacity by adjusting the sampling frequency, there is uncertainty about how fast the event-based data will overflow. In the building Gunay et al. (2016) studied, a data archiver was available. A data archiver device scans the control network and permanently stores the occupant data. The data gathered inside the data archiver can be accessed through a physical connection to the device. The data in the archiver is also duplicated in a cloud storage that can be accessed via the Internet.

Another example of a gateway application is Kleiminger et al. (2013) who detected occupancy from electricity consumption data. To collect data from a number of private households they installed a range of sensors in each home, including smart electricity meters and meters for plug-loads. Each sensor provided access to the measurements by different protocols and networking technologies. A gateway was installed to collect the locally data and then upload it over the Internet to a data storage platform hosted on a university server.

4.4.4 Internet-Enabled Sensors

In this setup, sensors are Internet-enabled, thus making direct communications between the sensors and the data storage platform possible. The platform might be hosted on a server or on a cloud platform, and the sensors might push the data to the platform or the platform might pull data from the sensors. The Internet-enabling of sensors is part of a trend targeting the development of Internet of Things (IoT) products and services. Notably, even though sensors are Internet-enabled they might not be accessible on the public Internet, but rather on a local subnet for security reasons. This creates some limitations on the physical placement of the data storage platform and might result in a need for a gateway that can access the local subnet and forward data over the public Internet as in the preceding setup.

The latency of Internet-enabled sensors can be low if the Internet connection from sensor node to storage platform has a low latency. The granularity of data in this setup is not limited if the Internet connection supports a high throughput. Moreover, the probability that all collected data arrive at the data platform (i.e., robustness) is generally high. The main issue is if the Internet connection gets disconnected either at the gateway or at the storage platform—in such situations, buffering at the sensor can help avoid data loss. Lastly, in terms of security, Internet communication is open to attacks by third parties that could result in interruptions in connectivity or copying of data. Again, this issue can be addressed by proper encryption of data.

As an example case, Kjærgaard et al. (2016) studied the total occupancy in commercial buildings through the deployment of a number of 3D stereovision cameras for counting people at all entrances to a building. Each camera is Internet-enabled and provides an application program interface (API) for either pulling or pushing data to a data storage platform.

Another example is the experimental setup built in two offices of the University of Calabria (Italy), where the device was designed to monitor presence and movement of the occupants in their offices. The thermo-physical properties of the internal environments and the electricity consumptions connected to the use of equipment were also collected. In this case, the necessity to work with remote input/output led to choosing devices that transmit data over hardware standard technology, Ethernet (hardware protocol) and standard HTTP (Hypertext Transfer Protocol). Furthermore, the system is able to gather, monitor, and archive analog and digital I/O values over the Internet/intranet. A master box is used to collect data from digital and analog web-IOs in a central embedded My SQL database. For this component, low energy consumption products were chosen.

4.5 Other Related Sensing Technologies

4.5.1 Indoor Environmental Sensing

There is a diverse set of sensing technologies beyond those discussed above, including those for measuring indoor environmental quantities of interest—for example, CO_2, temperature, illuminance, relative humidity, acoustic, and volatile organic compounds sensors. Sensor technologies are available to measure each of the four main elements of indoor environmental quality (IEQ): indoor air quality, thermal comfort, aural comfort, and visual comfort. In the context of occupant research, the primary reasons for measuring indoor environmental quality parameters are to establish relationships between triggering conditions and occupant actions and to provide additional proxies for presence. A consideration for selecting the indoor environmental quality-related parameters to measure is the eventual application of the data. For instance, measuring TVOCs as a predictor for window-opening behavior may have limited application in building simulation because TVOC is not commonly predicted in most mainstream simulation tools. Chapter 2 provides a comprehensive literature review on common occupant actions and the corresponding predictors that have been found to be significant or insignificant.

Also, as discussed at length in Chaps. 5, 6, and 7, a major challenge in deploying sensors is ensuring that they measure representative conditions as experienced by the occupant(s). Careful thought about position and orientation and avoiding obstruction are required. Alternatively, scientific-grade sensors connected to data acquisition systems can be installed, but these systems are frequently prohibitively expensive for widespread deployment and large sample sizes.

From this chapter's authors' experience, carbon dioxide concentration and acoustic sensors are considered among the most effective IEQ-related sensors at detecting occupant presence. For every sensor in this category, except acoustic sensors, there is a delay between occupants entering a space and occupant presence detection. This delay means that using these sensors as the sole source of occupant detection is often a poor choice. Acoustic sensors can relieve some of the errors induced by this approach, but due to the inherent variability in the amount of sound that occupants produce, they are rarely effective on their own. In total, the majority of the IEQ sensors reported in papers are used in conjunction with other sensors when detecting occupants. Temperature and relative humidity are critical for HVAC controls and CO_2 and total volatile organic compounds (TVOC) sensors are also sometimes used in that capacity, but not through their occupant detection abilities. Figure 4.6 provides an example of environmental sensor deployment.

4.5.2 Outdoor Environmental Sensing

Weather conditions have a great impact on buildings' energy consumption and indoor environmental quality (IEQ). Traditionally, historical weather data [e.g., typical meteorological year (TMY) or actual meteorological year (AMY) data] are used as input data for building energy modeling, lighting and shading modeling, and other performance analysis at the building design and commissioning stages for decades. Increasingly, real-time weather data is being used in building controls during building operation. For example, weather data can be directly fed into the BAS as control inputs to provide better IEQ and energy efficiency. In a recent study by Zhao et al. (2015), outdoor air temperature, relative humidity, wind speed and direction, and rainfall data were used to determine and optimize the natural ventilation control logic to provide comfort while reducing energy consumption. Weather data can also be used to create dashboards to increase awareness of sustainability for building occupants and visitors. Especially in museums, schools, and other educational buildings, weather data can be used as a tangible teaching tool to explain thermodynamics and other physical phenomena in the built environment.

There are two ways to acquire real-time weather data. One is using professional online weather services data, such as Weather Underground, a commercial weather service (https://www.wunderground.com/). These web services typically provide various APIs, which can be easily integrated with BASs and dashboards via web protocols. However, a major disadvantage is that the majority of the weather data in these services is from weather stations located at airports, and so the weather data may not accurately reflect the actual local climate of the building site.

The other way to acquire real-time weather data is to install an on-site weather station. These weather stations typically consist of various environmental sensors (solar radiation, wind speed and direction, air temperature, barometric pressure, relative humidity, CO_2, rainfall, etc.), a data logger, wired/wireless communication systems, and other supporting equipment. Some weather stations generate and store their own electrical power using photovoltaic and/or wind power systems. Wind characteristics tend to be very sensitive to sensor placement, particularly in the built environment where the airflow regime is turbulent. Measurement of solar radiation is similarly complex due to shading and reflections. Moreover, it is often necessary or convenient to have both direct and diffuse solar radiation separately. Direct measurement of direct (or beam) solar radiation requires a solar tracker, which can be prohibitively expensive (at least several tens of thousands of dollars). Alternatively, at least one available product measures global horizontal irradiance (i.e., total) and diffuse horizontal irradiance, from which direct normal irradiance can be estimated fairly accurately. This approach does not have moving parts and is thus considerably less costly and prone to failure than solar trackers. In many cases, it may be desirable to measure the solar irradiance that is incident on a façade. This can be achieved cost-effectively by mounting pyranometers on the weather station parallel to the façade of interest. CO_2, relative humidity, and air temperature sensors should be positioned away from building exhaust systems and should be ideally positioned upwind of them (based on the predominant wind direction). Local weather stations often provide accurate local climate information to be integrated with BAS systems as control inputs. The real-time weather data can also be uploaded to the database of online weather services in real-time to be used by other building operators and researchers.

In summary, researchers and building professionals have acknowledged the importance of real-time weather data to the building energy efficiency and IEQ. As more advanced building control systems are adopted by high performance buildings, it is likely that more weather stations and real-time web services will be implemented in the near future.

4.6 Conclusion

Over the last few decades, most occupant behavior field research has relied on a single motion or contact sensor to study the underlying motivations for human behavior. Recently, with vast development of the Internet of Things (IoT), occupant sensing and data acquisition are not limited to a single node. This chapter provided an overview of the current state-of-the-art on those technologies. In total, there were six listed categories of sensing technologies: image-based, threshold and mechanical, motion sensing, radio-based, human-in-the-loop, and consumption sensing. The applications of those technologies in occupant behavior research include: occupant presence, people counting, human building interactions such as turn on/off lights, thermostats and window blinds adjustment, energy consumption impacts of miscellaneous loads, and tracking movement. A number of examples of occupant sensing field test beds were given. In addition, nine performance metrics were developed and defined to evaluate existing sensing technologies. Such metrics provide a guideline for future occupant behavior researchers when field data collection is needed.

One future occupant sensing techbology could be "peel and stick" with minimum maintenance (DOE 2015); however, the challenges of such self-configured and long-lasting deployment of occupant sensing remain. Those challenges can be discussed in terms of four categories: (1) sensing element, (2) power consumption, (3) processing, and (4) communication. First, from this chapter, it is known that sensing elements fundamentally work based on physical laws. Current sensing elements still cannot satisfy certain building system requirements, for example, HVAC controls. The detection of the number of occupants is not accurate enough—even to detect if there is a human or not—for temperature setback or ventilation controls. There is thus a need for the development of advanced sensing elements that can capture occupants in a thermal zone accurately.

Second, most sensors depend on an external power source for a long-term experiment. This creates cost and wiring challenges for large-scale deployment, such as in a whole building. The power consumption mainly comes from three different sources on sensor board: processing units, communication, and the sensing element itself. There is a need to research low-power consumption mechanisms so that an occupant sensor can be self-sticking and self-contained over a long time. For example, how can sensors adaptively control their sensing power consumption as needed? Another concept with proven potential is energy harvesting, whereby sensors have on-board power collection and storage technologies (e.g., photovoltaic and piezoelectric generators).

Third, the current data processing unit is either on-board or cloud-based. An on-board processing unit consumes power and a cloud-based requires high data security. The question remains how to intelligently collect and process on-board data in a way that sensors will consume the least power possible. For example, there is no need to collect data if there is no occupant in a space for a certain period.

Fourth and finally, communication determines how frequent the data should be sent out for storage. Communication typically consumes the most power (>60%) of the whole sensor unit. The challenge is to determine how to perform communication as needed, and how to set up a communication network so that it minimizes the total data transmission power required.

Using this chapter, future researchers can choose the most appropriate technologies or technology paths to acquire and store occupancy data according to their particular study or applications. The next issue is how to utilize collected data for occupant behavior research—this is the focus of Chap. 5.

References

Agarwal Y, Balaji B, Dutta S, Gupta RK, Weng T (2011) Duty-cycling buildings aggressively: the next frontier in HVAC control. Proceedings of the 10th international conference on information processing in sensor networks (IPSN). ACM/IEEE, Chicago, pp 246–257

Agarwal Y, Balaji B, Gupta R, Lyles J, Wei M, Weng T (2010) Occupancy-driven energy management for smart building automation. Proceedings of the 2nd ACM workshop on embedded sensing systems for energy-efficiency in building (BuildSys'10), Zurich, November 2010. ACM, New York, pp 1–6

Aiello GR, Rogerson GD (2003) Ultra-wideband wireless systems. IEEE Microw Mag 4:36–47. doi:10.1109/MMW.2003.1201597

Brackney LJ, Florita AR, Swindler AC, Polese LG, Brunemann GA (2012) Design and performance of an image processing occupancy sensor. Proceedings of the second international conference on building energy and environment, Boulder, August 2012. COBEE, Boulder, pp 987–994

Caucheteux A, Es Sabar A, Boucher V (2013) Occupancy measurement in building: a literature review, application on an energy efficiency research demonstrated building. Int J Metrol Qual Eng 4:135–144. doi:10.1051/ijmqe/2013044

Chiesa M, Genz R, Heubler F, Mingo K, Noessel C, Sopieva N, Slocombe D, Tester J (2002) RFID: a week long survey on the technology and its potential. In: Harnessing Technology Project, Interaction Design Institute Ivrea. http://www.erasme.org/IMG/RFID_research.pdf. Accessed 15 Sept 2016

Deak G, Curran K, Condell J (2012) A survey of active and passive indoor localisation systems. Comput Commun 35:1939–1954. doi:10.1016/j.comcom.2012.06.004

Department of Energy (2015) DOE peer review for Building Technologies Office (BTO) sensors and controls technologies and emerging technologies R&D program. https://energy.gov/sites/prod/files/2015/05/f22/2015%20BTOpr%20Overview_Sensors%20and%20Controls.pdf

Ding D, Cooper RA, Pasquina PF, Fici-Pasquina L (2011) Sensor technology for smart homes. Maturitas 69:131–136. doi:10.1016/j.maturitas.2011.03.016

Dong B, Lam KP (2011) Building energy and comfort management through occupant behaviour pattern detection based on a large-scale environmental sensor network. J Build Perform Simul 4:359–369. doi:10.1080/19401493.2011.577810

D'Oca S, Fabi V, Corgnati SP, Andersen RK (2014) Effect of thermostat and window opening occupant behavior models on energy use in homes. Build Simul 7:683–694. doi:10.1007/s12273-014-0191-6

Dong B, Li Z, Mcfadden G (2015) An investigation on energy-related occupancy behavior for low-income residential buildings. Sci Techno Built Environ 21(6):892–901

Duarte C, Van Den Wymelenberg K, Rieger C (2013) Revealing occupancy patterns in an office building through the use of occupancy sensor data. Energy Build 67:587–595. doi:10.1016/j.enbuild.2013.08.062

Erickson VL, Lin Y, Kamthe A, Brahme R, Surana A, Cerpa AE, Sohn MD, Narayanan S (2009) Energy efficient building environment control strategies using real-time occupancy measurements. Proceedings of the 1st ACM workshop on embedded sensing systems for energy-efficiency in buildings (BuildSys 2009) in conjunction with ACM SenSys 2009, Berkeley, November 2009. ACM, New York, pp 19–24

Erickson VL, Carreira-Perpiñán MÁ, Cerpa AE (2014) Occupancy modeling and prediction for building energy management. ACM Trans Sens Netw. doi:10.1145/2594771

Faragher R, Harle R (2014) An analysis of the accuracy of bluetooth low energy for indoor positioning applications. Proceedings of the 27th international technical meeting of the satellite division of the Institute of Navigation (ION GNSS + 2014), Tampa, September 2014. ION publications, Folsom, pp 201–210

Gade R, Jørgensen A, Moeslund TB (2012) Occupancy analysis of sports arenas using thermal imaging. Proceedings of the international conference on computer vision theory and applications, Rome, February 2012. SCITEPRESS digital library, Setùbal, pp 277–283

Gade R, Jørgensen A, Moeslund TB (2013) Long-term occupancy analysis using graph-based optimisation in thermal imagery. Proceedings of the IEEE conference on computer vision and pattern recognition (CVPR'13), Portland, June 2013. IEEE Computer Society, Washington, pp 3698–3705

Ghai SK, Thanayankizil LV, Seetharam DP, Chakraborty D (2012) Occupancy detection in commercial buildings using opportunistic context sources. IEEE international conference on pervasive computing and communications Workshops (PERCOM Workshops), Lugano, March 2012. IEEE, Washington, pp 463–466

Gilani S, O'Brien W (2016) Review of current methods, opportunities, and challenges for in-situ monitoring to support occupant modelling in office spaces. J Build Perform Simul 0:1–27. doi:10.1080/19401493.2016.1255258

Gilani S, O'Brien W, Gunay HB, Carrizo JS (2016) Use of dynamic occupant behavior models in the building design and code compliance processes. Energy Build 117:260–271. doi:10.1016/j.enbuild.2015.10.044

Gomez C, Oller J, Paradells J (2012) Overview and evaluation of bluetooth low energy: an emerging low-power wireless technology. Sensors 12:11734–11753. doi:10.3390/s120911734

Goyal S, Ingley HA, Barooah P (2013) Occupancy-based zone-climate control for energy-efficient buildings: complexity versus performance. Appl Energy 106:209–221. doi:10.1016/j.apenergy.2013.01.039

Gunay HB, Fuller AF, O'Brien W, Beausoleil-Morrison I (2016) Detecting occupants' presence in office spaces: a case study. In: eSim 2016, McMaster University, Hamilton, 3–6 May 2016

Haldi F, Robinson D (2010) Adaptive actions on shading devices in response to local visual stimuli. J Build Perform Simul 3:135–153. doi:10.1080/19401490903580759

Hay S, Rice A (2009) The case for apportionment. Proceedings fo the first ACM workshop on embedded sensing systems for energy-efficiency in buildings (BuildSys'09), Berkeley, November 2009. ACM, New York, pp 13–18

Hnat TW, Griffiths E, Dawson R, Whitehouse K (2012) Doorjamb: unobtrusive room-level tracking of people in homes using doorway sensors. Proceedings of the 10th ACM conference on embedded network sensor systems (SenSys'12), Toronto, November 2012. ACM, New York, pp 309–322

Hutchins J, Ihler A, Smyth P (2007) Modeling count data from multiple sensors: a building occupancy model. 2nd IEEE international workshop on computational advances in multi-sensor adaptive processing (CAMPSAP), Virgin Islands, December 2007. IEEE, Washington, pp 241–244

Inkarojirit V (2005) Balancing comfort: occupants' control of window blinds in private offices. Dissertation, University of California

Jenkins CJ (2007) The weakly identifying system for doorway monitoring. Dissertation, Duke University

Kamthe A, Jiang L, Dudys M, Cerpa A (2009) SCOPES: Smart cameras object position estimation system. Proceedings of the 6th European conference on wireless sensor networks (EWSN'09), Cork, February 2009. Springer, Berlin, pp 279–295

Kapsis K, O'Brien W, Athienitis AK (2013) Time-lapse photography and image recognition to monitor occupant-controlled shade patterns: analysis and results. 13th conference of international building performance simulation association, Chambèry, August 2013. IBPSA, France, pp 3712–3719

Khoury HM, Kamat VR (2009) Evaluation of position tracking technologies for user localization in indoor construction environments. Autom Constr 18:444–457. doi:10.1016/j.autcon.2008.10.011

Kjærgaard MB (2007) A taxonomy for radio location fingerprinting. Third international symposium on location and context awareness (LoCA), Oberpfaffenhofen, September 2007. Springer, Berlin, pp 139–156

Kjærgaard MB, Johansen A, Sangogboye F, Holmegaard E (2016) Occure: an occupancy reasoning platform for occupancy-driven applications. 19th International ACM SIGSOFT symposium on component-based software engineering (CBSE), Venice, April 2016. IEEE, Washington, pp 39–48

Kleiminger W, Beckel C, Staake T, Santini S (2013) Occupancy detection from electricity consumption data. Proceedings of the 5th ACM workshop on embedded sensing systems for energy-efficiency in buildings, Rome, November 2013. ACM Publications, New York, pp 1–8

Konis K (2012) A method for measurement of transient discomfort glare conditions and occupant shade control behavior in the field using low-cost CCD cameras. In: American Solar Energy Society (ASES) National Solar Conference, Denver, 13–17 May 2012

Koyuncu H, Yang SH (2010) A survey of indoor positioning and object locating systems. Int J Comput Sci Netw Secur 10:121–128

Kumar S, Marks TK, Jones M (2014) Improving person tracking using an inexpensive thermal infrared sensor. In: Proceedings of the 2014 IEEE Conference on Computer Vision and Pattern Recognition Workshops (CVPRW'14), Columbus, June 2014. IEEE computer society Washington, pp 217–224

Lam KP, Höynck M, Dong B, Andrews B, Chiou Y-S, Zhang R, Benitez D, Choi J (2009) Occupancy detection through an extensive environmental sensor network in an open-plan office building. Proceedings of building simulation (IBPSA'2009), Glasgow, July 2009. IBPSA, England, pp 1452–1459

LaMarca A, de Lara E (2008) Location systems: an introduction to the technology behind location awareness. Synth Lect Mob pervasive Comput 3:1–122. doi:10.2200/S00115ED1V01Y200804MPC004

Li Z, Dong B (2017) A new modeling approach for short-term predictions of occupancy presence in residential buildings. Building and Environment 121:277–290

Li N, Calis G, Becerik-Gerber B (2012) Measuring and monitoring occupancy with an RFID based system for demand-driven HVAC operations. Autom Constr 24:89–99. doi:10.1016/j.autcon.2012.02.013

Martani C, Lee D, Robinson P, Britter R, Ratti C (2012) ENERNET: Studying the dynamic relationship between building occupancy and energy consumption. Energy Build 47:584–591. doi:10.1016/j.enbuild.2011.12.037

Meerbeek B, Gritti T, Aarts M, Van Loenen E, Aarts E (2014) Building automation and perceived control: a field study on motorized exterior blinds in Dutch offices. Build Environ 79:66–77. doi:10.1016/j.buildenv.2014.04.023

Meyn S, Surana A, Lin Y, Oggianu SM, Narayanan S, Frewen TA (2009) A sensor-utility-network method for estimation of occupancy in buildings. Proceedings of the 48th IEEE conference on decision and controll. doi:10.1109/CDC.2009.5400442

Misra P, Enge P (2011) Global positioning system: signals, measurements, and performance. Ganga-Jamuna Press, Lincoln

Narayana S, Prasad RV, Rao VS, Prabhakar TV, Kowshik SS, Iyer MS (2015) PIR sensors: characterization and novel localization technique. Proceedings of the 14th international conference on information processing in sensor networks (IPSN'15), Seattle, April 2015. ACM, New York, pp 142–153

O'Brien W, Gunay HB (2015) Mitigating office performance uncertainty of occupant use of window blinds and lighting using robust design. Build Simul 8:621–636. doi:10.1007/s12273-015-0239-2

Proulx G, Reid IMA (2006) Occupant behavior and evacuation during the Chicago cook county administration building fire. J Fire Prot Eng 16:283–309. doi:10.1177/1042391506065951

Ranjan J, Yao Y, Whitehouse K (2013) An RF doormat for tracking people's room locations. Proceedings of the 2013 ACM international joint conference on pervasive and ubiquitous computing, Zurick, September 2013. ACM, New York, pp 797–800

Ranjan J, Griffiths E, Whitehouse K (2014) Discerning electrical and water usage by individuals in homes. In: Proceedings of the 1st ACM Conference on Embedded Systems for Energy-Efficient Buildings (BuildSys'14). Memphis, November 2014. ACM, New York, pp 20–29

Rea MS (1984) Window blind occlusion: a pilot study. Build Environ 19:133–137. doi:10.1016/0360-1323(84)90038-6

Reinhart CF (2001) Daylight availability and manual lighting control in office buildings-simulation studies and analysis of measurements. Dissertation, University of Karlsruhe

Ruiz-Ruiz AJ, Blunck H, Prentow TS, Stisen A, Kjaergaard MB (2014) Analysis methods for extracting knowledge from large-scale WiFi monitoring to inform building facility planning. In: IEEE International Conference on Pervasive Computing and Communications (PerCom). doi:10.1109/PerCom.2014.6813953

Sabek I, Youssef M, Vasilakos AV (2015) ACE: an accurate and efficient multi-entity device-free WLAN localization system. IEEE Trans Mob Comput 14:261–273. doi:10.1109/TMC.2014.2320265

Seer S, Brändle N, Ratti C (2014) Kinects and human kinetics: a new approach for studying pedestrian behavior. Transp Res Part C Emerg Technol 48:212–228. doi:10.1016/j.trc.2014.08.012

Sutter Y, Dumortier D, Fontoynont M (2006) The use of shading systems in VDU task offices: a pilot study. Energy Build 38:780–789. doi:10.1016/j.enbuild.2006.03.010

Wang D, Fesenmaier DR (2013) Transforming the travel experience: the use of smartphones for travel. In: Cantoni L, Xiang Z (eds) Information and Communication Technologies in Tourism 2013. Springer, Berlin, pp 58–69

Yavari E, Jou H, Lubecke V, Boric-Lubecke O (2013) Doppler radar sensor for occupancy monitoring. IEEE topical conference on wireless sensors and sensors networks (WiSNet), Renaissance, January 2013. IEEE, Washington, pp 216–218

Zhao J, Lasternas B, Lam KP, Yun R, Loftness V (2014) Occupant behavior and schedule modeling for building energy simulation through office appliance power consumption data mining. Energy Build 82:341–355. doi:10.1016/j.enbuild.2014.07.033

Zhao J, Lam KP, Ydstie BE, Karaguzel OT (2015) EnergyPlus model-based predictive control within design–build–operate energy information modelling infrastructure. J Build Perform Simul 8:121–134. doi:10.1080/19401493.2014.891656

Chapter 5
Introduction to Occupant Research Approaches

William O'Brien, Andreas Wagner and Julia K. Day

Abstract There are numerous methods of collecting occupant-related data for the purpose of researching building occupants, each with its own strengths and weaknesses. The objective of this chapter is to guide the decision-making process for researchers who are about to embark on a new occupant data collection campaign. This chapter introduces Chaps. 6–8 by overviewing four methods for occupant research: in situ, laboratory, survey, and virtual reality. For each method, the advantages and disadvantages are laid out based on findings in the literature and the authors' experiences. Next, a comprehensive list of occupant-related phenomena of interest is provided, along with a qualitative discussion of the merits of each data collection method for studying them. Finally, mixed methods research approaches—whereby multiple, complementary approaches are adopted in a single study—are briefly discussed. Following this chapter, the reader is presented with three chapters that provide recommended best practice for each of in situ (Chap. 6), laboratory (Chap. 7), and survey (Chap. 8) methods to researching occupants in occupants.

5.1 Introduction

Background information has been provided in the previous chapters about the scope of occupant behavior, as well as research design and sensor technologies; now, specific methods for studying occupants and collecting data need to be discussed.

W. O'Brien (✉)
Department of Civil and Environmental Engineering, Carleton University,
3432 Mackenzie Building, 1125 Colonel By Drive, Ottawa, ON K1S 5B6, Canada
e-mail: liam.obrien@carleton.ca

A. Wagner
Building Science Group, Faculty of Architecture, Karlsruhe Institute of Technology,
Englerstr. 7, 76131 Karlsruhe, Germany
e-mail: wagner@kit.edu

J.K. Day
School of Design and Construction, Washington State University, PO Box 645815, Pullman,
WA 99164, USA
e-mail: julia_day@wsu.edu

A. Wagner et al. (eds.), *Exploring Occupant Behavior in Buildings*,
DOI 10.1007/978-3-319-61464-9_5

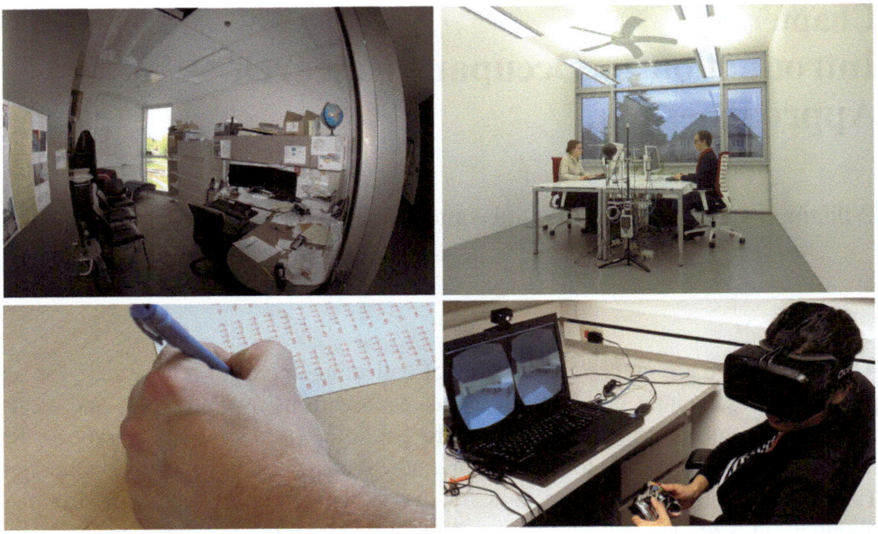

Fig. 5.1 Occupant measuring methods. Clockwise from *top-left*: In situ, laboratory, virtual reality (Heydarian et al. 2015b), and survey

This chapter introduces three major approaches to monitoring or otherwise studying occupants that will be further explored in upcoming chapters: in situ (Chap. 6), laboratory (Chap. 7), and survey questionnaire (or interview) studies (Chap. 8). It also briefly describes an emerging technique for studying occupant comfort and behavior: virtual reality. The chapter provides an overview of the methods to allow readers to decide on the best approach(es) for their specific research problem. After reading this chapter and the next three, readers will be able to answer the following questions:

- Which research methods are most suitable for understanding the occupant-related phenomena of interest?
- What research strategies and technologies have been used in the past and which have been successful?
- What is established best practice for applying the research methods?
- What practical challenges are sure to arise, and how can they be overcome?
- How should the research methods be documented and communicated to the audience?

First, each occupant research method is defined and discussed regarding their merits and shortcomings. Illustrative example applications of each method are provided in Fig. 5.1. Then, some commentary is provided on the suitability of each of the methods for measuring common occupant actions of interest and predictor variables. Finally, the notion of combining approaches in a mixed method study is also discussed.

5.2 Primary Occupant Research Approaches

This section briefly introduces each of the research approaches and then compares the primary three in Table 5.1.

Table 5.1 Summary of advantages and disadvantages of the three main methods for studying occupants

Method	Advantages	Disadvantages
In situ monitoring	• Suitable and cost-effective for long-term monitoring • Possibility to cover a great range of different scenarios (home, offices, etc.) • Relatively low-cost, particularly if existing sensors of the building automation system are adequate • Occupants are less aware that they are being monitored and thus behavior is more "pure" (natural)	• Some quantities cannot be measured using sensors • Sensors cannot explain all complexities of occupant behavior • Sensor position may be constrained because of interference with occupied spaces or concern for damage • Contextual information related to behavioral actions may be difficult or even impossible to retrieve • Sample size may be limited to the number of occupants in the studied building • Occupant awareness of monitoring may lead to bias • Significant degree of research ethics ambiguity
Laboratory experiments	• Full control over environment and indoor climate with possibility for reconfiguration • Fewer constraints on recruitment (relative to in situ studies) • Unconstrained access to equipment and the building space (e.g., for repairs) • Greater control over experiment, particularly to study adaptive opportunities • Possibility to use advanced sensors and other equipment because the same equipment can be used to measure all participants	• Occupants are aware that they are being monitored • Unfamiliar environment, technology, and fellow participants could introduce bias • Costly to build and operate • Costly for large sample sizes
Survey methods	• Insight into immeasurable (with sensors) phenomena (e.g., perception, attitudes, and personal characteristics) provided • Open-ended questions can reveal new insights that had not occurred to the researchers • Least costly for large sample sizes	• Subject to self-reporting bias • Low polling frequency makes it difficult to convert data to mathematical models for simulation • Sensors are not included—thus, corresponding contextual information and predictors are not available

In situ studies, described in detail in Chap. 6, involve monitoring occupants in their natural environments (workplace, home, school, etc.) and typically have a long data collection duration (weeks or years). Data are normally acquired passively through sensors, either built-in as part of the building automation system (BAS), or newly installed for research purposes that detect occupants, adaptive actions, energy use, and predictive variables, such as indoor environmental quality (Haldi and Robinson 2010; Pigg et al. 1996; Duarte et al. 2013). Because in situ studies use existing environments they are generally preferable for replicating reality (i.e., external validity) for the purposes of obtaining data for occupant modeling (De Dear 2004). In situ studies may also reduce the Hawthorne effect, the notion that knowledge of being studied affects occupants' behavior, as further discussed in Chap. 6. However, in situ monitoring does not necessarily provide detailed contextual insights about behavior, it has privacy implications, and it takes a considerable amount of time and effort to set up and collect data (O'Brien and Gunay 2014; Rea 1984; McLaughlin et al. 2011; Fogarty et al. 2006). Moreover, use of existing occupied spaces limits the flexibility of experiments, while research visits to the space can be invasive for occupants. In contrast to the other occupant research methods, in situ methods often limit the sample size to the number of willing participants in the subject building. Lack of flexibility in sensor placement because of the constraint of not interfering with occupants' activities or avoiding disturbance of measurements by occupants' interference (e.g., covering of illuminance sensors on a desk) can pose some measurement challenges that reduce accuracy and may introduce errors (Reinhart and Voss 2003; Andersen et al. 2013). While existing built-in sensors can provide a cost-effective method for collecting data, adding, maintaining, and removing additional sensors and related infrastructure—and the labor for doing so—can become costly for large sample sizes (e.g., more than ten). Ethics, participant recruitment, and informed consent remain fundamental challenges for this approach (Gilani and O'Brien 2016). Research ethics and privacy implications for in situ studies are discussed at length in Chap. 11.

Laboratory studies, covered at length in Chap. 7, involve recruiting participants to spend time and interact with a fabricated environment that is specifically intended for scientific study of building performance and occupant behavior and/or comfort (Meinke et al. 2017; Clausen 2004). In the past several decades, numerous laboratory environments have been built, mostly for studying comfort, and more recently also for investigating occupant behavior. Many of them look like real indoor environments, but are heavily equipped with sensors and have greater control over layout, technologies, and environmental conditions. This degree of control offers a significant experimental advantage over in situ studies, where spaces are less flexible and are not designed for studying occupants. Basically, a wide range of indoor environmental scenarios can be simulated according to an experimental design. For example, the impact of specific indoor environmental variables on occupants' perceptions or reactions by interacting with different devices can be systematically tested (e.g., Wyon and Sandberg 1996; Schweiker et al. 2013). Moreover, the social impact of the presence of other occupants on occupants' adaptive actions can be measured very efficiently (Schweiker and

Wagner 2016). Furthermore, laboratory studies offer greater flexibility over recruiting participants because subjects do not have to be occupying a specific building (as for in situ studies) and can be selected based on pre-defined criteria. A briefing of the participants prior to an experiment with specific instructions ensures an "as-planned" procedure, including formal entrance and exit surveys or interviews. A disadvantage of laboratory studies is that facilities for occupant research are typically costly to build and operate. Likewise, the experiments themselves are significantly more expensive compared to in situ studies, mainly due to human resources (supervising research assistants) and, to a much lesser extent, to recruitment and compensation of participants. Another downside is that the short-term and potentially unnatural characteristics of some laboratory environments may influence occupants in complex ways. For instance, an occupant in a laboratory study may perceive their environment differently compared to someone under stress from work in a real office. Schweiker and Wagner (2016) addressed this issue by having study participants work on their regular work tasks during a one-day test. Similarly, sensor equipment that is visible to participants reminds them that they are being monitored, which may constrain their behavior. Another issue with laboratory studies can be the presence of unknown persons in an experimental setting, which may influence their perceived sense of control over the indoor environment (Hawighorst et al. 2016). Finally, participating occupants may operate windows, blinds, fans, etc. not according to the experimental design, even if instructed beforehand, thus leading to undesired or unexpected results.

Surveys differ considerably from the previous two occupant research methods and are covered in detail in Chap. 8. Unlike the other methods, they rely on self-reporting of personal behaviors (Vine 1986), either by filling out questionnaires or through interviews and focus groups. This method can reveal the logic and rationale behind habits and behaviors in ways that sensor-based methods do not (Day et al. 2012). Oftentimes, post-occupancy evaluation (POE) studies rely on surveys to understand how well a building is functioning, including occupant comfort and satisfaction (Cohen et al. 1999; Wagner et al. 2012). Occupants seem to detect and report failures of systems which lead to discomfort (Gossauer and Wagner 2008; Wagner et al. 2012). Surveys are a cost-effective means to achieving a large sample size and can measure phenomena that would be difficult or impossible to measure with sensors (e.g., thermal comfort sensation and clothing level). Several recent studies (Becerik-Gerber et al. 2011; Konis 2013; Haldi and Robinson 2008) have relied on custom technological survey solutions for polling occupants more frequently than what a telephone, paper, or online survey would allow. Surveys have also been used to develop models (e.g., Haldi and Robinson 2008). While there are many benefits to using surveys in occupant research, a number of established psychological biases, including the Hawthorne effect and social desirability bias (i.e., propensity for participants to answer according to what they think is socially desirable, rather than truthfully), suggest that self-reported behaviors may not always match observed ones (McCambridge et al. 2014). In addition, it is possible that lack of understanding of different building services systems or misinterpretation of questions may cause occupants to unknowingly report things incorrectly; researchers should thoughtfully consider their approach to both survey design, particularly the formulation of questions, and occupant responses. A final disadvantage of survey studies is that relative to in situ and laboratory monitoring approaches they typically do not facilitate frequent sampling because they rely on occupants' active input and, therefore, may be less suited for longitudinal studies. Despite these limitations, surveys are an effective tool for improved understanding of occupant behaviors, and can be used to narrow down predictors for in situ and laboratory studies.

An emerging approach in the study of occupant behavior is virtual reality-based immersive environments (Heydarian et al. 2015a). Rather than traditional laboratory or in situ approaches, the virtual reality approach immerses study participants in interactive building environments. This approach offers the advantages of relatively low costs (a real environment is not required) and greater control over environmental conditions, room layout, and user interfaces. At the same time, however, the present equipment is primarily limited to the visual or acoustic domains (not thermal or indoor air quality). In addition, many of the disadvantages mentioned under laboratory studies are also true here; particularly, virtual scenes do not represent the subjects' natural environment. An unfamiliar environment without standard tasks may cause occupants to perceive less control and greater discomfort than in familiar environments with everyday stresses and distractions.

5.3 Objective Comparison of Approaches

The three primary occupant research methods—in situ, laboratory, and survey—vary significantly with regards to capability and domains of interest. For instance, clothing cannot be easily measured using sensors, but temperature cannot be measured using surveys. Table 5.2 provides a summary of the merits of measuring commonly studied variables using the three methods.

5.4 Mixed Methods Research Design

Oftentimes, it may be appropriate or necessary to exploit the benefits of several of the above methods to achieve research goals. Indeed, there is considerable strength in combining, or mixing, methods. For instance, one method may have insurmountable obstacles that can be solved effectively using a second method. Or, multiple methods may be used in the same study for triangulation purposes, i.e., to form greater confidence and approach a problem from several directions (Creswell and Clark 2007). There may also be strategic benefits: for example, a low-cost, short-term survey may be used to screen out insignificant factors, which can then help narrow the focus of a laboratory or in situ study. While any research design requires careful thought and planning, mixed methods designs demand additional

Table 5.2 Summary of variables and the practicality of measuring them using the three main study methods

Variable	In situ	Laboratory	Survey/questionnaire
Adaptive comfort measures			
Clothing level	Difficult because there are no available sensors apart from cameras, which are associated with significant privacy concerns.	Can be monitored by a questionnaire filled in either by subjects or study assistant (including changes over the period of the experiment). Alternatively, clothing level can be estimated by inspecting photo/video-based observations from inside (subjects have to be informed beforehand). Can also be regulated by experimental design.	Due to privacy concerns, surveys are most suitable for clothing level measurement, but results are subjective and may be biased Except for longitudinal surveys, they cannot provide much insight about long-term clothing level.
Drinking hot or cold beverages	Difficult as there are no available sensors apart from cameras, which are associated with significant privacy concerns.	Can be monitored by counting the number of cups or bottles having been consumed during the experiment. Also, balance between consumed and secreted fluids can be measured using scales at the beginning and the end of the experiment.	Surveys can be used to ask occupants for self-reported beverage consumption, but results may be inaccurate.
Changing posture	Several commercial posture sensors exist and typically are built into chairs, mounted near the occupant, or are worn by the occupant; video is also possible but could violate privacy; see discussion of Microsoft Kinect in Chap. 4.		Surveys can be used to ask occupants for self-reported posture, but direct observations or measurements would likely be more accurate.
Window opening/closing	Can be measured using contact sensors or by inspecting photo/video-based observations from outside the building. It is difficult to detect the extent of opening for hinged windows and difficult to detect window opening for sliding windows using contact sensors. If motorized operable windows are used, manual adjustments can be logged via the BAS, but the extent of motorized movement may have to be determined via timing of motor runtime.		Survey questions can be used to ask occupants if: (a) they have control over operable windows, (b) how often windows are adjusted, and (c) how the windows are used and (d) what factors trigger window use behavior. Longitudinal window movement data relies on self-reporting.

(continued)

Table 5.2 (continued)

Variable	In situ	Laboratory	Survey/questionnaire
Blinds/shades adjustment	Controlled and motorized blind positions (if integrated to the BAS) can be detected via BAS. Otherwise, they can be detected by inspection of photos/video from outside. It is difficult to detect venetian blind slat angle and shade position, but measurement could be achieved using proximity sensors or rotary encoders that are mounted inside the blind shaft or on the cord.		Survey questions can be used to ask occupants if: (a) they have control over blinds, (b) how often blinds are adjusted over time, (c) to what level they are adjusted, and (d) what factors trigger blind use behavior. Longitudinal blind movement data relies on self-reporting.
Light switching/dimming	For advanced BAS, light switch events can be logged if lights are integrated with the BAS. Lights can be inferred from predetermined step changes measured by illuminance sensors or possibly power use profiles (likely with some error).		Survey questions can be used to ask occupants if: (a) they have control over lighting, (b) how often lights are adjusted, (c) to what level lights are adjusted, and (d) what factors trigger light use behavior. Longitudinal light use data relies on self-reporting.
Use of fans (desk, ceiling, and ventilation system)	Ceiling fans and ventilation systems may be integrated into BAS, which can be used to log occupant inputs. Desk fans could be metered for experimental purposes.	Use of desk and ceiling fans can be monitored via the control and data acquisition system or BAS. Air velocity near workplaces can be measured with air speed sensors (which are part of the comfort meters used for determining thermal comfort parameters).	Survey questions can ask individual occupants if and how they use fans in their workspace or home.
Door opening	Can be measured with contact sensors, but must be combined with occupancy sensing to predict occupant travel direction.	Can be measured with contact sensors, but must be combined with occupancy sensing to predict occupant travel direction. Door opening behavior may not needed as experiments are normally designed to study behavior in the designated space. Leaving and re-entering the room (e.g. for going to the toilet or lunch) is normally registered by the research assistant.	Survey questions can ask about door use, but observation would be best because results would depend on self-reported behavior. Asking occupants to record departure events requires them to anticipate this action if the recording medium is in the studied space.

(continued)

Table 5.2 (continued)

Variable	In situ	Laboratory	Survey/questionnaire
Thermostat adjustments for central heating and cooling systems	If integrated with the BAS, manual thermostat adjustments can be detected and logged. New smart thermostats provide considerable data and can be accessed via the internet.		Survey questions can be used to ask occupants if: (a) they have control over thermostats, (b) how often controls are adjusted, and (c) to what level they are adjusted.
Use of independent space heaters and cooling (single room)	Can be measured using meters, current transducers, or electricity load disaggregation.		Survey questions can ask individual occupants if they use space heaters or personal fans in their workspace or home.
Non-adaptive behaviors			
Presence/number of occupants	Sensors for space monitoring passive-infrared PIR), CO_2 concentration, door position, occupant interactions with other monitored control interfaces. Sensors for individual tracking (RFID and access cards) are also available. Refer to Chap. 4 for details on technologies.	The number of subjects in the designated test space is normally fixed by the experimental design and does not change; otherwise, the research assistant normally logs occupancy data.	In a commercial building, an individual occupant would likely not know the total number of occupants in a building; in a residential scenario, the surveyed occupant would likely know the total number of occcupants.
Use of electrical/gas appliances	Can be identified via the control and data acquisition system or BAS if the devices are connected; otherwise, they can be measured by submetering or load disaggregation methods.	Can be identified via the control and data acquisition system or BAS if the devices are connected; otherwise, they can be measured or by submetering or load disaggregation methods. Gas appliances are not normally used in labs.	Survey questions can ask individual occupants if and when they use electrical/gas appliances (and which ones).
Hot and cold water consumption and water heating energy use	Most water meters are not equipped to provide a high temporal or spatial resolution. Adding water or hot water heater power meters is costly.	Water is not normally provided in labs; otherwise, as above.	Survey questions can ask individual occupants to estimate total hot and cold water consumption, but it is unlikely that an accurate volume would be reported.

(continued)

Table 5.2 (continued)

Variable	In situ	Laboratory	Survey/questionnaire
Indoor environmental parameters			
Air temperature	Air temperature sensors are one of the most common types in living spaces, as part of thermostats. Single sensor points are more valid if air is well mixed (e.g., for entrainment ventilation; not displacement ventilation); regardless, position is critical (e.g., must not be near heat sources or long-term sun exposure). If they are not part of the BAS, temperature sensing and logging capabilities can be provided by wide variety of available technologies.	Labs allow various options for measuring air temperature, ranging from single sensor points to a spatial sensor distribution (x, y, z), depending on the purpose of a study and the spatial dimensions of the lab. For determining thermal comfort indices air temperature is measured close to a person (or its assumed position) according to ASHRAE Standard 55. Sensors are mounted at three different heights which differ for sedentary or standing occupants. Temperature sensors should generally not disturb occupants and interfere with their movements and actions.	Occupants cannot be expected to reliably report indoor environmental quality-related values and may be biased by confounding factors, but questions can be asked to assess the occupants' overall perception of and satisfaction with comfort conditions.
Relative humidity (RH)	Same as above, but less common. RH sensors must not be near humidity sources (e.g., occupants, kettle); they must also not be near heating or cooling sources, since local heating or cooling of the sensor will affect the RH measurements.	Same as above; necessity of high spatial resolution is minimal if no specific experiments with regards to humidification/dehumidification are performed.	Same as above.

(continued)

Table 5.2 (continued)

Variable	In situ	Laboratory	Survey/questionnaire
Mean radiant temperature (MRT)	MRT can be very position-sensitive, especially if cold or warm surfaces are present (e.g., radiant heating or large windows). Ideally MRT can be measured from center of occupied space. MRT can be estimated using black bulb thermometers.	MRT can be estimated using black bulb thermometers. For determining thermal comfort indices the sensor is positioned close to a person (or its assumed position) at a specific height taking into account sedentary and standing occupants, according to ASHRAE Standard 55. In addition, all surface temperatures of the space enclosure can be measured with sensors fixed directly beneath the surface (except windows), normally at several points of a surface if it is large.	Same as above.
Carbon dioxide (CO_2) concentration	CO_2 loggers are available, but at a considerable cost and have similar constraints as other IEQ sensors. Calibration becomes an issue over time (e.g., months), though self-calibrating CO_2 sensors are available.	Same as above; CO_2 sensor distribution depends on the purpose of a study and the spatial dimensions of the lab; normally little necessity of high spatial resolution.	Same as above.
Other air contaminants and odor	Sensors to measure VOCs and other common contaminants are available, though not commonly included in BAS except in specialized environments (e.g., chemical laboratories and parking lots).	Same as above; VOC sensor distribution depends on the purpose of a study and the spatial dimensions of the lab; normally necessity of high spatial resolution is little. As air quality is	Same as above.

(continued)

Table 5.2 (continued)

Variable	In situ	Laboratory	Survey/questionnaire
	Odors cannot be directly measured, only the concentration of gas causing them.	normally controlled in lab experiments, there is no necessity to monitor VOC and other pollutants in the lab (unless a specific experiment, e.g., on intended pollution, would require it). Contaminants can be monitored directly in the inlet airflow of the ventilation/air-conditioning system. Perception of odors by subjects can also be part of an experiment in a lab.	
Local air speed and direction	Air speed and direction, as sensed by occupants, are difficult to measure in situ because of the challenges of locating sensors near occupants.	Same as above, sensor distribution depends on the purpose of a study and the spatial dimensions of the lab. For determining thermal comfort indices omnidirectional air speed is measured close to a person (or its assumed position) according to ASHRAE Standard 55. Sensors are mounted at three different heights which differ for sedentary and standing occupants. A high responsivity (<0.2 m/s) is necessary to approach draft phenomena. Sensors should not disturb occupants and interfere with their movements and actions.	Same as above.
Illuminance	Workplane illuminance is difficult to measure in situ due to the risk of occupants obstructing them (e.g., with desktop items). It is possible to estimate	Labs permit the measurement of illuminance levels in any plane or surface. For desktop illuminance, subjects should to be instructed to not	Same as above.

(continued)

Table 5.2 (continued)

Variable	In situ	Laboratory	Survey/questionnaire
	workplane illuminance from ceiling illuminance (where sensors are often located for daylighting-based lighting control), but this can introduce uncertainty, especially in rooms with specular surfaces.	obstruct sensors. For visual comfort evaluation, the cylindrical illuminance level is needed which can be derived from measuring the vertical illuminance in the four main directions at the workplace (and the height of the subjects' eye position). In principle, illuminance levels at the eye of the subjects can be measured with special devices (e.g., mounted on eyeglasses), as well. Additionally, RGB (red green blue) sensors (calibrated photo-diodes) can be used for identifying the light color in a space. Sensors should not disturb occupants and interfere with their movements and actions.	
Glare-related quantities	Luminance-based glare metrics (e.g., DGP, DGI) can be calculated using cameras in spaces that are oriented the same direction as occupants, though there may be little certainty about occupant position and orientation.	Luminance levels can be derived from camera images. Cameras are either oriented towards facades with the highest probability of glare or (preferably) into the view angle of a subject. Participants must to be informed about the use of cameras, as movements/actions can be tracked, as well. Alternatively, high dynamic range (HDR) cameras mounted on the subject's head allow determining luminance levels at a subject's eyes. Cameras should not disturb occupants or interfere with their movements and actions.	Same as above.

(continued)

Table 5.2 (continued)

Variable	In situ	Laboratory	Survey/questionnaire
Noise and other acoustic properties	Sound level meters may be installed in buildings where their location is central to avoid isolation from noise. Signal processing could be used to estimate intentional noise (e.g., voices) versus unwanted background noise (e.g., HVAC, transmission from neighbouring zones). Care to respect privacy is required for indoor spaces (i.e., only sound pressure level should be recorded rather than the entire audio signal).	Sound levels, or preferably sound pressure levels due to privacy issues, are measured with microphones. Sensor distribution depends on the purpose of a study and the spatial dimensions of the lab. Signal processing could be used to estimate intentional noise (e.g., voices) versus unwanted background noise (e.g., HVAC, transmission from outdoors). Reverberation time of the lab can be measured at the beginning of the experiment (with subjects in the space). Very specific requirements regarding sound absorption/reflection from the space enclosure occur in acoustic labs.	Same as above.
Exterior environmental parameters			
Air temperature	Air temperature is normally relatively uniform in outdoor environments as long as temperature sensors are shielded from solar radiation and not located near objects that emit heat (e.g., exhaust air, vehicles, cooling towers, and dark-colored facades that are exposed to solar radiation). It is often reasonably accurate for most occupant monitoring campaigns to use local weather station data, though sound engineering judgment is required.	Same as in situ; however, air temperature should be measured on site to include site-specific influences (built environment, etc.) on outdoor temperature.	With knowledge of survey participant location, weather conditions can be estimated using local weather stations to improve contextual insights.

(continued)

Table 5.2 (continued)

Variable	In situ	Laboratory	Survey/questionnaire
Relative humidity	Same as for air temperature, but care must be taken to avoid sources of humidity or dry air (e.g., cooling towers).		Same as above.
Local wind speed and direction	Can be measured using an anemometer and wind vane, though wind speed and direction are highly variable in turbulent environments such as in urban environments.		Same as above.
Solar radiation, direct and diffuse	At least global horizontal radiation should be measured on site. Additionally, diffuse radiation can be measured with a shadow ring pyranometer. If the lab has window facades, total vertical solar radiation in the directions of the windows should be measured, as well. Pyranometers should be placed to minimize shading from adjacent buildings and other objects.		Same as above.
Sun position	Solar geometry can be derived with knowledge of time/date/building site; building orientation is needed for relative sun position (e.g., surface-solar azimuth and solar profile angle).		For digital surveys (e.g., online or smart phone-based), the sun position could be calculated if location is provided and time is logged.
Cloud cover	Cloud cover data is often available from meteorological stations.		Occupants can be asked to report the exterior sky conditions at the time the survey is taken so that responses can be measured/evaluated appropriately.
Rain	Rainfall data is often available from meteorological stations or on-site weather stations.		Occupants can be asked to report the exterior sky conditions at the time the survey is taken so that responses can be measured/evaluated appropriately.
Outdoor noise	Can be measured using sensors (microphones). The main objective for collecting sound data is likely for it to serve as a predictor for window opening position. Thus, sound should be measured near each façade with care to aim directional sensors normal to facades. Properties of noise may be of interest, including periodicity, sound pressure level, and frequency-dependence of sound levels.		Occupants can be asked to assess overall perception of and satisfaction with the acoustic environment (exterior noise entering the building).
Pollutants/air quality	CO_2 and other common contaminants should preferably be measured near the building's/lab's windows/facades in order to relate data to indoor concentrations.		N/A.

consideration to ensure methods are compatible and do not interfere with each other.

At this point it should be noted that the authors of this chapter use the term "mixed methods" as a broad term that is not limited to the combination of different research paradigms, i.e., qualitative and quantitative, as it is often intended. Rather, "mixed methods" is used in a way that might be referred to elsewhere as "combined methods" or "multiple methods", whereby multiple approaches to data collection and analysis—whether qualitative, quantitative, or both—are used in a single research study. In other words, "mixed methods" here is a question of methods (data collection and analysis), and not one of methodology (ontology, epistemology, axiology—i.e., the "worldview" that underpins research design). This application of the term intentionally circumvents the extensive discussions of research paradigm that have been known to dominate mixed methods discussions (e.g., in social sciences). These discussions are important and touched upon in Chap. 2, but are far beyond the scope of this chapter. For a more in-depth discussion, the reader is encouraged to refer to texts on mixed methods research (e.g., Creswell and Clark 2007; Teddlie and Tashakkori 2009).

Mixed methods studies can be designed in a number of ways, all with the common feature of combining multiple methods (qualitative, quantitative, or both) in a single study. If qualitative ("qual") and quantitative ("quan") methods are combined, the combination can have greater weight on one or the other. Alternatively, both parts might have equal weight in the final results. For example, a quantitative survey might be followed by qualitative interviews that provide additional insight into survey results, but ultimately the research question demands a quantitative result (a "what" question, not a "how" question). In this case, the survey results would be featured more prominently than the interview results and study would be considered quantitative-dominant—or QUAN-qual. If the same study were conducted with a qualitative research question, however, it would be considered a quan-QUAL design, where the order of phases remains the same, but the emphasis is on the qualitative element (e.g., interviews). Recognizing that mixed methods is adaptable to a variety of research questions and needs makes it a potentially very fruitful choice.

A common classification system for mixed methods is: convergent parallel, exploratory sequential, explanatory sequential, and embedded (Creswell and Clark 2007). Briefly, a convergent parallel design involves quantitative data collection and analysis and qualitative data collection and analysis to be performed in parallel, and then compared and interpreted together. An exploratory sequential design involves using qualitative methods (first phase) to help inform quantitative methods (second phase). An explanatory sequential design starts with quantitative data methods (first phase) followed by qualitative methods (second phase) that help explain the quantitative data. Finally, an embedded design involves applying a quantitative or qualitative method, but embedding a lesser amount of the opposite (e.g., a quantitative survey with a few qualitative questions within in). In contrast to convergent parallel studies, embedded studies involve analyzing the qualitative and quantitative data together, rather than performing separate analyses. Beyond the four basic mixed methods designs, there is also the more advanced multiphase design, which involves a combination of qualitative and quantitative methods performed either in series or in parallel, with each phase informing the next one. Further details of the four basic designs and illustrative examples from the field of occupant research are provided below. Note that the studies presented may not perfectly match the approaches as discussed above, but nonetheless serve to introduce the reader to some of the possibilities of mixed methods designs.

Convergent parallel research designs allow researchers to quantify occupant actions and obtain a better understanding of cause and effect (e.g., performing a survey (QUAL) and measuring behaviors in situ (QUAN) for the same group of participants). In this way, a convergent parallel design can play a critical role in supporting methodological research endeavors in building occupancy studies. For instance, this design can be used in future research that aims to determine whether occupants behave similarly in laboratories and their natural environments (e.g., offices, classrooms, homes) (Schweiker and Wagner 2016). Likewise, the research design can be used to determine whether self-reporting is a reliable means to collect longitudinal behavior data relative to sensor-based methods (e.g., Fabbri 2016).

There are several relevant examples of occupant behavior studies that use a convergent parallel design. Gunay et al. (2014) measured temperatures in 40 apartments for four months in the heating season to understand occupants' thermostat-related behavior. The occupants also performed an extensive survey during this time to better understand their attitudes and behavior towards control of heating. As per the definition of convergent parallel design, the qualitative and quantitative analyses were performed in parallel, and then compared for final interpretation. Building upon this work, Bennet and O'Brien (2017) combined six months of apartment temperature and relative humidity measurements with a survey at both the beginning and end of the measurement period. This allowed participants to be surveyed with the same comfort-related questions in both the summer and winter, while enabling logistical efficiency because the equipment was set up during the first survey and retrieved during the second survey. Day et al. (2012) distributed a questionnaire to 35 office participants to understand how and why they use their window blinds and their corresponding satisfaction with them. The researchers also photographed the exterior of the building over three 10-day periods to quantify actual window blind use.

There are several other notable mixed methods studies that employ a convergent parallel design using predominantly quantitative methods. For example, Konis (2013) used detailed measurements about indoor environmental quality (e.g., air temperature, mean radiant temperature, and window luminance) coupled with a quantitative survey on occupant satisfaction with comfort. Haldi and Robinson (2008) conducted a longitudinal study using pop-up computer surveys with office workers about things that would be impractical to measure in another way (e.g., clothing level, drinking actions); meanwhile, indoor air temperature and weather conditions were simultaneously measured and recorded. This study is not strictly parallel convergent because the data from each measurement instrument were correlated immediately rather than in parallel, but it is nonetheless is a valuable example of parallel data collection. Wagner et al. (2007) conducted a four-week monitoring campaign in an office building where they recorded indoor and outdoor temperatures, as well as relative humidity continuously in several rooms. This was accompanied by a survey conducted twice a week, in the morning and afternoon, to study thermal adaptation by retrieving comfort temperatures of the occupants under varying outdoor conditions.

Explanatory sequential mixed methods designs are appropriate for situations where the quantitative data that was collected cannot be fully explained by the data alone, and qualitative methods may offer more insight. Meerbeek et al. (2014) conducted a study whereby they monitored office workers' window blind usage, and then asked select participants to keep a diary to help explain the rationale behind their blind movement actions. Day and Gunderson (2015) similarly applied an explanatory design to study the relationship between occupant knowledge of passive building systems and behavior, comfort, and satisfaction. In their study, first, a survey was conducted across ten high-performance buildings (n = 118), and then follow-up interviews were conducted with several of the survey participants (n = 41) to better understand the results of the survey.

Exploratory sequential designs are particularly well suited for researching building occupants because qualitative methods (e.g., focus groups) can be used to identify the most important phenomena to measure in follow-up quantitative laboratory or in situ studies. Given the cost to conduct laboratory and in situ studies, as discussed in Chaps. 6 and 7, identifying the most important measurement equipment is critical. An exploratory sequential design is not as common as previous methods in the occupant behavior literature; however, as observed by O'Brien et al. (2013), there has been a trend over the past decades from qualitative and exploratory research to quantitative research. Undoubtedly, the quantitative research has benefitted tremendously from foundational work of the last three decades of the 20th century.

Lastly, an example of embedded research design is Gilani and O'Brien (2017), where the primary researcher took the opportunity to converse with occupants to better understand comfort in 25 private offices as she configured and placed the sensors. The primary goal of the study was to quantify how behavior affected building energy, but these informal discussions yielded interesting and unexpected insights (e.g., that a few occupants reported that fritted glass was causing them to have headaches).

5.5 Conclusion

The introductory chapters in this book dealt with fundamental occupant research concepts and data collection technologies. This chapter introduced and compared data collection approaches for occupant research campaigns. In all, four approaches were described: in situ, laboratory, survey, and virtual reality. The first three are addressed in more depth in Chaps. 6–8. Finally, following a discussion of each of the above methods, this chapter provided some guidance on combining or mixing methods to yield greater insights and provide validation.

Overall, the literature suggests that in situ methods yield greater external validity and the potential for long-term studies, but are constrained by sample size, the characteristics of the building at hand, and sensor location. In contrast, laboratory studies provide greater flexibility with regards to sensing equipment, better control over indoor environment, and more straightforward experiments, but are costly in terms of human resources and may suffer from biases stemming from placing occupants in unfamiliar environments and performing unnatural activities. Survey approaches yield insights into occupant behaviors and allow phenomena to be measured that sensors may be incapable of measuring, but they rely on self-reporting, which may be subject to significant bias. Lastly, virtual reality is an emerging method to study occupants, but for now is limited to the visual and acoustic domains (i.e., not thermal comfort or indoor air quality). A large table of occupant-related phenomena versus research methods was provided to yield insights into the merits of using each method to collect data on that topic. The next three chapters provide much greater detail on best practice, research strategies, available technologies, and challenges associated with in situ, laboratory, and survey methods.

References

Andersen R, Fabi V, Toftum J, Corgnati SP, Olesen BW (2013) Window opening behaviour modelled from measurements in Danish dwellings. Build Environ 69:101–113

Becerik-Gerber B, Jazizadeh F, Li N, Calis G (2011) Application areas and data requirements for BIM-enabled facilities management. J Constr Eng Manage 138(3):431–442

Bennet I, O'Brien W (2017) Field study of thermal comfort and occupant satisfaction in Canadian condominiums. Archit Sci Rev 60(1):27–39. doi:10.1080/00038628.2016.1205179

Clausen G (2004) Ventilation filters and indoor air quality: a review of research from the International Centre for Indoor Environment and Energy. Indoor Air 14:202–207. doi:10.1111/j.1600-0668.2004.00289.x

Cohen R, Ruyssevelt P, Standeven M, Bordass W, Leaman A (1999) Building intelligence in use: lessons from the probe project. Usable Buildings, UK

Creswell JW, Clark VLP (2007) Designing and conducting mixed methods research. Sage, Thousand Oaks

Day JK, Gunderson DE (2015) Understanding high performance buildings: the link between occupant knowledge of passive design systems, corresponding behaviors, occupant comfort and environmental satisfaction. Build Environ 84:114–124. doi:10.1016/j.enbuild.2013.08.062

Day J, Theodorson J, Van Den Wymelenberg K (2012) Understanding controls, behaviors and satisfaction in the daylit perimeter office: a daylight design case study. J Inter Des 37(1):17–34

De Dear R (2004) Thermal comfort in practice. Indoor Air 14(s7):32–39

Duarte C, Van Den Wymelenberg K, Rieger C (2013) Revealing occupancy patterns in an office building through the use of occupancy sensor data. Energy Build 67:587–595. doi:10.1016/j.enbuild.2013.08.062

Fabbri K (2016) How to use online surveys to understand human behaviour concerning window opening in terms of building energy performance. Adv Build Energy Res 10(2):213–239

Fogarty J, Au C, Hudson SE (2006) Sensing from the basement: a feasibility study of unobtrusive and low-cost home activity recognition. In: Proceedings of the 19th annual ACM Symposium on User Interface Software and Technology. ACM, pp 91–100
Gilani S, O'Brien W (2016) Review of current methods, opportunities, and challenges for in-situ monitoring to support occupant modelling in office spaces. J Build Perform Simul 1–27. doi:10.1080/19401493.2016.1255258
Gilani S, O'Brien W (2017) Modelling and simulation of lighting use patterns in office spaces. Paper presented at the Building Simulation, San Francisco, CA
Gossauer E, Wagner A (2008) Occupant satisfaction at workplaces—a Field study in office buildings. In: Proceedings of Windsor Conference on Air-Conditioning and the Low Carbon Cooling Challenge, London Metropolitan University, Windsor, 27–29 July 2008
Gunay HB, O'Brien W, Beausoleil-Morrison I, Perna A (2014) On behavioral effects of residential electricity submetering. Build Environ 81:396–403
Haldi F, Robinson D (2008) On the behaviour and adaptation of office occupants. Build Environ 43(12):2163–2177. doi:10.1016/j.buildenv.2008.01.003
Haldi F, Robinson D (2010) Adaptive actions on shading devices in response to local visual stimuli. J Build Perform Simul 3(2):135–153. doi:10.1080/19401490903580759
Hawighorst M, Schweiker M, Wagner A (2016) Thermo-specific self-efficacy (specSE) in relation to perceived comfort and control. Build Environ 102:193–206
Heydarian A, Carneiro JP, Gerber D, Becerik-Gerber B (2015a) Immersive virtual environments, understanding the impact of design features and occupant choice upon lighting for building performance. Build Environ 89:217–228. doi:10.1016/j.buildenv.2015.02.038
Heydarian A, Carneiro JP, Gerber D, Becerik-Gerber B, Hayes T, Wood W (2015b) Immersive virtual environments versus physical built environments: a benchmarking study for building design and user-built environment explorations. Autom Constr 54:116–126
Konis K (2013) Evaluating daylighting effectiveness and occupant visual comfort in a side-lit open-plan office building in San Francisco, California. Build Environ 59:662–677
McCambridge J, Kypri K, Elbourne D (2014) Research participation effects: a skeleton in the methodological cupboard. J Clin Epidemiol 67(8):845–849. doi:10.1016/j.jclinepi.2014.03.002
McLaughlin S, McDaniel P, Aiello W (2011) Protecting consumer privacy from electric load monitoring. In: Proceedings of the 18th ACM Conference on Computer and Communications Security. ACM, 87–98
Meerbeek B, te Kulve M, Gritti T, Aarts M, van Loenen E, Aarts E (2014) Building automation and perceived control: a field study on motorized exterior blinds in Dutch offices. Build Environ 79:66–77
Meinke A, Hawighorst M, Wagner A, Trojan J, Schweiker M (2017) Comfort-related feedforward information: occupants' choice of cooling strategy and perceived comfort. Build Res Inform 45 (1–2):222–238. doi:10.1080/09613218.2017.1233774
O'Brien W, Gunay HB (2014) The contextual factors contributing to occupants' adaptive comfort behaviors in offices—a review and proposed modeling framework. Build Environ 77:77–87
O'Brien W, Kapsis K, Athienitis AK (2013) Manually-operated window shade patterns in office buildings: a critical review. Build Environ 60:319–338
Pigg S, Eilers M, Reed J (1996) Behavioral aspects of lighting and occupancy sensors in private offices: a case study of a university office building. Paper presented at the ACEEE 1996 Summer Study on Energy Efficiency in Buildings Pacific Grove
Rea MS (1984) Window blind occlusion: a pilot study. Build Environ 19(2):133–137
Reinhart CF, Voss K (2003) Monitoring manual control of electric lighting and blinds. Light Res Technol 35(3):243–260
Schweiker M, Wagner A (2016) The effect of occupancy on perceived control, neutral temperature, and behavioral patterns. Energy Build 117(1):246–259. doi:10.1016/j.enbuild.2015.10.051
Schweiker M, Brasche S, Bischof W, Hawighorst M, Wagner A (2013) Explaining the individual processes leading to adaptive comfort: exploring physiological, behavioural and psychological reactions to thermal stimuli. J Build Phys 36(4):438–463

Teddlie C, Tashakkori A (2009) Foundations of mixed methods research: integrating quantitative and qualitative approaches in the social and behavioral sciences. Sage, Los Angeles

Vine EL (1986) Saving energy the easy way: an analysis of thermostat management. Energy 11 (8):811–820

Wagner A, Gossauer E, Moosmann C, Gropp T, Leonhart R (2007) Thermal comfort and workplace occupant satisfaction—results of field studies in German low energy office buildings. Energy Build 39(7):758–769

Wagner A, Schakib-Ekbatan K, Schweizer-Ries P (2012) The influence of building energy performance standards on user satisfaction in office buildings, Germany

Wyon DP, Sandberg M (1996) Discomfort due to vertical thermal gradients. Indoor Air 6(1): 48–54

Chapter 6
In Situ Approaches to Studying Occupants

William O'Brien, Sara Gilani and H. Burak Gunay

Abstract This chapter provides an overview of in situ methods to study occupant behavior and presence. The aim of the chapter is to provide new and established researchers with a systematic approach to in situ occupant monitoring studies, while also providing illustrative examples to demonstrate the complexities and solutions for navigating this method. The chapter begins with a recommended systematic procedure for designing, conducting, and publishing in situ occupant studies. Following that, in situ-specific sensor technologies and sensing strategies are discussed in detail, with numerous real examples. This chapter devotes considerable discussion on nuances and practical issues that are frequently encountered during in situ studies, including: sensor placement, validation, access to studied spaces, monitoring spaces with multiple occupants, biases such as the Hawthorne effect, participant recruitment, and ethical considerations. Next, recommendations are provided for the level of documentation that should be provided when publishing in situ studies, with particular attention to the contextual factors that could influence the results. Finally, the use of surveys to complement in situ sensor-based methods is discussed.

6.1 Introduction

"If you want to understand how animals live, you go to the jungle, not the zoo."—Jim Stengel

In situ monitoring of occupants and indoor environments offers an opportunity to obtain realistic data about building occupants at a relatively low cost. In contrast to other occupant research methods, in situ studies have relatively high ecological

W. O'Brien (✉) · S. Gilani · H. Burak Gunay
Department of Civil and Environmental Engineering, Carleton University,
3432 Mackenzie Building, 1125 Colonel By Drive, Ottawa, ON, Canada
e-mail: liam.obrien@carleton.ca

S. Gilani
e-mail: SaraGilani@cmail.carleton.ca

H. Burak Gunay
e-mail: burak.gunay@carleton.ca

A. Wagner et al. (eds.), *Exploring Occupant Behavior in Buildings*,
DOI 10.1007/978-3-319-61464-9_6

validity (Nestor and Schutt 2014) since they are performed in occupants' real environments.

In situ monitoring studies do not require a laboratory because they use existing buildings as living laboratories. Occupancy, occupant actions, and their predictors (e.g., indoor environmental conditions and time of day) can be estimated using a combination of built-in sensors (e.g., controls and energy meters) and additional sensors with central or distributed data-logging capabilities. The expenditure for equipment is typically up-front (rather than on-going) and fixed for in situ monitoring, meaning that monitoring campaigns can last months or ideally for an entire year or more (see discussion of sample size in Chap. 3). Furthermore, electronic sensing can be a relatively nonintrusive approach to monitor occupants, though privacy, security, and ethics are important considerations. Unlike many in situ building study approaches (e.g., post-occupancy evaluations) that require spot checks, longitudinal occupant monitoring campaigns require robust sensors to be positioned such that they do not interfere with occupant activities—and likewise that occupants do not interfere with them.

In situ monitoring campaigns are not without drawbacks and challenges. Deploying, checking, and retrieving sensors can be very labor-intensive, particularly for large sample sizes and studies involving multiple distributed buildings. Purely sensor-based approaches may provide hints about cause and effect, but these cannot be confirmed without interacting with the occupants. Numerous studies have revealed that occupant actions are far more complex than can be captured by sensors alone (O'Brien and Gunay 2014; Day et al. 2012). For example, occupants may open window blinds to provide light to their house plants rather than for themselves (Veitch et al. 2013).

Sensors for in situ monitoring are constrained by: location (e.g., sensors cannot be suspended in the center of a space); availability (sensors may not be available to measure all desired quantities accurately—or at all); cost (sensors and the associated logging infrastructure and labor may be costly); privacy (occupants may not wish to be recorded, e.g., concerns about employers tracking office worker presence); and aesthetics (occupants may not like the appearance of sensors in their space).

Unlike in laboratory studies, researchers may have limited access to sensors and other equipment used for in situ monitoring. Adjustment or replacement of monitoring equipment can be invasive and time-consuming, while frequent visits may remind occupants that they are being monitored. Thus, a methodological question arises: are occupants and their behavior significantly affected by the knowledge that they are being monitored? This chapter addresses this concern and explores measures to mitigate the Hawthorne effect.

The objective of this chapter is to provide researchers with a comprehensive guide on best practice and overcoming challenges of in situ occupant research methods. First, this chapter proposes a generalized in situ monitoring framework, and then discusses the complexities and nuances of in situ occupant monitoring. A combination of the authors' experience and the literature are drawn upon to provide a practical grounding.

6.2 In Situ Monitoring Approaches

This section describes an ideal approach and outcomes of in situ occupant monitoring studies. The four major phases of in situ monitoring studies are: (1) investigation and design of experiment; (2) participant recruitment and equipment installation; (3) study; and (4) publishing. The recommended procedure is presented below, and then certain steps are expanded in the sections that follow. Note that the exact order of steps will vary greatly from study to study. For instance, the researcher can likely not enter occupants' private spaces (e.g., private offices or homes) prior to obtaining their informed consent. Thus, an iterative approach involving several visits to assess the space, install sensors, and interview the occupants may be required.

1. **Investigation and design of experiment phase**

This preparatory planning phase involves designing the research project, selecting and investigating the space, assessing steps required to prepare the spaces, getting research ethics approval, and budgeting.

Step 1. Determine the occupant behaviors of interest to be studied, including presence (see Chap. 2 for discussion on key behaviors).

Step 2. Determine whether the in situ monitoring will be accompanied by a survey or laboratory-based methods in order to obtain greater insights on the phenomena of interest. Mixed method occupant studies were discussed in Chap. 5, laboratory studies are discussed in Chap. 7, and survey studies are discussed in Chap. 8.

Step 3. Review Chap. 3 to determine the adequate sample size (number of occupants and study duration) that applies to the behaviors to be studied. A major consideration for the extensiveness of an in situ monitoring campaign is budget. A sample budget for an in situ study is shown in Table 6.1. Note that there can be significant variation between the cost of sensing equipment, depending on accuracy, battery life, memory capacity, etc., as discussed in Chap. 4. To some extent, economies of scale can be realized because of the fixed cost and time for activities such as ethics review, travel to the subject building(s), and data analysis (if automated).

Table 6.1 Sample in situ budget for a study that involved 20 homes for a three-month period

Item	Approximate cost ($US)
Window contact sensors (20)	$2,500
Air temperature loggers (20)	$500
Incentives for participation (20 participants)	$400
Salary for research assistant to design and conduct study and compile results	$7,000
Total	$10,400

Research design is likely to be an iterative process and thus new insights (e.g., importance of measuring an additional item) mean that the budget may evolve over time.

Step 4. Review Chap. 2 to assess the influencing factors (e.g., indoor environmental parameters) that are known to affect or not affect the behaviors of interest. If the literature has not set a precedent for whether a particular influencing factor has a statistically significant contribution to predicting occupant actions, the researcher is advised to consider including it in the monitoring campaign.

Step 5. Obtain research ethics clearance, as discussed in Chap. 11. Note that permission from occupants is likely mandated by the local ethics board for visits to and photography of private spaces.

Step 6. Inspect the building(s) and spaces to be monitored via a walkthrough, drawings, and/or building facility management to develop an inventory of: the current space layout and equipment; potential built-in sensors (e.g., those connected to the building automation system); control interfaces; heating, cooling, and ventilation equipment; failed or broken equipment; and occupant interventions to equipment and user interfaces. Record this information and sketch the spaces. Note that for studies involving homes and other private spaces, this step will likely occur after recruitment, as participants in these spaces would normally have to consent to researchers performing this investigation.

Step 7. Assess the study's needs for weather data (e.g., type of data, temporal resolution, and spatial resolution). Many occupant research studies and modeling efforts attempt to correlate occupant actions with weather phenomena; if this is the intention, available weather data sources should be surveyed. Frequently—and unbeknownst to the researcher—seemingly rich weather data streams could be based on a combination of manual observations and models. Given the spatial nature of weather, airport weather stations may be too far away to be representative. Terrain and urban impacts (e.g., urban heat island effect and solar reflection or shading from neighbouring buildings) could also justify the need for a new weather station to be installed. Chap. 4 discusses weather stations in greater detail.

Step 8. Determine the sampling frequency/frequencies for measurements and data logging. Ideally, the frequencies of all systems should match and sampling should be synchronized. Previous studies have used sampling periods ranging from minutes to hours. Electrical load measurements may require even higher frequency if they are rapidly fluctuating and an objective is to disaggregate the load (see Chap. 4). Ideally, sampling frequency should be at least as frequent as commonly used in building simulation timesteps (i.e., 5–15 min). Researchers should be aware of expected frequency of occupant actions and rate of change of states and evaluate a practical sampling frequency accordingly. For modeling occupant actions, it is important to measure the time of actions so that their triggers can be reliably identified. If sampling frequencies are insufficient and unsynchronized, error can be introduced into models through time averaging or interpolation (see Chap. 10). If local data storage capacity is limited, sampling frequency may have to be compromised to reduce the number of data retrieval visits, which may disturb occupants or invoke the Hawthorne effect. Event-based logging is more appropriate than time interval sampling for discrete events, such as window openings and occupancy. Event-based logging is also much more memory-efficient, as only events are recorded (see Chap. 4 for details).

Step 9. Determine the most suitable sensors and data-logging infrastructure for the measured parameters of interest, using advice from Chap. 4. Note that some of these may already exist in the space as part of the building automation system (BAS). Reliance on BAS sensors is discussed at length later in this chapter. Other proxies for occupancy and occupant actions may be available using existing infrastructure and data sources (e.g., security card systems, Wi-Fi devices).

Step 10. Assess the BAS and energy and water meters to determine the availability of data that could be used to study occupants. The accuracy of the sensors/meters should be assessed for adequacy. Sample data should be inspected to ensure results are within the expected range and being stored. Ideally, the data from meters should be validated (e.g., using portable equipment for spot checks).

Step 11. Based on existing sensor/meter availability, determine which additional sensors are required, based on discussion in Chaps. 3 and 4.

Step 12. Procure sensors and data-logging infrastructure, if applicable. Such equipment can be sourced from scientific supply companies and building controls equipment suppliers, but may also come from companies that manufacture or supply equipment for entirely different purposes than the one at hand. For instance, the authors regularly use packaged temperature sensors and loggers that are nominally intended to log the temperature of shipped goods.

Step 13. Test all sensors for at least several days or weeks prior to study to ensure proper functionality. Ideally, the sensors used to measure the same conditions (e.g., temperature sensors immersed in the same air) can be compared to a sensor with a known high accuracy. Similarly, contact sensors (to measure door and window state) and occupancy sensors can be tested in a research space prior to deployment. The diligent researcher should refer to ASTM (American Society for Testing and Materials) or other calibration standards. Key practical questions that the researcher should determine through sensor testing include:

- How easily are the sensors dislodged if they are bumped or jostled by closing doors and windows?
- How sensitive are the sensors to orientation and location? What are the most suitable placement or mounting strategies to be used in the occupant spaces?

- What are the failure modes caused by occupant interference (e.g., permanent manual overrides such as covering sensors with tape) and what corresponding instructions must occupants be given?
- How sensitive are the sensors? For instance, if a door is left ajar, does the contact sensor measure the state as open or as closed?
- For distributed sensors that transmit wireless signals, what is the possible range and impact of walls and floors?
- How sensitive are indoor environmental sensors to sources of heat, moisture, and CO_2?
- How sensitive are occupancy sensors, considering mounting position, orientation, field of view, obstructions, delay time, and type (e.g., passive infrared or ultrasonic)?

Step 14. Test sensors mid-study (or at regular intervals for long-term studies), if possible, to ensure that they are functioning properly and readings have not drifted significantly. Sensor drift should be assessed and reported at the end of the study. Details on ground truth and data validation are provided in Chap. 9.

An overview of the investigation and design of experiment phase in a monitoring study is shown in Fig. 6.1.

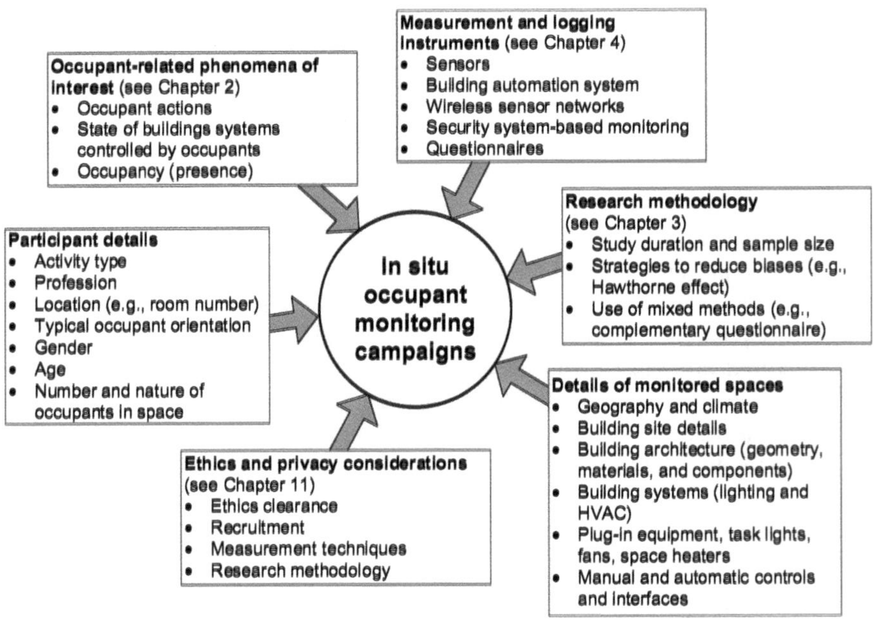

Fig. 6.1 Overview of the elements of in situ occupant monitoring studies

2. **Occupant recruitment and equipment installation phase**

The occupant recruitment and equipment installation phase normally occurs immediately prior to the study phase. The researcher should be aware that this seemingly straightforward phase can take many weeks, largely because of the uncertainties associated with recruiting and interacting with participants.

Step 1. Recruit participants as per the procedure laid out in the research ethics proposal (e.g., see Chaps. 8 and 11), with a detailed explanation of the experiment, including, but not limited to:

- Duration of study
- Expected timing of visits (e.g., for installation, data retrieval, and removal of sensors)
- Type of sensors and what they measure
- Clear instructions to the participants on how to relocate sensors if relocation is absolutely necessary
- Details on data storage, security, publication, and anonymity
- Availability of data and final results if occupants wish to obtain them
- Collection and publication of other information (e.g., planned questionnaires or photographs)
- Terms for ceasing participation of study
- Compensation for participating in the study, if applicable.

Step 2. Obtain permission and informed consent from occupants for monitoring private spaces (see Chap. 11).

Step 3. Provide a record of occupants via walkthroughs of monitored spaces to investigate, including but not limited to: their profession (for workplaces), gender, number of occupants, locations, and type of activities.

Step 4. Repair failed equipment (e.g., broken blinds, operable window cranks, poor automatic light controls logic), if possible; otherwise the data are tainted by these anomalies.

Step 5. Visit the occupants to discuss the study, check the space(s), and install sensors. For commercial buildings, it may be possible to gain access to spaces with the assistance of the building managers or operators without the presence of occupants. However, occupant/participant permission should be sought regardless, as per the terms of the ethics application.

Step 6. Inform the occupants of sensor locations and any specific instructions to reduce the likelihood of obstruction, disconnection, or damage. Researchers should remind participants to contact them if they are leaving the space (e.g., changing offices or moving homes) so that the equipment is not lost and the data are not misinterpreted as having minimal occupant presence and actions. Providing researcher contact information on all distributed sensors is a wise strategy.

Step 7. Photograph and take notes about the spaces and sensors locations. Sensors should be labeled so that there is no risk of mixing them up after retrieval. Many purpose-built packaged sensor and data logging systems also allow digital naming via software. This extent of documentation is critical for retrieval at the end of the study and to help explain any unexpected measurements. Best practice for documentation is discussed later in this chapter.

3. **Study phase**

This phase is the time—weeks to years—when the actual data is collected.

Step 1. Plan regular data checks, if possible, to ensure that sensors and data storage are functioning. If data storage is local and requires site visits, the researcher should avoid frequent visits to minimize effort and avoid disrupting occupants. Note that the amount of lost data could be as high as the time between visits. For instance, monthly visits will help ensure that at most only one month of data is lost. For low-cost sensors, redundancy is strongly encouraged. In studies involving multiple sensors and where correlations between concurrent data are sought, the importance of ensuring reliability is further increased. If possible, back-up sensors, batteries, and other equipment and tools should be brought to site visits in the event that sensor failure has occurred. Data should be backed-up on multiple storage devices, while abiding by data security regulations laid out in the ethics application.

Step 2. Perform scheduled intermediate surveys, if applicable.

Step 3. Ensure secure data storage and occupant anonymity, as per the details in the research ethics application, to protect their identity and measured data. Coding schemes can be used to disassociate occupant names from data (i.e., pseudonyms, as described in Chap. 11). This is particularly sensitive for occupancy data, which could be used by thieves or employers. Normally, ethics clearance requires thorough planning for these matters.

4. **Publishing phase**

Given the great effort required to conduct in situ occupant monitoring campaigns, the resulting data and analysis are of tremendous value to the research community. Thus, the importance of such studies having high attention to detail, scientific rigor, and transparency cannot be underestimated.

Step 1. Provide a significant level of detail about the equipment specifications, spaces, participants, occupant behaviors of interest, and details of the procedures listed above. Sufficient detail to allow readers to repeat the experiment is best scientific practice. Recommendations for providing contextual information are provided later in this chapter.

Step 2. Publish anonymized data in raw or aggregated form, where possible, such that other researchers and stakeholders can verify the published results. Chap. 10 covers details on data management and reporting protocols.

6.3 Sensors and Data Acquisition Architecture: Practical Considerations

At the heart of in situ occupant monitoring studies are the sensors that detect occupancy, occupant actions, and the predictors for occupant actions. Sensor technologies are discussed at length in Chap. 4; the current chapter is focused on in situ-specific matters. This section discusses centralized and decentralized system architectures, while also exploring novel sensing methods and exploiting BAS.

6.3.1 Building Automation Systems

Often, the most practical and economical method to detect occupant actions, presence, and environmental conditions is to use electronic sensors that are connected to a central BAS. Once such systems are configured to record and store data, minimal maintenance is required and faults or unexpected readings can be detected and possibly resolved remotely without intruding on occupant spaces. Furthermore, data storage is virtually limitless with regards to cost—except perhaps for video or audio storage, which is significantly more memory-intensive.

Newer building automation systems tend to include data archiving systems with data access through a web browser. It is particularly helpful if the building is controlled by a central system with an open protocol such as BACnet (Building Automation and Control network). Presence of digital controls and sensing capabilities does not necessarily translate to straightforward data collection. Most current building controllers have on the order of days' worth of data storage capacity; thus, a higher-level archiving system is essential to facilitate research. Figure 6.2

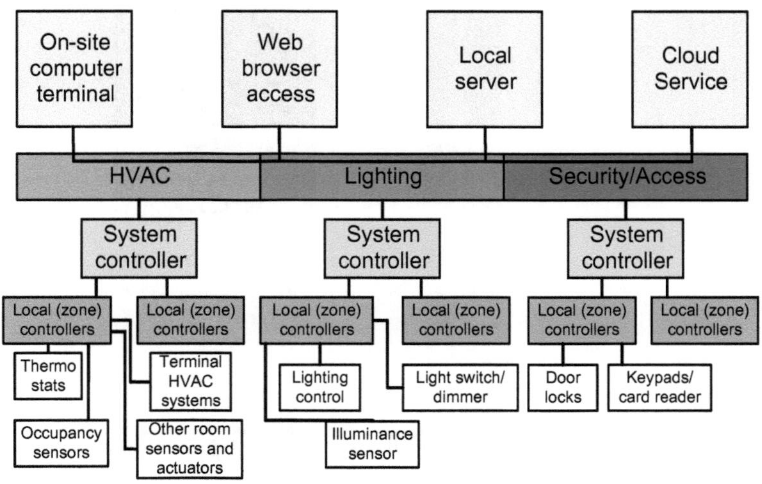

Fig. 6.2 An exemplary building automation system architecture

shows a generic BAS architecture, whereby all controlled domains (HVAC, lighting, and access) are connected and available via a variety of local and remote portals.

In contrast to the system illustrated below, some buildings have a separate lighting network that uses a closed and proprietary communications protocol that prevents light use data from being collected. In the case of a building that the authors studied, the cost to convert the lighting controls system to the central BAS was prohibitive. Thus, researcher input into the control system architecture of new buildings is critical to future cost-effective research. An overview of relevant BAS sensors and data acquisition systems is provided in Chap. 4.

Of the configurations like those described above, the most ideal circumstance is to have extensive sensing and logging capabilities in the BAS. Modern zone-level controls components can measure and log temperature, relative humidity, CO_2, occupant presence, illuminance, other indoor environmental metrics, and occupant interactions with building systems (e.g., Figs. 6.3 and 6.4). This hardware is often used for routine controls applications, but can also be used to study occupants. It is

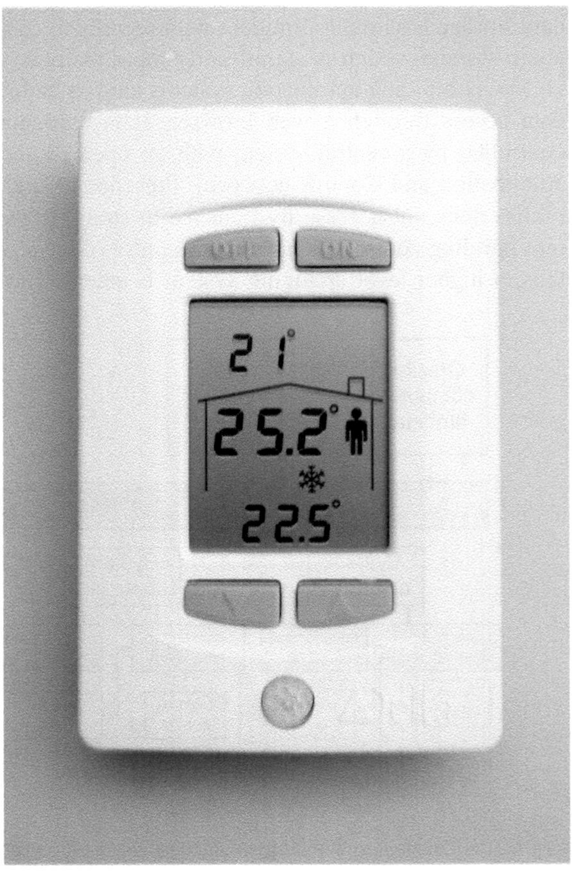

Fig. 6.3 Example thermostat with buttons for light switching and temperature setpoints adjustment, which also measures occupancy, CO_2 concentration, temperature, and relative humidity

Fig. 6.4 Ceiling-mounted illuminance sensor for controls that can also be used to measure daylight levels in the room as part of occupant monitoring studies

even more ideal for live weather data to be available to the BAS, either from a web-based service or by an integrated weather station mounted on the building. Modern BASs tend to measure outdoor air temperature and relative humidity because these variables are used in HVAC controls.

However, there are some caveats to relying on built-in BAS-related sensors. First, they may not have the same requirements for accuracy and range as those needed for scientific occupant studies. For instance, temperature sensors may have less accuracy than scientific-grade sensors, data might be logged at a lower precision or frequency than desired, or the location of sensors may not be suitable for scientific studies. In a building that the authors studied, they found that measurements by the room supply air temperature sensors were about 2 °C higher than those measured by a scientific-grade temperature sensor.

On the subject of sensor location, an example to consider is ceiling-mounted illuminance sensors, which are often adequate for lighting controls applications (e.g., to determine whether conditions are too bright or too dark), but less adequate for estimating workplane illuminance for occupant model development. Illuminance sensors may have a limited range (e.g., 0–1000 lx) because their primary purpose is to detect whether conditions are too dark. On the other hand, one wireless illuminance sensor product that the authors tested is solar-powered and it only measures and transmits illuminance readings during daylit (>100 lx) conditions. Again, this limitation is due to its primary purpose to detect adequate daylight levels rather than fulfill scientific objectives.

BAS-integrated sensors may be obstructed by furniture, located near heat or humidity sources, or even intentionally obstructed by occupants to override undesirable controls. These challenges are discussed at length in the next section. Finally, researchers should be very cautious about conclusions drawn when occupant actions are inferred from occupant interactions with motorized/automated systems; the fact that automatic building systems may be easier to operate relative to manual systems means that such systems may encounter greater occupant interaction frequency. For instance, Sutter et al. (2006) found that occupants adjust blinds three times more frequently if they are motorized with a switch than if they are manually controlled via mechanical means. However, adaptive system states are less likely to be monitored by the BAS unless they are directly automated or motorized. As discussed later in this chapter and in O'Brien and Gunay (2014), care must be taken to document occupant monitoring studies so that results are not inappropriately extrapolated to a different context.

6.3.2 Adding Additional Sensors to BASs

In the frequent event that existing sensing capabilities in a space are inadequate for the extent of the desired monitoring, a viable option is to add additional sensors. The preferable option is to integrate wired or wireless commercial sensors into the BAS such that data can be reliably collected and stored with minimal impact on occupants. Modern building automation systems often integrate with wireless sensor networks or can be expanded to do so. For post-construction upgrades, wireless sensors are preferable for occupant monitoring because of the ease of integration, lack of requirement for wiring, and minimal disruption to occupants. Some new BAS-integrated wireless sensors and user interfaces also use environmental energy harvesting (e.g., from integrated photovoltaics or kinetic energy from light switching) and thus do not require battery replacement (e.g., Fig. 6.5). Despite the ease of wireless sensors, they tend not to be as reliable as wired networks. Example problems include: communication range issues, depleted batteries, and infrequent sampling, which may lead to a time delay between a measured value (e.g., illuminance) and occupant action (e.g., blind closing caused by glare).

Regardless of the relatively low hard cost of BAS-integrated sensors, the soft cost of sensor installation, configuration, and calibration can still be quite high—at least as much as the hardware itself. The purpose of BAS sensors is somewhat different from scientific studies and thus considerable manual effort may be required. For instance, there is typically no need for controls to store more than tens of readings about historical room air temperature, whereas researchers normally expect to collect months or years of data.

Fig. 6.5 Example of a wireless contact sensor with integrated photovoltaic energy harvesting that communicates with the BAS via a wireless gateway

Some common challenges that may be encountered upon attempting to add additional sensors to an existing BAS include: (1) the number of unused input/output (I/O) ports on controllers may not allow for more sensors to be added to the BAS network, (2) the communication protocol of the BAS and sensor may not match, and (3) installation costs may be prohibitive if a controls contractor is hired. For the first challenge, it is probably possible to add a controller with more I/O ports, but the cost to use the BAS network may no longer be attractive compared to a new network. For the second challenge, a gateway may fulfill the purpose of bridging communication protocols. To keep costs under control, the authors have relied heavily on students for equipment installation and configuration, as well as, close collaboration with facilities managers and equipment manufacturers.

6.3.3 Obtaining BAS Data

Building automation systems represent an invaluable approach to cost-effectively collecting vast amounts of building performance data (Zibin et al. 2016). While the sensor availability within a BAS may be adequate for conducting extensive occupant monitoring studies, contemporary BASs are typically not aimed at research. For instance, controllers have limited memory—often enough for only days or weeks of data. The novice researcher will quickly discover that regularly downloading short datasets from multiple controls is very tedious. Thus, it is highly desirable to have an integrated server that archives data that is easily accessible and convenient to download. Some newer buildings have a dedicated BAS-integrated computer that stores data locally but also sends it to a cloud. These data can be automatically retrieved remotely via an API (application program interface). Connecting networks of systems via cloud services [e.g., Internet of Things (IoT)] can economize on the infrastructure costs (e.g., Kovatsch et al. 2012).

6.3.4 New Sensor Networks

Suitable BAS-integrated sensors may not exist off-the-shelf, particularly for unique applications related to occupant action sensing (e.g., Venetian blind slat angle adjustment). Furthermore, BAS-grade sensors may be more expensive due to their packaging, communication protocol, electrical safety rating, and design for long-term reliability. Finally, existing BAS infrastructure may not easily allow for expansion. For instance, a building may still have pneumatic controls or decentralized digital controls. Thus, two major options remain: the addition of sensors with centralized data acquisition or packaged systems with decentralized logging. As discussed previously here and in Chap. 4, centralized data acquisition is preferable for detecting sensing faults and obtaining data unobtrusively.

Aside from the BAS-grade sensors, there are many wired and wireless sensors and data acquisition systems available commercially. Generally, these are not

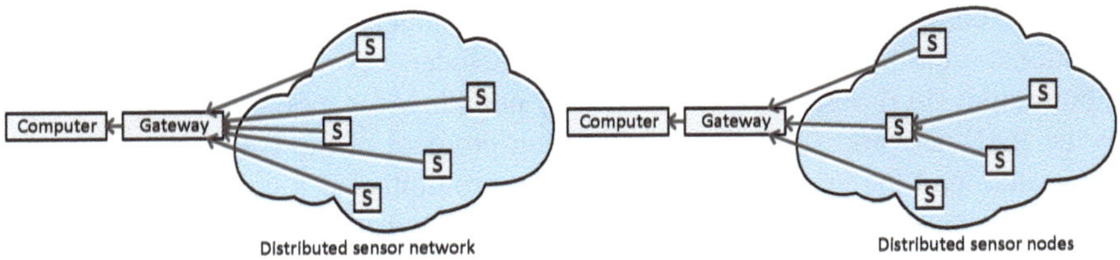

Fig. 6.6 Sensor network configurations. In the left system, all sensor modules must communicate with the gateway and are thus limited in range. In the right system, the sensor modules also act as nodes and extend the network range

Fig. 6.7 Example plug load meter that integrates into a building's wireless network

designed for occupant research or even building applications, but rather general scientific research. Some sensor modules integrate into building computer networks using Internet protocol, while others rely on stand-alone networks. Two illustrative sensor network configurations are shown in Fig. 6.6, where the second configuration has sensor nodes that help to increase the range for wireless networks.

Figure 6.7 shows an example product that acts as both a sensor and a node to extend the range of the network, so long as all nodes are within range of another node. This system communicates with standard network routers.

The viability of using sensor networks for in situ occupant monitoring studies is dependent on the building geometry and construction and the density of occupants. For instance, if a sample of offices in a single office building is studied, it may be practical to have one or more dense sensor networks. For a network of illuminance sensors, one gateway is likely adequate for each floor and wing of a larger office building. However, if apartments or detached houses are being studied, it is likely that the distance between sensors is greater than the feasible network range.

6.3.5 Distributed Stand-Alone Sensors and Data Loggers

Off-the-shelf packaged sensors and logging systems are often low-cost and easiest to install because they are designed as an integrated product with minimal user configuration required. The benefits of packaged sensor-logger systems are balanced by the disadvantages of data and battery capacities (and hence the required frequency of downloading data and replacing batteries) and risk of theft or misplacement, especially in public places. Typically, available products have batteries that last on the order of months or years and have sufficient on-board memory for several thousand recordings with user-defined recording frequency. Thus, researchers may have to compromise between sampling frequency and data downloading frequency for extended studies. Stand-alone sensors can often be mounted in more suitable locations than built-in sensors, which are difficult to move, and may be placed out of the way, in remote locations. However even if theft is not an issue, occupants may choose to relocate packaged sensors for their convenience. The researcher cannot be sure whether relocation, tampering, or exposure to unexpected conditions occurred during the study. In contrast, built-in sensors provide greater certainty in this regard.

Some example stand-alone sensors are shown in Fig. 6.8. Note the care that was taken to mount the sensors in secure and convenient ways in these examples. For instance, the CO_2 sensor was mounted on a piece of wood to minimize damage to the wall. Many stand-alone sensors are provided with magnetic or adhesive mounting systems.

One innovative project that was run by Cornell University (Chong 2016) further exploited stand-alone sensors by mailing temperature loggers out to participants. In all, they reached over 500 participants and collected 15 days of high-resolution household air temperature data. Participants were provided with an instructional video and were asked to activate the sensor upon its arrival to avoid it depleting memory and measuring temperature in transit. However, there is no way to verify that the participants properly configured their sensors or watched the video. Since that study, Toronto, Canada-based smart thermostat company ecobee has developed an opt-in scheme for homeowners to share their thermostat and occupancy data with researchers (ecobee 2017).

Fig. 6.8 Example stand-alone sensor systems. Clockwise from *top-left* CO_2 concentration sensor connected to data logger; custom-built thermostatic radiator valve position sensor connected to a data logger; light state logger; and operable window state logger (photos from Rune Andersen)

6.3.6 Image-Based Sensing for in Situ Occupant Monitoring Studies

A relatively unexploited source of information for existing in situ occupant monitoring studies is image- and video-based recordings. First introduced in Chap. 4, image-based sensing is further discussed here specifically with practical aspects of in situ monitoring. Ideally, images from existing cameras (e.g., security cameras) can be obtained. There is also a large number of available commercial camera products that are suitable for building applications. Many of these cameras can be integrated into a building's existing network infrastructure or have built-in batteries and memory. An example experimental set-up of a time-lapse camera array and corresponding sample image are shown in Figs. 6.9 and 6.10.

Rather than manual interpretation of images—a very time consuming process (Rea 1984)—advanced computer vision can be used to assess occupant location, window blind position, and operable window state (Kapsis et al. 2013; Benezeth et al. 2011; Shih 2014) (e.g., Fig. 6.11).

Image-based techniques can be further used to identify occupant numbers, actions, position, and posture. Jalal et al. (2012) demonstrated that occupants' household activities can be identified through computer interpretation of silhouettes. Others (e.g., Davis III and Nutter 2010; Erickson et al. 2009) have used both

Fig. 6.9 Array of time-lapse cameras, configured to capture an entire façade. Another set of cameras aimed at an oblique angle with respect to the facade was used to detect window opening state

Fig. 6.10 Image from one of the time-lapse cameras shown in Fig. 6.9

manual and automated methods to count occupants either in a space or passing into a space.

In addition to being the primary means of obtaining occupant data, photography can also be used to verify sensor readings. For instance, if a time-lapse camera were

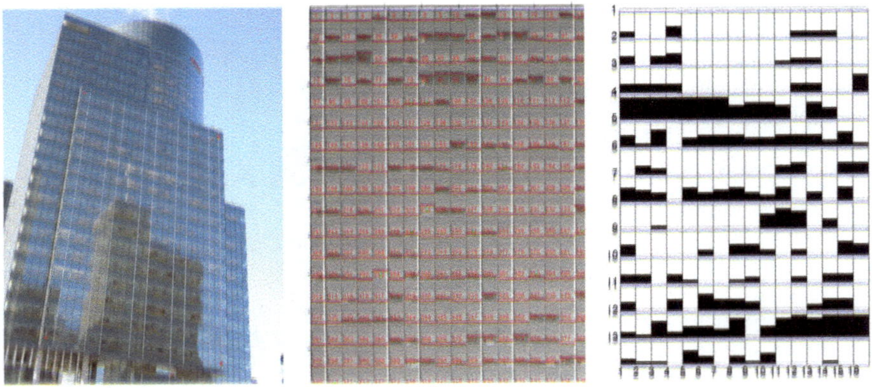

Fig. 6.11 Example of time-lapse photography and computer vision algorithm to convert photos of a façade into window shade position data (O'Brien et al. 2010)

set up to face a façade, suspicious readings from blind or window state sensors could be verified using photos at the same time.

Privacy remains a major concern for camera-based occupant and behavior sensing. In general, researchers should adhere to the policy that cameras should not capture images in places where there is an expectation of privacy (e.g., homes, offices, washrooms). All restrictions imposed by the ethics board hold and it is ideal for images to be converted to anonymized data (e.g., number of occupants in a space), and then deleted to minimize consequences of data breaches.

6.3.7 Virtual Sensors for in Situ Occupant Monitoring Studies

A newer and promising approach to estimating occupancy, occupant actions, and indoor environmental conditions is virtual sensors and sensor fusion. Virtual sensors are model/software-based sensors that infer states and events from a combination of one or more real sensors or data inputs rather than relying on direct measurements.

Virtual sensors provide the benefit that parameters that are difficult to directly measure can be estimated. Thus, virtual sensors have particularly high potential for in situ monitoring studies where the addition of physical sensors in a space may be impractical. Moreover, virtual sensors are low- or no-cost once they are developed because they are software-based. A few examples of virtual sensors include the following:

- Solar angles (e.g., solar altitude and daylight penetration depth) can be estimated with a high degree of accuracy if the time of year, geographical location, and façade orientation are known.

- With partial electrical load metering, it may be possible to infer loads in one space using simple accounting. For instance, it is possible to estimate common area electrical loads in an office building through subtraction if both the entire building and each suite are metered. A more advanced approach to electrical power metering is to use load disaggregation, whereby the signal is analyzed to extract data for individual electrical loads (Marceau and Zmeureanu 2000).
- Operable window state can be predicted by CO_2 concentration or relative humidity measurements if supply air rate, CO_2 concentration, and occupancy are estimated. Confidence about window state could be further increased by also performing a water vapor balance or energy balance.
- The number of occupants in a space may be estimated if a complete CO_2 mass balance is performed, such that ventilation rate and CO_2 concentration are known (Dong and Andrews 2009; Lam et al. 2009; Zoha et al. 2012).
- A radiator's surface temperature can be used to estimate a radiator's heat output (Tahmasebi and Mahdavi 2012).
- Long periods of vacancy may be estimated by use of CO_2 concentration (Andersen et al. 2013).

However, virtual sensors introduce error into the data because they are not direct measurements, and some assumptions need to be made. As the number of physical sensors involved in a virtual sensor increases, the error associated with the virtual sensor output will compound. In order to develop and apply virtual sensors, they should be thoroughly validated against ground truth values—ideally for the space(s) being studied, or at least for several similar spaces. The field of virtual sensing in buildings and for occupant monitoring is quite new and not widespread; further research is required for their development and field validation.

6.3.8 *Future Sensing Technologies for in Situ Occupant Monitoring Studies*

Numerous new and anticipated technologies will substantially broaden the types of occupant actions that can be sensed. Given the relatively small field of occupant research, researchers should look to other domains for research methods and technologies. Notable examples include:

- Accelerometers and furniture-integrated sensors can be used to estimate occupant posture, orientation and adaptation to daylight glare, and occupant presence (Labeodan et al. 2015; Coen 1998; Mathie et al. 2004; Godfrey et al. 2008; Yang and Hsu 2010; Zhao et al. 2015);
- Wearable sensors can be used to measure occupant metabolic rate and interaction with building systems (Yang and Hsu 2010; Butte et al. 2012);
- Computer keyboard and mouse data can be used to assess productivity (George 1996);

- Wireless computer networks can be used to estimate the number of personal wireless devices (e.g., smart phones, laptops) as a proxy for occupancy (Balaji et al. 2013);
- Smart phones and Global Positioning System (GPS) can be used to estimate occupant location with respect to buildings (Gupta et al. 2009); and,
- Printable sensors that can be embedded in construction materials, furniture, and other building components (Brown et al. 2016).

6.4 Practical and Methodological in Situ Monitoring Challenges

The above discussion provided brief descriptions of possible sensor network configurations. However, many additional challenges remain for in situ occupant monitoring studies since researchers have little control over the space in which the experiment is being conducted.

6.4.1 Sensor Placement and Obstruction

Sensor positioning is critical to accurately sensing and detecting environmental conditions, occupancy, and occupant actions. In situ studies often prevent placement of sensors in optimal locations because of practical constraints. When measuring indoor environmental conditions, the ideal situation is to sense them at the location of the occupant, as this is most representative of the environment that the occupant is immersed in. For instance, laboratory-based thermal comfort studies involve measurement of temperature at foot, waist, and head height and skin temperature at numerous locations. Because experiments are typically quite short, one set of reusable sensors is able to serve a large sample of participants. In situ studies are less likely to afford this resolution of measurement.

Visual comfort is the most sensitive to occupant orientation and position (Wienold and Christoffersen 2006). With current technology, it is simply not practical for in situ studies to measure all environmental conditions at the location of the occupant, as the occupant would sense them, because of the interference with occupants. To address this, Wienold and Christoffersen (2006) used an experiment to develop a better daylight glare metric. Using two identical offices—one occupied and one unoccupied—they positioned a camera in the unoccupied office in the equivalent location and orientation as in the office with occupants. But this is clearly impractical for most in situ studies, as an empty office would be required and there is little certainty about occupant orientation. Konis (2013) performed a small-scale in situ study using high dynamic range (HDR) photography to predict glare. An example experimental set-up for measuring and recording glare is shown in Fig. 6.12.

Fig. 6.12 HDR camera configuration for capturing glare data in a classroom (photo from Julia Day)

While daylight glare is highly direction- and location-specific, occupant orientation and location are difficult to measure in situ and virtually impossible to predict in simulation. Thus, proxies are required for both monitoring and modeling studies. Workplane illuminance is often used because it is relatively low cost to measure and is readily estimated in many modeling tools. Placing illuminance sensors directly on the workplane risks that they will be covered by occupant belongings (e.g., papers). A method to avoid this includes mounting sensors above the workplane on a pedestal (e.g., Fig. 6.13).

Similar to illuminance, other major indoor environment parameters (air temperature, mean radiant temperature, airspeed, acoustic noise, and CO_2 concentration) likely cannot be measured near the occupant for the sake of practicality (i.e., uncertainty of their location and nuisance of hardware and wiring). Moreover, locating indoor environmental quality sensors near the occupant—and their associated heat, moisture, and CO_2 generation—to measure these parameters would risk tainting the results. Similarly, temperature, relative humidity, and illuminance sensors may be significantly affected by the proximity of equipment, appliances, supply air diffusers, electric lights, and solar radiation. For instance, consider the situation in Fig. 6.14, where solar radiation directly hits the thermostat in an office for the months of April and August at approximately 6 PM.

Fig. 6.13 Illuminance configuration to reduce obstruction by desktop objects (photo from Ardeshir Mahdavi)

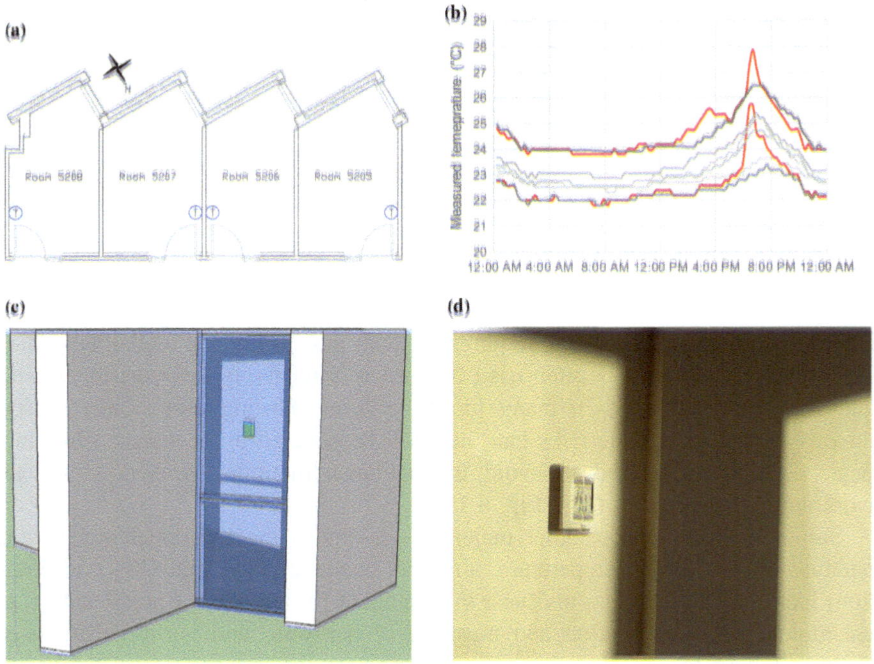

Fig. 6.14 An example of thermostats in offices that suffer from periodic incident solar radiation such that they read 2–3 °C higher than in the identical adjacent offices. (**a**) Office floor plans and window orientation; (**b**) a graph showing sensed thermostat temperature for eight identical offices on a sunny day in August, including the four shown in the drawing; (**c**) the 3D model of the same office from the west in late afternoon in mid-August; and (**d**) a photo of the thermostat in direct sun (figures from Justin Berquist)

In small spaces, such as offices with well-mixed air (e.g., with entrainment ventilation), it would be reasonable to assume that measurements of key properties of indoor air at the wall are representative of those near the occupant. Air-related environmental conditions should not be measured at the ceiling because conditions there may be significantly different than in the occupied zone, particularly in rooms with displacement ventilation or radiant heating and cooling. Built-in sensors (e.g., digital thermostats) are usually mounted to ensure usability and so that they can be easily accessed and viewed. The typical thermostat mounting height is approximately 1.5 m (Webster et al. 2002); whereas ASHRAE Standard 55 (ASHRAE 2010) requires that temperature be measured at 0.1, 0.6, and 1.1 meters above the floor, for ankles, waists, and heads of seated occupants.

Occupancy sensing using traditional means (i.e., passive infrared (PIR) or ultrasonic sensors) is directionally sensitive and requires occupant movement to be frequent. Occupancy sensors should be aimed at the region in the room where occupants are most likely to be for the majority of the time. For instance, in offices they should be aimed at the center of the room near where the desk chair is likely to be located. Often, insufficient thought is invested in selecting the location and orientation of occupancy sensors that are used for building controls. For instance, there are some instances of sensors being pointed around a corner from the main part of an office (Gilani and O'Brien 2016). As a result, the occupant may only be detected when they arrive and leave an office, with no direct methods to predict whether the motion is associated with an arrival or departure event. Advanced methods to detect and even count occupants in a space are discussed at length in Chap. 4. Some potential proxies to use for in situ occupant detection include: wireless networks or Bluetooth data associated with wireless device, plug-in appliances state or power, and CO_2 concentration.

A major issue for in situ sensing is sensor obstruction, whether intentional or unintentional. Wall-mounted sensors may be covered by furniture, while workplane-mounted or horizontally-mounted sensors may be obstructed by occupant possessions. For instance, Reinhart and Voss (2003) opted to use virtual sensing of illuminance in place of direct illuminance measurement because they found that occupants frequently covered or obstructed the sensors with their office equipment and supplies. Ceiling-mounted sensors are a good approach to avoid obstruction, though this is more suitable for occupancy detection than for daylight illuminance. Figure 6.15 shows some typical occupancy sensor locations from a real shared office. Details about the respective PIR sensors' effectiveness for this office can be found in Gunay et al. (2016).

To improve data quality, redundant sensors can be used for in situ occupant monitoring studies. For instance, multiple PIR sensors could be used in different locations in a space to improve the chances of occupancy detection. Redundancy has the added benefit that data are less likely to be lost if one sensor or its data logging infrastructure fails.

Sensors that are connected to building controls are prone to intentional obstruction by occupants if the controls logic is undesirable. For instance, occupancy and daylight-controlled light can be particularly annoying to occupants;

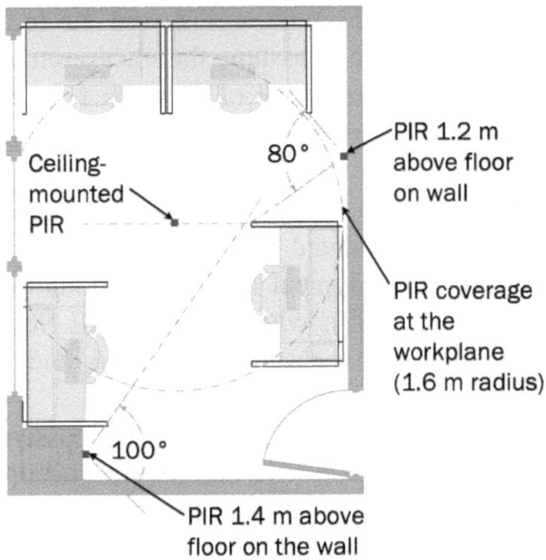

Fig. 6.15 Example occupant (PIR) sensor fields of view. The ceiling-mounted sensor is preferable over the other options shown because it covers all seated occupants. However, if knowledge of arrival times is critical, the sensor that is located across from the door is a worthwhile addition. The sensor on the wall between two desks has a significantly reduced field of view due to the desk beside it. Note that the PIR sensors shown also include a temperature sensor; they were intentionally installed away from windows and on interior surfaces to avoid bias from solar radiation or outdoor temperature

occupants have been observed on numerous occasions covering sensors with tape to prevent lights from turning on or off unexpectedly (Heerwagen 2000). An example of sensor obstruction is shown in Fig. 6.16. Figure 6.17 shows a CO_2 sensor that was covered by coats after installation. This could be problematic because poorly mixed air will cause a significant delay between CO_2 concentration at the center of the room and what this sensor measures.

Methods to avoid obstruction include: talking to (and educating) occupants to convince them to keep sensors uncovered, resolving poorly conceived controls that prompted occupant interference, mounting ceiling sensors, and grouping user interfaces with sensors such that occupants cannot obstruct sensors with large objects such as furniture without losing access to interfaces. For instance, the device in Fig. 6.16 includes lighting and thermostat push buttons and a PIR motion sensor.

A major consideration for sensor placement is occupant satisfaction. When installing sensors in private spaces, the location of the sensors is often a negotiation. The best location of a sensor from a scientific viewpoint may not be acceptable to the occupants. It is very important that the sensors be placed in a location that the residents find acceptable; if this is not the case, it is likely that the residents will remove the sensors after the installation visit. In one anecdote, a researcher lost two CO_2 sensors (valued at $1000 each) because the participant did not like the way they looked and disposed of them. Another research collaborator experienced

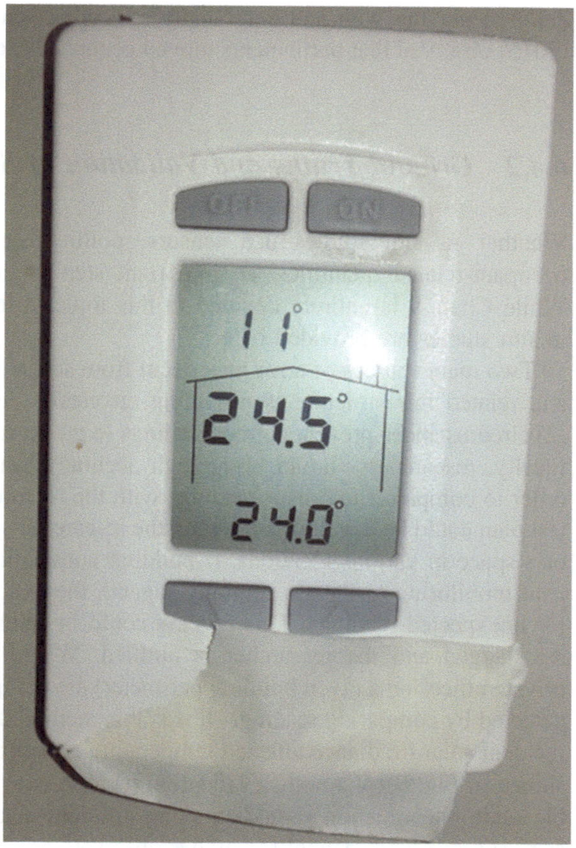

Fig. 6.16 Example of intentional sensor obstruction, where the occupant covered the PIR sensor with masking tape

Fig. 6.17 A packaged temperature, relative humidity and CO_2 concentration sensor that was subsequently covered by coats (Andersen 2013)

children playing with and continually relocating sensors. Meanwhile, Dong et al. (2015) observed that participants moved occupancy sensors to protect their privacy.

6.4.2 Ground Truths and Validation of Sensor Readings

Whether in situ studies use sensors, polling, or other methods to measure occupant-related quantities, an important step is to validate the collected data. While Chap. 9 is entirely devoted to this topic, a few details that are specific to in situ studies are provided here.

Two main sources of error may occur from sensor readings: (1) the actual sensor and related infrastructure is providing erroneous outputs or is uncalibrated, and (2) circumstances prevent correct readings (e.g., sensor obstruction or interference). Ideally, researchers should bring high accuracy sensors to initial space visits in order to compare site sensor readings with the reference. For occupancy sensors, a test plan could be developed whereby the researcher systematically walks in and out of a space in various locations. If building automation systems are used for occupant monitoring or data is centrally logged, they should be continually monitored for unexpected readings. This process could be automated whereby outlying data are flagged and the researcher is notified. Where nearly-identical spaces (e.g., private offices on a given building perimeter) are being monitored, anomalies can be detected by comparing readings. It was this method that alerted the researchers that incident solar irradiance affected temperature sensor readings in the above example shown in Fig. 6.14. Another validation method is to use mixed methods, whereby the same phenomenon is studied using different measurement methods (e.g., sensors, surveys, and manual researcher observation).

6.4.3 Limited Access to Spaces

In contrast to laboratory-based studies, researchers normally have limited access to spaces to use for in situ studies. Moreover, study participants may have limited patience for invasive and frequent researcher visits. Moreover, frequent visits increase occupants' cognizance that they are being studied and the associated Hawthorne effect (discussed later in this chapter). Thus, it is critical to plan visits carefully to maximize efficiency. For instance, sensor positioning, photography and note taking, and brief occupant surveys could be completed in a single visit. Bennet and O'Brien (2017) performed a six-month residential study whereby two visits were made to each of the 20 participants—one in the summer and one in the winter. In the first visit, sensors were deployed and a survey with questions oriented towards summer thermal comfort was distributed. In the second visit, the sensors were picked up and the participants were asked about winter thermal comfort in a second survey (surveys are discussed in Chap. 8). Some non-comfort related questions were asked in both surveys to help validate responses (i.e., to quantify consistency of responses).

The time required to visit occupants cannot be underestimated when budgeting for a planned in situ study. For instance, gaining access to apartments or single-family houses normally requires separate appointments for every single participant. Even if the apartments or houses are located close to each other, it is often not possible to get access to all dwellings in the same day. As a consequence, several visits to the site are required—often in the evening to accommodate occupants' work schedules. In some circumstances, safety considerations may require paired researchers to make the visits, which may further complicate finding a time for an appointment.

6.4.4 Monitoring Spaces with Multiple Occupants

A largely unresolved issue, which extends to the occupant modeling domain, is monitoring multiple occupants using in situ methods. For multi-occupant spaces (e.g., homes, classrooms, hospitals, and shared offices), robust methodologies to distinguish between occupants for in situ monitoring are still emerging (Dong et al. 2010). For instance, we cannot be sure which occupant turned on the light in a double-occupancy office. The existing methods for identification (e.g., camera or security passes) pose considerable threats to privacy and the required technology could be costly. Still, the question of whether this data is necessary for modeling and simulation efforts remains. The majority of existing studies on spaces with multiple occupants do not distinguish occupants and merely quantify states and actions of the population (e.g., Haldi and Robinson 2010; Zhang and Barrett 2012). However, presence of multiple occupants is known to play a major role in likelihood of adaptive actions being taken (Haldi and Robinson 2010; Peffer et al. 2011). Until we are able to reliably distinguish between occupants, understanding the social dynamics and hierarchies may be best performed using laboratory or survey-based studies (Schweiker and Wagner 2016).

6.4.5 Hawthorne Effect

A major advantage to in situ studies is that they take place in the natural environments where occupants engage in their everyday activities, rather than an artificial environment (e.g., laboratory) with constant reminders that their actions and environment are being monitored. However, most in situ studies have some minor contact between the researcher and participant wherein consent is obtained and possibly additional equipment is installed or surveys answered. Therefore, in situ study participants may also be affected by the knowledge that they are being studied, and consequently alter their natural behavior to please the researcher or society (i.e., social desirability bias) (McCambridge et al. 2014). For energy consumption-related studies where occupants are responsible for energy bills, frequent reminders of the study may prompt occupants to conserve energy.

A methodological question for any new in situ study arises: does occupants' knowledge of the study affect their behavior? It is not clear that the occupant monitoring research field has rigorously answered this question, as most prominent papers do not discuss the Hawthorne effect as a possible source of bias. Some researchers (e.g., Van Dam et al. 2010; Vassileva et al. 2012) have acknowledged the possibility of the Hawthorne effect biasing their study, but have not quantified its impact. Other studies have intentionally minimized the Hawthorne effect by minimizing researcher-participant contact. For instance, Meerbeek et al. (2014), who studied office window blinds, avoided informing occupants that they were being studied to prevent influencing the results. The occupants were contacted near the end of the study period when they were asked to participate in the survey. It should be noted that ethical protocol must be upheld.

Some established methods to reduce the Hawthorne effect include:

- Do not reveal the full purpose of the study, or develop an artificial task or motive that is distinct from the actual research (e.g., Meerbeek et al. 2014; Boyce et al. 2006);
- Do not instruct participants how to behave (Wood and Newborough 2003);
- Minimize site visit frequency and elongate study durations so that participants may forget about the study;
- Minimize the visibility and inconvenience of sensors in occupied spaces; and
- Rely only on built-in sensors or use the cover of building management (Mahdavi 2011).

Chapter 8 discusses the Hawthorne effect in the context of surveys. For greater detail and approaches to navigate biases, the reader is directed to introductory psychology textbooks (e.g., Elmes et al. 2011; Breakwell et al. 2012). Future research is needed to conclusively determine the impact of in situ study participants' knowledge of monitoring. Ideally, two adequately large samples could be tested: one control whose participants are not informed that they are being monitored and the other who are informed (ethics clearance must be obtained in both cases). A second approach would be to observe the frequency of occupant actions as a function of time after researcher-participant interaction events. Based on the latter approach, Tiefenbeck (2016) found that studied households increased their resource use by 5–20% after the first few weeks of studies.

6.4.6 Participant Recruitment

One of the most challenging aspects of in situ occupant studies is recruiting participants. Typical recruitment methods include unsolicited visits, posters, email, and social media. Unlike recruitment for laboratory studies and surveys, recruitment for in situ studies may, by necessity, require more intensive advertising for occupants of a single building. If a particular sample size is being sought in just one building, there is extra incentive to reach all occupants.

It is advisable to offer a modest honorarium to compensate participants for their time. Such honoraria can be divided such that some of it is provided at the end of the study. This strategy helps to ensure that temporarily installed sensors can be retrieved. The authors have typically offered $20 in cash or gift cards to compensate participants for about one hour of sensor installation and surveying. Regardless, the authors have found that about half of a contacted population does not respond, even if there is some sense of obligation to the researchers (e.g., the population works for the same university as the researchers).

Care should be taken to attempt to randomly sample a population of occupants in order to increase external validity. Specific biases that often arise for in situ occupant studies include:

- Recruiting only low-income participants who find the incentive particularly attractive, but are sensitive to energy costs;
- Recruiting particularly environmentally-conscious occupants who are interested in the study, but are not representative with regards to energy-related behaviors;
- Recruiting occupants from a particular façade or building orientation (e.g., north-facing apartments) or room elevation (e.g., top floor residents who suffer from stack effect); and,
- Recruiting occupants who are particularly dissatisfied with the space and believe that complaining and exaggerating about discomfort may result in building improvements.

Another challenge that researchers should be aware of is participants who are new arrivals to an unfamiliar space, will be absent for extended periods, or are planning to leave before the study period is complete. It is not unusual for occupants to have long-term absences or to move out entirely—for example, professors may go away on sabbatical for a year, apartment tenants may change, or employees may retire or switch offices. While study participants should be asked about their intention to stay in their space for the long term, unexpected events can arise. In such cases, equipment may be lost or future occupants may not give the researcher access to the space. Thus, it is important to build a margin of 10% or more into participant sample sizes, as discussed elsewhere in this book (e.g., Chaps. 3 and 8). The reader is also referred to the numerous available texts on research methods in social science and psychology (e.g., Gliner et al. 2011) for sampling and participant recruitment methods.

6.4.7 *Ethical Obligations and Implications of Performing in Situ Monitoring*

Ethics play a major guiding role in all occupant studies, as discussed at length in Chap. 11. Some additional ethical considerations particular to the context of in situ studies are briefly summarized here.

Observation of occupants can be covert or overt: while occupants are unaware of being observed in covert observations, they are notified of the study in an overt observation. As a general rule, if occupants should expect privacy and are being monitored or observed, then ethics clearance is required. For instance, pedestrians in public spaces are aware that they are in the presence of a large group of people, and so the researcher likely does not have to obtain individual consent. However, occupants in private offices or homes likely do not expect that they are being observed; thus, informed consent is required. Regardless, confirmation with the ethics board is always required.

Ethical obligations to occupants and building owners can obstruct in situ studies in two ways: (1) they may require occupants to be informed that they are being observed, thus invoking Hawthorne effect risks; and (2) they may require informed consent, which can reduce study sample sizes, as some targeted occupants may not be willing to participate in the study. Anecdotal evidence of the authors suggests that approximately 20% of contacted potential participants express concern for due diligence over ethics clearance. In general, the more invasive the monitoring, the more likely the occupants are to feel uncomfortable. For instance, in one study involving university professor and administrator offices, participants were quite concerned that confidential records and student information would be visible in photographs. Thus, ethics clearance should be obtained early in the research process and the population of occupants that are initially contacted should ideally be at least double the desired sample size (on the basis that about half do not respond to invitations to be participate).

6.5 Qualitative Aspects of in Situ Monitoring

While electronic sensors greatly improve the efficiency and accuracy of quantifying occupants and their actions, these sensed values should not be used to blindly develop models or draw conclusions because many confounding and contextual factors are likely to exist. Inventories should be performed before consideration of sensor-based monitoring to assess the available occupant interfaces, equipment, space geometry, and envelope features. Table 6.2 summarizes the suggested observations that should be noted and recorded during walkthroughs. These details should also be included with published datasets and research papers.

Furthermore, photographs and diagrams with all notable features labeled should be provided, where permission is obtained. Examples are shown in Figs. 6.18, 6.19 and 6.20.

Table 6.2 Features that should be recorded from walkthroughs and/or building drawings

Space characteristics
• Major geometry and approximate location of occupants, where applicable
• Nominal number of occupants in room/unit/building
• Ease of access to adaptive systems relative to typical occupant positions (e.g., obstructions and required effort to reach interfaces)
• Plug-in equipment type
• Type and position of furniture
• Location of computer monitors and televisions
• Typical activities expected in the space as it is currently used based on walkthrough or discussion with occupants (e.g., reading, computer work, sleeping, cooking, television watching, exercise, teaching)
• If pets (e.g., dogs, cats) are present, they should be noted, as their presence and actions could be mistaken for human occupants. Pets may also pose constraints on the indoor environmental conditions maintained by the occupants
Envelope properties
• Envelope properties (U-factor, window-to-wall area ratio, window visible transmittance, window solar heat gain coefficient, type/size of operable window section)
• Thermal and optical properties of window blinds/shades, if applicable
• Approximate surface finishes and reflectance (i.e., specularity and visible reflectance)
Building systems
• Detailed description of adaptive systems on envelope (e.g., window blinds, operable windows) and means for control (e.g., crank, lever, chain, motorized)
• Presence and specification of fixed shading
• Light switch type and controls (e.g., daylight controlled, occupancy controlled, vacancy controlled, dimming possibility, presence of other lamps)
• Thermostat type and control logic (e.g., ease of use and accessibility, setpoint schedule, deadband, override/reset period before switching to automation); map of rooms served by thermostat; responsiveness of thermostats (e.g., is there immediate feedback?); and systems controlled by the thermostat (supply water temperature, supply water flow, electric heating/cooling)
• Heating and cooling system types and location
Building site
• View type/quality of view to outdoors (both geometry and content)
• External solar obstructions and reflective objects (e.g., reflective facades of neighborhood buildings) surround the building being monitored

6.6 Use of Surveys to Complement Monitoring

As discussed in Chap. 8, surveys are an important method for understanding intricacies in occupant decision-making, as sensors alone cannot necessarily capture the cause and effect of occupant actions and presence. For instance, building systems may be difficult to use or access, causing significantly diminished frequency of use (Day et al. 2012). Meanwhile, corporate policy may affect clothing choices, which in turn affects thermostat and operable window use (Morgan and de Dear 2003). Or perhaps there is local discomfort (e.g., drafts or a small beam of sunlight that hits the computer monitor) that is too subtle for sensors to detect.

Fig. 6.18 Sample image showing key building systems and occupant interfaces in a private office

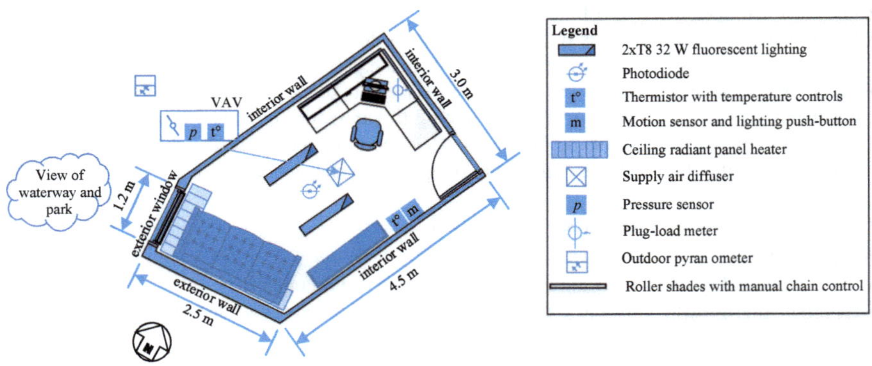

Fig. 6.19 Floor plan of monitored office

Two longitudinal survey approaches that could complement in situ methods and have been applied extensively in the past include: periodic surveys and frequent automated polling. Periodic surveys reduce the risk of a significant Hawthorne effect. Frequent automated polling increases regularity and the amount of data collected, but at the cost of potentially influencing results. Issues that are specific to these for in situ studies are discussed below.

Frequent polling of occupants for in situ studies is emerging as a potentially effective means to obtain longitudinal data. These draw from survey or laboratory methodologies, but are used for in situ environments and tend to be longer term than the other methods. Two methods that have been used in practice are to have a

Fig. 6.20 West-facing façade of monitored building

computer-based survey that pops up periodically to ask questions, and small consoles with simple user interfaces. The advantages of either approach are that a large longitudinal sample is obtained and the frequency of response reduces the error from recall bias (i.e., participants are less likely to forget events or sensations that occurred in the past several hours). Notable recent examples in the literature include Konis' (2013) polling station and Haldi and Robinson's (2008) computer-based survey.

Konis' device included an interface for occupants to provide subjective visual comfort, as well as sensors to measure illuminance and black globe temperature. This approach is quite elegant because it does not require a computer to be on at the workstation—or a computer at all—and it allows environmental variables to be measured.

Haldi and Robinson's (2008) questionnaire was installed on occupants' desktops and the occupant was able to choose the frequency that it popped up. The researchers reported that generally a two-hour frequency was selected, such that the questionnaire was answered three to four times per day. Using this method, they were able to obtain data about immeasurable quantities—clothing changes, hot or cold drink consumption, and activity level—as well as variables that are more easily measured, such as blinds, window opening, and fan (Fig. 6.21). Potential drawbacks of the approach include fatigue bias (the participants responded to the questionnaire on the order of 100 times), concern over privacy (particularly related to clothing level), recall bias, and the possibility that an occupant is reminded of adaptive opportunities that may not have occurred to them because of the frequent questionnaire.

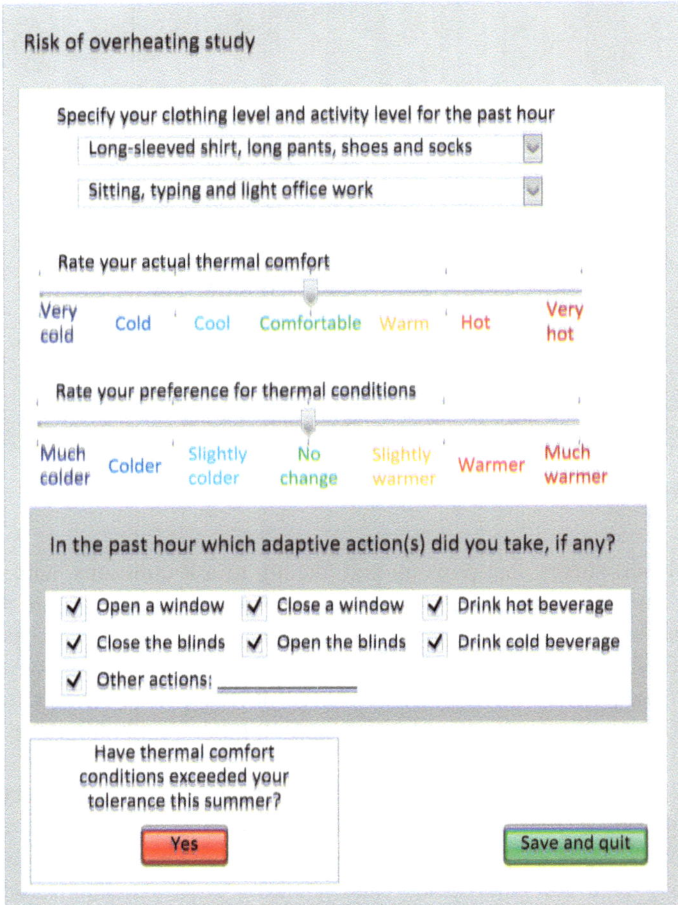

Fig. 6.21 Example pop-up computer-based poll [inspired by Haldi and Robinson (2008)]

Another relatively unexploited research method is to provide a user interface through which the building occupants can register their comfort state at the building level (e.g., Fig. 6.22).

Further research is required to reduce biases from sampling frequency. For instance, occupants may respond differently if they recently arrived and have an elevated metabolic rate than if they have been in the space for an extended period. Moreover, weighting of responses requires careful attention. Some occupants may be much more responsive about reported comfort merely because of they have more time to devote to the research. Lastly, language barriers may alter survey effects. For instance, Europeans often refer to cool air as "fresh", whereas in English it usually implies outdoor air that is free of detectable contaminants.

Fig. 6.22 Example interface that could be used to assess occupant comfort of buildings (photos from Carolyn Wayne)

6.7 Conclusion

This chapter proposed a generalized procedure for performing in situ occupant monitoring studies. Following that, it discussed a variety of practical solutions and complications to overcome in studying occupants in situ. Finally, a variety of case studies from the literature were briefly explored to illustrate challenges and solutions that international researchers have applied.

In situ occupant monitoring studies are likely to be the most reliable method to obtain data for occupant models that are suitable for building simulation tools. They are naturalistic and do not rely on occupants' memory or their willingness to frequently respond to surveys or other polls. They are also relatively resistant to the Hawthorne effect due to their long-term and discreet nature. Their duration, which can last months or years, allows a relatively large temporal sample to be obtained.

Nonetheless, the cost to collect data—both hardware and researcher effort—may limit sample sizes to tens of occupants. Furthermore, purely sensor-based measurements limit researchers' ability to explain cause and effect relationships. In addition, the researchers generally cannot control major building and activity parameters, such as orientation, window size, and occupant traits. Thus, contrary to laboratory-based occupant studies, researchers using in situ studies are at the mercy of having a suitable building—and willing participants—to perform in situ studies.

Regardless of the recent popularity of performing in situ occupant monitoring campaigns, many fundamental and technological challenges and questions remain, such as:

- Accurately counting occupants in spaces
- Measuring clothing level and other adaptive actions that do not relate to building systems
- Determining which occupant acted (in shared spaces)
- Accurately measuring representative indoor environmental quality parameters, particularly spatially sensitive quantities, such as illuminance, glare, and mean radiant temperature
- Determining adequate sample size (number of occupants and duration), depending on the purpose of the study
- Quantifying the impact of biases, such as the Hawthorne effect
- Cost-effectively obtaining large sample sizes.

The next two chapters discuss two complementary occupant study methods: laboratories and surveys. These methods—particularly surveys—should not be considered in isolation, but rather as critical tools to forming a complete picture of our understanding of occupants in buildings.

References

Andersen R (2013) Erfaringer fra projekt Dynamisk varmeregnskab (trans: Energi CfIo). Danmarks Tekniske Universitet

Andersen R, Fabi V, Toftum J, Corgnati SP, Olesen BW (2013) Window opening behaviour modelled from measurements in Danish dwellings. Build Environ 69:101–113

ASHRAE (2010) Thermal environmental conditions for human occupancy. Standard 55. Atlanta

Balaji B, Xu J, Nwokafor A, Gupta R, Agarwal Y (2013) Sentinel: occupancy based HVAC actuation using existing WiFi infrastructure within commercial buildings. Paper presented at the proceedings of the 11th ACM conference on embedded networked sensor systems. Roma, Italy

Benezeth Y, Laurent H, Emile B, Rosenberger C (2011) Towards a sensor for detecting human presence and characterizing activity. Energ Build 43(2):305–314

Bennet I, O'Brien W (2017) Field study of thermal comfort and occupant satisfaction in Canadian condominiums. Architectural Sci Rev 60(1):27–39. doi:10.1080/00038628.2016.1205179

Boyce PR, Veitch JA, Newsham GR, Jones C, Heerwagen J, Myer M, Hunter C (2006) Lighting quality and office work: two field simulation experiments. Lighting Res Technol 38(3):191–223

Breakwell GM, Smith JA, Wright DB (2012) Research methods in psychology. SAGE Publications, Beverley Hills

Brown S, Freas B, Goldstein N, Kallai P, Tischler M, Valiquette L (2016) Printable and flexible electronics enabled intelligent buildings. Canadian Printed Electronics Industry Association/Continental Automated Buildings Association (CPEIA/CABA) White Paper. https://www.navigant.com/-/media/www/site/insights/energy/2016/energycpeiawpib.pdf

Butte NF, Ekelund U, Westerterp KR (2012) Assessing physical activity using wearable monitors: measures of physical activity. Med Sci Sports Exerc 44(1 Suppl 1):S5–12

Chong H (2016) Temperature dataloggers in single family homes in winter. http://www.dataloggerproject.com/

Coen MH (1998) Design principles for intelligent environments. In: AAAI/IAAI, pp 547–554

Davis JA III, Nutter DW (2010) Occupancy diversity factors for common university building types. Energ Build 42(9):1543–1551. doi:10.1016/j.enbuild.2010.03.025

Day J, Theodorson J, Van Den Wymelenberg K (2012) Understanding controls, behaviors and satisfaction in the daylit perimeter office: a daylight design case study. J Inter Des 37(1):17–34

Dong B, Andrews B (2009) Sensor-based occupancy behavioral pattern recognition for energy and comfort management in intelligent buildings. In: Proceedings of International IBPSA Conference

Dong B, Andrews B, Lam KP, Höynck M, Zhang R, Chiou Y-S, Benitez D (2010) An information technology enabled sustainability test-bed (ITEST) for occupancy detection through an environmental sensing network. Energ Build 42(7):1038–1046. doi:10.1016/j.enbuild.2010.01.016

Dong B, Li Z, McFadden G (2015) An investigation on energy-related occupancy behavior for low-income residential buildings. Sci Technol Built Environ 21(6):892–901. doi:10.1080/23744731.2015.1040321

ecobee (2017) Donate your data. https://www.ecobee.com/donateyourdata/

Elmes DG, Kantowitz BH, Roediger III HL (2011) Research methods in psychology. Nelson Education, Canada

Erickson VL, Lin Y, Kamthe A, Brahme R, Surana A, Cerpa AE, Sohn MD, Narayanan S (2009) Energy efficient building environment control strategies using real-time occupancy measurements. In: Proceedings of the First ACM Workshop on Embedded Sensing Systems for Energy-Efficiency in Buildings. ACM, pp 19–24

George JF (1996) Computer-based monitoring: common perceptions and empirical results. Mis Q 459–480

Gilani S, O'Brien W (2016) Review of current methods, opportunities, and challenges for in-situ monitoring to support occupant modelling in office spaces. J Build Perform Simul 1–27. doi:10.1080/19401493.2016.1255258

Gliner JA, Morgan GA, Leech NL (2011) Research methods in applied settings: an integrated approach to design and analysis. Routledge, London

Godfrey A, Conway R, Meagher D, ÓLaighin G (2008) Direct measurement of human movement by accelerometry. Med Eng Phys 30(10):1364–1386. doi:10.1016/j.medengphy.2008.09.005

Gunay HB, Fuller A, O'Brien W, Beausoleil-Morrison I (2016) Detecting occupants' presence in office spaces: a case study. Paper presented at the eSim 2016. Hamilton, Ontario

Gupta M, Intille SS, Larson K (2009) Adding gps-control to traditional thermostats: an exploration of potential energy savings and design challenges. In: Pervasive Computing. Springer, Berlin, pp 95–114

Haldi F, Robinson D (2008) On the behaviour and adaptation of office occupants. Build Environ 43(12):2163–2177. doi:10.1016/j.buildenv.2008.01.003

Haldi F, Robinson D (2010) Adaptive actions on shading devices in response to local visual stimuli. J Build Performan Simul 3(2):135–153. doi:10.1080/19401490903580759

Heerwagen J (2000) Green buildings, organizational success and occupant productivity. Build Res Inf 28(5–6):353–367

Jalal A, Uddin MZ, Kim JT, Kim T-S (2012) Recognition of human home activities via depth silhouettes and $\Re$ transformation for Smart Homes. Indoor Built Environ 21(1):184–190

Kapsis K, O'Brien W, Athienitis AK (2013) Time-lapse photography and image recognition to monitor occupant-controlled shade patterns: analysis and results. Paper presented at Building Simulation, Chambery, France, 25–28 Aug 2013

Konis K (2013) Evaluating daylighting effectiveness and occupant visual comfort in a side-lit open-plan office building in San Francisco, California. Build Environ 59:662–677

Kovatsch M, Lanter M, Duquennoy S (2012) Actinium: A RESTful runtime container for scriptable internet of things applications. In: Internet of Things (IOT), 2012 3rd International Conference on the Internet of Things, 2012. IEEE, pp 135–142

Labeodan T, Zeiler W, Boxem G, Zhao Y (2015) Occupancy measurement in commercial office buildings for demand-driven control applications—a survey and detection system evaluation. Energ Build 93:303–314. doi:10.1016/j.enbuild.2015.02.028

Lam KP, Höynck M, Dong B, Andrews B, Chiou Y-S, Zhang R, Benitez D, Choi J (2009) Occupancy detection through an extensive environmental sensor network in an open-plan office building. IBPSA Build Simul 1452–1459

Mahdavi A (2011) The human dimension of building performance simulation. In: Building Simulation-12th Conference of International Building Performance Simulation Association

Marceau ML, Zmeureanu R (2000) Nonintrusive load disaggregation computer program to estimate the energy consumption of major end uses in residential buildings. Energ Convers Manag 41(13):1389–1403

Mathie M, Celler BG, Lovell NH, Coster A (2004) Classification of basic daily movements using a triaxial accelerometer. Med Biol Eng Compu 42(5):679–687

McCambridge J, Kypri K, Elbourne D (2014) Research participation effects: a skeleton in the methodological cupboard. J Clin Epidemiol 67(8):845–849. doi:10.1016/j.jclinepi.2014.03.002

Meerbeek B, te Kulve M, Gritti T, Aarts M, van Loenen E, Aarts E (2014) Building automation and perceived control: a field study on motorized exterior blinds in Dutch offices. Build Environ 79:66–77

Morgan C, de Dear R (2003) Weather, clothing and thermal adaptation to indoor climate. Clim Res 24(3):267–284

Nestor PG, Schutt RK (2014) Research methods in psychology: investigating human behavior. Sage Publications, Beverley Hills

O'Brien W, Gunay HB (2014) The contextual factors contributing to occupants' adaptive comfort behaviors in offices—a review and proposed modeling framework. Build Environ 77:77–87

O'Brien W, Kapsis K, Athienitis AK, Kesik T (2010) Methodology for quantifying the performance implications of intelligent shade control in existing buildings in an urban context. Paper presented at the SimBuild 2010, New York City, 11–13 Aug 2010

Peffer T, Pritoni M, Meier A, Aragon C, Perry D (2011) How people use thermostats in homes: a review. Build Environ 46(12):2529–2541

Rea MS (1984) Window blind occlusion: a pilot study. Build Environ 19(2):133–137

Reinhart CF, Voss K (2003) Monitoring manual control of electric lighting and blinds. Lighting Res Technol 35(3):243–260

Schweiker M, Wagner A (2016) The effect of occupancy on perceived control, neutral temperature, and behavioral patterns. Energ Build 117(1):246–259. doi:10.1016/j.enbuild.2015.10.051

Shih H-C (2014) A robust occupancy detection and tracking algorithm for the automatic monitoring and commissioning of a building. Energ Build 77:270–280. doi:10.1016/j.enbuild.2014.03.069

Sutter Y, Dumortier D, Fontoynont M (2006) The use of shading systems in VDU task offices: a pilot study. Energ Build 38(7):780–789

Tahmasebi F, Mahdavi A (2012) Optimization-based simulation model calibration using sensitivity analysis. Paper presented at the 7th conference of IBPSA-CZ, Brno, Czech Republic

Tiefenbeck V (2016) On the magnitude and persistence of the Hawthorne effect—evidence from four field studies. In: 4th European Conference on Behaviour and Energy Efficiency, Coimbra, Portugal, 8–9 Sept 2016

Van Dam S, Bakker C, Van Hal J (2010) Home energy monitors: impact over the medium-term. Build Res Inf 38(5):458–469

Vassileva I, Wallin F, Dahlquist E (2012) Analytical comparison between electricity consumption and behavioral characteristics of Swedish households in rented apartments. Appl Energ 90 (1):182–188

Veitch JA, Mancini S, Galasiu AD, Laouadi A (2013) Survey on Canadian households' control of indoor climate. Research Report RR-344. National Research Council (NRC) Construction. http://nparc.cisti-icist.nrccnrc.gc.ca/eng/view/fulltext/?id=e13b92f0-4044-4643-9bd3-2728e06c3aaa

Webster T, Bauman F, Shi M, Reese J (2002) Thermal stratification performance of underfloor air distribution (UFAD) systems. Paper presented at the Indoor Air, Monterey, CA, June 30–July 5

Wienold J, Christoffersen J (2006) Evaluation methods and development of a new glare prediction model for daylight environments with the use of CCD cameras. Energ Build 38(7):743–757. doi:10.1016/j.enbuild.2006.03.017

Wood G, Newborough M (2003) Dynamic energy-consumption indicators for domestic appliances: environment, behaviour and design. Energ Build 35(8):821–841

Yang C-C, Hsu Y-L (2010) A review of accelerometry-based wearable motion detectors for physical activity monitoring. Sensors 10(8):7772–7788

Zhang Y, Barrett P (2012) Factors influencing occupants' blind-control behaviour in a naturally ventilated office building. Build Environ 54:137–147

Zhao Y, Zeiler W, Boxem G, Labeodan T (2015) Virtual occupancy sensors for real-time occupancy information in buildings. Build Environ 93(Part 2):9–20. doi:10.1016/j.buildenv.2015.06.019

Zibin N, Zmeureanu R, Love J (2016) Automatic assisted calibration tool for coupling building automation system trend data with commissioning. Automation in Construction 61:124–133

Zoha A, Gluhak A, Imran MA, Rajasegarar S (2012) Non-intrusive load monitoring approaches for disaggregated energy sensing: a survey. Sensors 12(12):16838–16866

Chapter 7
Laboratory Approaches to Studying Occupants

Andreas Wagner, Rune Korsholm Andersen, Hui Zhang, Richard de Dear, Marcel Schweiker, Edwin Goh, Wouter van Marken Lichtenbelt, Rita Streblow, Francesco Goia and Sumee Park

Abstract Laboratories offer the possibility to study occupant behavior in a very detailed manner. A wide range of indoor environmental scenarios can be simulated under precisely controlled conditions, and human subjects can be selected based on pre-defined criteria. The degree of control over experiments is high and a large number of physical, physiological, and psychological quantities can be monitored. This chapter gives an overview of various types of test facilities in the world and their main features in terms of experimental opportunities. It then presents typical technical equipment and sensor technologies used in laboratory environments. Finally, questions on appropriate laboratory design and experimental set-ups are

A. Wagner (✉)
Building Science Group, Faculty of Architecture, Karlsruhe Institute of Technology, Englerstr. 7, 76131 Karlsruhe, Germany
e-mail: wagner@kit.edu

R.K. Andersen
Department of Civil Engineering, Technical University of Denmark, Nils Koppels Allé, Building 402, Office 218, 2800 Kongens Lyngby, Denmark
e-mail: rva@byg.dtu.dk

H. Zhang
Center for the Built Environment, University of California, 390 Wurster Hall #1839, Berkeley, 94720-1839 CA, USA
e-mail: zhanghui@berkeley.edu

R. de Dear
Indoor Environmental Quality Laboratory, School of Architecture, Design and Planning, The University of Sydney, 2006 Sydney, NSW, Australia
e-mail: richard.dedear@sydney.edu.au

M. Schweiker
Building Science Group, Faculty of Architecture, Karlsruhe Institute of Technology, Englerstr. 7, 76131 Karlsruhe, Germany
e-mail: marcel.schweiker@kit.edu

E. Goh
Berkeley Education Alliance for Research in Singapore Limited, 1 Create Way, #11-02, CREATE Tower, 138602 Singapore, Singapore
e-mail: edwingoh@bears-berkeley.sg

A. Wagner et al. (eds.), *Exploring Occupant Behavior in Buildings*,
DOI 10.1007/978-3-319-61464-9_7

discussed. One conclusion is that, in spite of many advantages, there are limits to investigating occupant behavior in a laboratory's "artificial" environment, in part due to the fact that subjects always feel observed to some extent. However, valuable results can be achieved if the specific opportunities of laboratories are utilized both by appropriate design and precise experiments during operation.

7.1 Laboratories in Indoor Environmental Quality Research

In a general sense, a laboratory (i.e., "lab") offers the possibility to study occupant behavior in an environment which can be specifically designed and set up for certain purposes and study objectives. In contrast to in situ studies, the participating persons can be selected based on pre-defined criteria (e.g., sample size, age group, gender). Further, the degree of control over the experimental procedure, as well as the variety of applicable sensor equipment and technologies, is much higher compared to a field study. This chapter focuses on lab environments for behavioral studies to maintain or improve comfort in indoor spaces and the implications for energy consumption, but will exclude all other set-ups for perceptional-behavioral (e.g., color preferences and noise levels), sociological, or psychological studies. The main focus of this chapter is on experimental settings for office spaces, but many of the aspects addressed are also true for residential environments, or can be transformed for this purpose.

There is a long tradition in research on indoor environmental quality (IEQ)—referring to thermal, visual, and aural comfort, as well as indoor air quality—to conduct experiments in labs. According to the different research questions from

W. van Marken Lichtenbelt
Department of Human Biology and Human Movement Sciences, NUTRIM School for Nutrition and Translational Research in Metabolism, Maastricht University Medical Center+, 6200 MD PO Box 616, Maastricht, The Netherlands
e-mail: markenlichtenbelt@maastrichtuniversity.nl

R. Streblow
Institute for Energy Efficient Buildings and Indoor Climate, E.ON Energy Research Center, RWTH Aachen University, Mathieustr 10, 52074 Aachen, Germany
e-mail: rstreblow@eonerc.rwth-aachen.de

F. Goia
Department of Architecture and Technology, Faculty of Architecture and Design, Norwegian University of Science and Technology, Alfred Getz vei 3, 7491 Trondheim, Norway
e-mail: francesco.goia@ntnu.no

S. Park
Department of Energy Efficiency and Indoor Climate, Fraunhofer Institute for Building Physics, Fraunhoferstr. 10, 83626 Valley, Germany
e-mail: sumee.park@ibp.fraunhofer.de

each field of comfort, the designs of lab facilities can differ substantially. For example, with a focus on thermal comfort, different climate chambers are used for exposing subjects to a well-defined indoor climate which also might include certain disturbances (e.g., temperature asymmetry, draft) or changes over time. Test facilities for air quality add options for changing air change rates and introducing fresh or polluted air to the subjects through different technical means. Parameters of visual comfort are tested by providing an exposure of subjects to different (artificial and natural) light sources and levels, as well as control options, mostly under defined (and comfortable) thermal conditions. Aural comfort experiments include exposure of subjects to different noise sources (e.g., ventilation systems, other persons) and acoustic qualities of spaces. In contrast to this, we can also find very specific labs which try to simulate well-defined environments and situations (e.g., office rooms, class rooms, living rooms, car interiors, airplane cabins). In these tailored settings, the perception of single comfort parameters is investigated with a larger range of other influencing parameters—or, the subjects' overall comfort and the weights or interdependencies of thermal, aural, olfactory or visual perceptions are addressed, often also under varying indoor conditions and over different time periods.

In recent years, interest in subjects' reaction to their given environmental conditions has risen, particularly with regard to comfort and the related energy consumption. To address this topic on a lab scale, experimental environments have to provide a surrounding as realistic and familiar as possible for subjects, as experiments have to be performed over longer periods (half to a whole day, or even longer), and the lab's influence on the subjects' general perception should be kept to a minimum. As the number of such behavioral experiments in labs starts to increase, several questions have to be answered—for example, whether or to which extent experiments on occupant behavior can be performed in a lab environment at all, and what are the topics with regard to behavior to be addressed preferably in lab experiments? In this context, the chapter mainly addresses occupant behavior related to thermal comfort in building indoor environments.

7.2 Examples of Typical Laboratory Designs and Their Technical Equipment

With the following examples (summarized in Table 7.1), the range of existing types of test facilities for IEQ research, as well as their experimental opportunities shall be illustrated as a basis for further discussion of behavioral studies in lab environments. There are numerous other climate chambers and lab facilities at different universities, research institutions, and private companies in the world which are not mentioned in this section. The authors attempted to foreground the historic development from the seventies until today and the variety of test facilities, rather than drawing a complete picture. They are aware that they presumably missed facilities worthy of description.

Table 7.1 Overview of test facilities described in this chapter

Sections	Name of the laboratory	Location	Key features
7.2.1	International Centre for Indoor Environment and Energy (ICIEE)	Danish Technical University (DTU), Denmark	Wide variety of different climate chambers and field laboratories, mainly for experiments on thermal comfort, air quality, air distribution, ventilation systems, and combined effects of indoor environmental variables
7.2.2	Controlled Environmental Chamber	Center for the Built Environment (CBE), University of California at Berkeley, USA	Chamber in real office design for thermal comfort experiments and reproducing the effect of different air distribution systems
7.2.3	Indoor Environmental Quality Laboratory (IEQ Lab)	University of Sydney, Australia	Two connected chambers, designed to be as realistic as possible to any setting (bedroom, office, etc.), to examine how combinations of the key IEQ factors relate to comfort, productivity, and health of occupants
7.2.4	Laboratory for Occupant Behavior, Satisfaction, Thermal comfort and Environmental Research (LOBSTER)	Karlsruhe Institute of Technology (KIT), Germany	Facility hosting two office-like rooms with real windows to the outdoors and providing different adaptive opportunities for occupants, to study adaptation and behavioral actions
7.2.5	SinBerBEST Test Bed	Berkeley Education Alliance for Research in Singapore (BEARS) Limited, Singapore	Fully configurable space with moveable and interchangeable wall panels, designed for experiments on air quality, in combination with thermal and visual comfort
7.2.6	Metabolic Research Unit Maastricht (MRUM)	University of Maastricht, The Netherlands	Facility with a variety of chambers to experimentally investigate human energy and substrate metabolism

(continued)

Table 7.1 (continued)

Sections	Name of the laboratory	Location	Key features
7.2.7	Institute for Energy Efficient Buildings and Indoor Climate, E.ON Energy Research Center	RWTH Aachen University, Germany	Various labs from climate chambers over generic constructions of indoor environments (vehicle test facilities) to a living lab office building
7.2.8	The ZEB Living Laboratory	Norwegian University of Science and Technology (NTNU) and SINTEF, Norway	Single-family house built as a living lab, fully equipped to monitor occupants' interactions with the building and technical services
7.2.9	High Performance Indoor Environment Laboratory (HiPIE-Lab), Indoor Air Test Center (IATC), Modular Test Facility for Energy and Indoor Environments (VERU)	Fraunhofer Institute for Building Physics (IBP), Germany	Laboratories for IEQ tests, including impact of acoustics, lighting, and indoor climate on human beings; investigation of air quality, airflow, and effectiveness of active and passive air purification systems Test bed for façade systems and their effects on indoor environments and energy consumption
7.2.10	Flight Test Facilities	Fraunhofer IBP and RWTH Aachen University, Germany	Air plane mock-ups to study cabin air quality and thermal comfort, as well as technical equipment

7.2.1 International Centre for Indoor Environment and Energy (ICIEE), Technical University of Denmark, Denmark

(ICIEE 2017)

The ICIEE operates a wide variety of chambers (see Fig. 7.1) and is the largest test facility of its kind in the world. It is situated at the Technical University of Denmark (DTU) in Lyngby. Chambers 1 and 2, constructed in 1988, were primarily designed for air quality experiments. They both have a floor area of 9 m^2 and a height of 2.5 m, and are connected with a door for easy access in comparative studies. The chambers can provide ranges of temperature from 10 to 40°C and relative humidity from 10 to 90%. A particularly high air exchange rate—up to 70 air changes per hour (ACH)—can be adjusted by displacement ventilation. The stainless steel

Fig. 7.1 Climate chambers 3 (*left*) and 2 (*middle*), and field lab (*right*) at the ICIEE (photos by Rune Korsholm Andersen (*left* and *middle*) and Pawel Wargocki)

material minimizes sorption processes by indoor surfaces. Chamber 3, built in 1972 for thermal comfort experiments, has a floor area of 28 m^2 and a height of 2.5 m (volume: 70 m^3). It is worthwhile mentioning that this is the chamber of the famous comfort researcher P.Ole Fanger used for many of his experiments with regard to the Predicted Mean Vote (PMV) model. Up to 3500 m^3/h of supply air can be provided, which means an air change rate of 50 ACH. The conditioned air is supplied uniformly from floor to ceiling and also passes through thin textile layers of the walls to control air temperature and radiant temperature. The chamber can provide the same range of temperature and relative humidity as above. No specific means are foreseen for subjects to interact with the indoor climate. In 2016, Chambers 1, 2, and 3 were upgraded with a new control system and many parts of the ventilation systems were replaced. Chamber 3 was equipped with facilities for experiments with personalized ventilation systems.

Chambers 4, 5, and 6, put into operation in 2001, have a size of 23 $m^2 \times 2.5$ m (volume: 57 m^3). Chambers 4 and 5 are multipurpose chambers with the possibility of mixing or displacement ventilation; the maximum airflow rate is 610 m^3/h (11 ACH). The temperature can vary between 10 and 40 °C and the relative humidity between 10 and 90%. Chamber 6 is an air quality chamber with an available airflow rate double that of the other two chambers. It is designed with displacement ventilation only. In contrast to Chambers 1, 2, and 3, Chambers 4, 5, and 6 appear like normal offices and are suited for long-term exposures. External pollution chambers can be connected in the ventilation system upstream, which enables double blind experiments to be run (having the source of pollution hidden within the ventilation system). Chamber 7 is designed for testing different air distribution principles (mixing or displacement ventilation) and can be changed in height. Its maximum size is 35 $m^2 \times 6$ m (volume: 208 m^3). The ceiling is divided into three parts which can be adjusted to different heights, each part separately or as a whole. The chamber size can also be modified with a floor-to-ceiling partition, which splits the chamber into two sections.

In contrast to these more traditional climate chambers, ICIEE also operates three field laboratories which present a flexible room set-up to be used as one large room or divided with partitions into three separate and fully independent rooms with no leakage between. Each room has a floor area of 18 m^2 and a volume of 55 m^3 and can be supplied with a maximum of approximately 400 m^3/h of outdoor air (7 ACH). Each of the rooms has an independently controlled ventilation system and a separate temperature control. Humidifiers can be additionally installed in each room. The separate rooms are suitable for testing a simulated office environment or can be used for full scale sensory assessments of pollution sources. The so-called "pollution box" in each room provides space for placing sources of pollution invisible for subjects. Fans on the top of the box draw air through the source and ensure mixing of the polluted air in the room.

The personalized ventilation field lab with a size of 36 m^2 × 3 m (volume: 108 m^3) was especially designed for personalized ventilation studies under conditions as close as possible to reality. The lab is equipped with two different and separately controlled ventilation systems, a total volume system for generating different background environments in the room, and a personalized ventilation system. The total volume system can generate either mixing or displacement air distribution in the lab with up to 7.5 ACH and a room air temperature between 20 and 28 °C. It can be operated under different modes: from only recirculation of the ventilation air to 100% outdoor air. The personalized ventilation system supplies 648 m^3/h of outdoor air in total to six workplaces. Individual control is provided at each workplace in terms of flow rate (0–108 m^3/h), supply air temperature (18 to 28 °C), and relative humidity of the supply air (30–70%), either for the subjects themselves at the workplaces or through the building management system.

Another field lab is a classroom (73.7 m^2) designed for research, education, and demonstration of different ventilation and air-conditioning systems. A number of ventilation systems are integrated in the room: mixing, displacement, automatically controlled natural, hybrid, and personalized. The temperature of both ceiling and floor is adjustable (both heating and cooling). The advanced control system makes it possible to combine different ventilation and heating/cooling systems to establish any required indoor air environment. The system has the flexibility to integrate many kinds of air processing equipments for testing their performance. The classroom can also be used as a field lab to study the effect of different ventilation principles in a space with high population density.

None of the ICIEE chambers—except for the field labs and the classroom—has operable windows with views to the outside but three of the chambers have fixed windows to an internal space where subjects are thermally pre-conditioned for their experiments.

7.2.2 Controlled Environmental Chamber, Center for the Built Environment (CBE); University of California at Berkeley, USA

(Bauman et al. 1988; Zhang et. al. 2009; Zhai et al. 2013; Pasut et al. 2015)

Located at a corner of the Architecture Faculty building, with windows on both sides to provide views to the outside, the Controlled Environmental Chamber at UC

Berkeley provides a “real” office appearance. It has a dimension of 5.5 m × 5.5 m × 2.5 m (75.6 m^3 in volume). The triple-glazed windows, which cannot be opened by the subjects, are well shaded by fixed external shades and internal Venetian blinds. The chamber is designed in such a way that it can reproduce the effects of many types of heating, ventilation, and air-conditioning (HVAC) distribution systems, as well as allow for performing pure tests of human physiology and comfort. For example, the chamber has been used extensively in comfort and efficiency studies of underfloor air distribution (UFAD), displacement ventilation, personal comfort systems (also called “task ambient systems”), overhead diffuser design, ceiling fans, and other types of air circulation devices. It has also been used for creating fundamental thermal distributions around the human body for developing multi-segment thermal comfort models and highly efficient personal comfort systems for offices and vehicles.

The raised floor system consists of 0.6 m^2 square panels. Floor registers, diffusers, desktop and partition-based supply nozzles (see Fig. 7.2, photo on the left), and other spot cooling air flow connections are installed directly into the floor panels, permitting maximum flexibility in the selection of supply and return locations. The 0.6 m high sub-floor area serves as a supply or return plenum, while also providing adequate space for connecting ducted floor registers and running instrumentation, power, and communication cables. A 0.5 m high ceiling plenum is provided for a similar purpose above the suspended ceiling, made up of 0.6 m^2 square acoustical ceiling tiles. The air supply and return can be from overhead or from below. The air handling system has been designed to provide three separately controllable air supplies to maintain desired conditions in the main chamber (temperature, humidity, and supply air volume), in the annular space between the window glazing layers, and in flexible spot heating/cooling supplies within the main chamber. Under normal steady state operation, the annular space temperature is controlled to be equal to the average temperature in the chamber, eliminating any asymmetrical thermal conditions during the experiments. The spot cooling is separated from the main supply to the chamber, allowing the test condition simulating cool air from a personal ventilation system to be tested with a warm ambient temperature in the test chamber.

The chamber supply air volume can be varied from a minimum of 40 m^3/h (0.5 ACH) to a maximum of 1520 m^3/h (10 ACH). Outside ventilation air can be provided over a full range of 0 to 100%. The minimum and maximum steady state

Fig. 7.2 Different experimental set-ups of the climate chamber at CBE (all photos by CBE)

temperatures can be maintained at a range from 13 to 35 °C and relative humidity over the range of 10–90%. Depending on the experiment, occupants may have control of the spot heating/cooling conditions, and of local air speed and radiation sources. Typically, experiments involve one to four subjects with some office equipment (chair, desk, and partitions).

7.2.3 Indoor Environmental Quality Laboratory (IEQ Lab), University of Sydney, Australia

(Nathwani et al. 2012; de Dear et al. 2013)

The IEQ Lab is situated in the School of Architecture building of the University of Sydney. It was designed to examine how combinations of the key IEQ factors influence comfort, productivity, and health of occupants. The facility hosts two purpose-designed climate chambers with approximately 60 m^2 (Chamber 1) and 25 m^2 (Chamber 2), both with an accessible raised floor and Chamber 1 also with a suspended ceiling (height of the room: 2.6 m). Chamber 1 provides space for 8 to 12 persons in a typical office fit-out, whereas Chamber 2 can host 4 to 6 persons in an office-type workstation layout (see Fig. 7.3). The chambers are connected directly by an internal door which allows moving directly from one conditioned space to another. The northern perimeter zone of both labs directly connects to an “environmental corridor” which can simulate “sunshine” with solar lamps and a large range of “outdoor” conditions with temperatures from around 4 to 40 °C. Removable insulating panels which cover the external single-pane windows in the corridor allow for natural ventilation and daylight in the chambers. This is also true for the external (also single-pane) windows on the southern façade of Chamber 1.

Building services have been designed to apply different conditioning strategies in the IEQ Lab, i.e., full air conditioning, natural ventilation or mixed-mode regimes. In Chamber 1, a constant or variable air volume (VAV) system allows flexible supply air temperature configurations. Air diffusion is realized with linear diffusers in the ceiling grid and the return air leaves the chamber through slots in the light fittings into the ceiling plenum. Alternatively, Chamber 1 can be operated with displacement under-floor ventilation; in this case conditioned air is supplied through the plenum under the raised floor, into the appropriate zone. The air is diffused into the chamber through swirl type diffusers in the floor. Operable windows on the southern and northern side of Chamber 1 also allow for cross-flow ventilation and daylighting.

Chamber 2 is equipped with a passive as well as an active chilled beam air conditioning system. The passive chilled beams provide sensible cooling by natural convection only, which depends on the difference between the zone and the beam temperatures. Latent cooling is provided by conditioned outside or fresh air which is delivered via the under-floor air distribution (UFAD) system. With the active chilled beams, sensible cooling can be increased by inducing an air flow over the beams’ coils with conditioned outside air. Latent cooling is again provided as

Fig. 7.3 Chamber 1 (*left*) with view towards the "environmental corridor", Chamber 2 (*middle*) with subjects, and the "environmental corridor" (*right*) of the IEQ Lab at the University of Sydney (all photos by Richard de Dear, University of Sydney)

described above. Similar to Chamber 1, there is the option of displacement ventilation via an under-floor plenum for Chamber 2. Electric duct heaters offer the possibility of heating the supply air, when necessary.

All indoor environmental parameters can be precisely controlled over a broad range of values, in any combination. The range for air temperatures in the chambers reaches from 16 to 38 °C and fresh outdoor airflow can be adjusted between 36 and 108 m^3/h per person. Depending on the ventilation mode, supply airflows reach from 1440 to 3960 m^3/h in Chamber 1 and from 1080 to 1512 m^3/h in Chamber 2. All indoor and "outdoor" conditions and scenarios foreseen by the experimental design can be programmed and controlled with a building management and control system which also logs all relevant data during an experiment. Effort went into making the IEQ Lab's interior as realistic as possible, to give it the look and feel of normal rooms, rather than experimental chambers. Initially the chambers were equipped and furnished as grade-A commercial office spaces, but they can easily be fitted-out to resemble residential, industrial, retail, cinema/theatre, leisure facility, and even vehicular (car, bus, train, plane) interiors for modest reconfiguration costs. The IEQ Lab also includes audio infrastructure for simulating acoustic environments within the experimental chambers. An integrated high-fidelity 24-channel audio system can present soundscapes appropriate to the research design in question (such as external environmental noise, including aircraft and road, as well as internal sources, such as co-worker "talkers" and office equipment).

7.2.4 Laboratory for Occupant Behavior, Satisfaction, Thermal Comfort and Environmental Research (LOBSTER), Karlsruhe Institute of Technology (KIT), Germany

(Schweiker et al. 2014)

The intention of building the LOBSTER was to study various aspects of human adaptation including behavioral reactions to variations in the indoor and outdoor environment in depth. Therefore, the main objectives were to create an environment which comes as close as possible to a real-world office environment and to provide different typical adaptive opportunities for the occupants to react on the (changing) outdoor and indoor climate. The LOBSTER is situated on a mostly unobstructed site on the western campus of KIT. The facility houses two identical office rooms (24 m^2 floor area, 3 m height), which both directly connect to the outdoor environment with windows (total area: 12.6 m^2) in a post and beam façade. Each room has two windows (0.9 m width by 1.5 m height) which can be opened and tilted, and two top light windows (0.9 m width by 0.5 m height), which can only be tilted. All windows are triple glazed (U_g = 0.7 W/m^2K, total solar transmittance = 0.5) and the opaque balustrade is equipped with vacuum insulation panels (U_{panel} = 0.2 W/m^2K). The framing is of insulated aluminum (U_f = 1.3 W/m^2K). Shading is provided by electrically driven venetian blinds with daylight guidance through the upper section of the blinds (Fig. 7.4).

The LOBSTER is designed as a timber frame construction. For insulation, wood fibers have been blown into the spaces between the frames, resulting in a layer of 18 cm which is topped by another 10 cm wood fiber insulation board. The internal walls of the test rooms have 5 cm of insulation, and between the two test rooms is a cavity which serves for technical services infrastructure and maintenance. The roof and the floor of the test facility are insulated with 26 cm of wood fibers. The resulting U-values are 0.13 W/m^2K for the exterior walls, 0.12 W/m^2K for the roof, and 0.12 W/m^2K for the floor, which is elevated from the ground due to a rotating assembly underneath. The insulation quality of the whole envelope is close to Passive House standard.

All interior surfaces of the test rooms—except the post and beam façade—can be thermally activated with a capillary tube system, which allows surface temperatures of each wall, the floor, and the ceiling to be controlled individually. Heat and cold is provided by two separate heat pumps with water storages connected to 13 water circuits. Indoor air temperature is additionally influenced by the ventilation strategy. In addition to having operable windows for ventilation, the ventilation concept includes two decentralized under floor convectors able to heat or cool the inlet air for each room and a central fan driven exhaust system for both rooms. The adjustable indoor temperature range reaches from around 19 to 32 °C, with higher temperatures possible by heating up the hot water storage to 60 °C with an electric auxiliary heater; the relative humidity cannot be modified without extra devices.

Indoor and outdoor climate quantities, as well as relevant lab and system parameters (e.g., surface temperatures in the offices) are continuously monitored during an experiment by either stationary sensors or mobile sensing equipment. Depending on the experimental setup, adaptive opportunities for the participants may include operating the windows, blinds, ventilation system, heating and cooling system, or a ceiling fan. On the other hand, all those interactions—except a full opening of the windows—can be done by the researcher or an algorithm through the building control system accessible through LabView®. The central control/ building management system allows the researchers to run various indoor climate

Fig. 7.4 View of the main façade with shadings closed (*left*), interior of the test office spaces with subjects and comfort meter (*middle*) and with shadings closed (*right*) of the LOBSTER at KIT (photos by Moritz M. Karl (*left*) and Marcel Schweiker, KIT)

scenarios over the length of an experiment, e.g., ascending temperature ramps over a day in order to simulate the thermal performance of a real office room. The rooms are normally furnished as grade-A commercial office spaces.

For visual or thermal comfort studies with regard to direct solar incident, the whole facility can be rotated 355° (limited by the electricity supply cable), allowing different azimuth angles for the window façade.

7.2.5 SinBerBEST Test Bed, CREATE Tower, Berkeley Education Alliance for Research in Singapore (BEARS) Limited, Singapore

The SinBerBEST test facility resides in the CREATE Tower in Singapore and provides a fully configurable space of approximately 100 m^2. The structure features moveable and interchangeable wall panels, and it can be subdivided using a fixed raised floor, fixed suspended ceiling, and wall modules on a rail system. On the north side, a modular steel frame panel is used to facilitate different configurations of passive façades (Fig. 7.5). Each room has a fully controlled air handling system and sufficient thermal insulation to allow varying its interior parameters: room temperature from −4 to +8 °C around the nominal 24 °C office environment; humidity levels from 30 to 90%; and CO_2 levels from 400 to 1200 ppm. These features enable to simulate various built environments, from aggressively chilled up to naturally ventilated spaces in tropical climates. With regards to lighting, an array of LED lights is placed in front of the north side of the test facility enabling to evaluate the daylight performance of façade systems. It can be programmed and controlled to emulate daylight from sunrise to sunset with changing colour temperature ranging from 2400 to 10,000 K. The maximum power consumption is about 50 kW, providing 40,000 lx at a distance of 0.5 m. Although the entire test bed is situated in the existing conditioned environment of the CREATE Tower, various lighting, heating/cooling, and air handling conditions can be emulated. A central data control unit allows full control of the test bed.

The physical structure is of a hybrid design, divided into four equal subspaces of approximately 25 m^2. Two rooms on the north side consist of fully movable wall panels (hanging from rails), and the other two rooms on the south side are made up of fixed panels using clean room quality and low emission components. The moveable panels are designed for sound insulation, but they will provide sufficient environmental isolation as well, given that they "lock" in place with gaskets and there are no significant pressure differentials. Façade panels on the north side can be easily interchanged using the ceiling mounted rail system. As the two chambers on the south side are made for clean room use, they have strict air tightness standards (smaller than 0.6 l/s at 50 Pa), and low out-gassing surfaces, including stainless steel. This allows for serious experimentation on air quality, in combination with thermal and visual comfort.

The air conditioning and mechanical ventilation system consists of an air and a water stream to attain the desired settings for experiments. The system is also configured with two air channels; one with a conventional type where the air handling unit is handling both latent and sensible load, and another one with an advanced desiccant dehumidification system where each load is tackled by a specific component.

All rooms are equipped with a controllable LED lighting system together with smart sensing. The modular interchangeable 0.6 m × 0.6 m ceiling grid in all four rooms also allows for different types of lighting fixtures through normal placement on the ceiling grid, suspended from the ceiling grid, and other types of lighting fixtures. Especially the rooms located on the north side provide various possibilities for visual comfort studies with different types of blinds, façade systems, the daylight emulator, and internal lighting.

Fig. 7.5 Instrumented test room (*left*) and flexible façade system (*right*) of SinBerBEST Test Bed at BEARS (photos by BEARS)

7.2.6 Respiration Chambers, Metabolic Research Unit Maastricht (MRUM), University of Maastricht, the Netherlands

(MRUM 2017; Schoffelen et al. 1997)

The MRUM at the University of Maastricht is a facility to experimentally investigate human energy and substrate metabolism in depth. The whole lab hosts twenty so-called “metabolic rooms” providing state-of-the-art equipment and infrastructure for a wide range of experiments with regard to indoor environmental conditions and physical setups for the subjects. Five of the rooms are specially designed as climate-controlled respiration chambers, called “room calorimeters”, which allow to determine human energy expenditure by monitoring oxygen consumption and CO_2 production, as well as nitrogen loss in urine under strictly standardized conditions.

The respiration chambers, with a floor area of 3.1 m × 2.25 m and a gross volume of 14 m^3, are designed to appear like small hotel rooms: they are equipped with a desk, bed, sink, and toilet. Emphasis was put on creating a friendly atmosphere for the subjects, including social contact with other persons during their stay in the rooms. For example, windows in the door provide contact with the researchers, while windows in the wall and between the chambers are for outside view and for visual contact between the subjects, respectively (see Fig. 7.6). Curtains ensure privacy when needed. Although enclosed in the chamber during an experiment, the subjects can engage in normal daily activities, such as sleeping, eating, reading, office work, etc., which is supported by a computer, TV, and DVD player. Physical exercise of the subjects is possible with a cycle ergometer, a treadmill, or a stepping platform. This means that human energy metabolism can be measured under defined “living conditions”. Experiments may run over a period of 12 h up to seven days.

The construction of the chambers was realized with prefabricated steel panels, assembled completely airtight. Three air locks with flexible seals allow the exchange of food, the sampling of blood, and the collection of feces and urine. For this purpose, the rooms are equipped with a deep-freeze toilet for collecting feces; urine is collected separately in bottles.

The rooms are equipped with an ultrasonic system for the registration of the physical activity, and with a SkyRibbon® LED light system. The latter allows for tunable white light with correlated color temperatures ranging from 2000 to 10,000 K, offering a maximum intensity of 1600 lx (under 4000 K) and provides the possibility to deviate from the black body curve. The temperature range reaches from 10 to 45 °C and the relative humidity from 20 to 80%. Indoor climate is controlled by the climate control system and continuously monitored by an automated information system. Both temperature and light conditions can also be modified by the participants within pre-set ranges. Air is re-circulated through the internal air conditioning within the enclosed compartment. This allows a ventilation rate in the range of 200–800 m^3/h (14–57 ACH). The air can be supplied in two different ways: through mixing with the supply air entering the room via an inlet

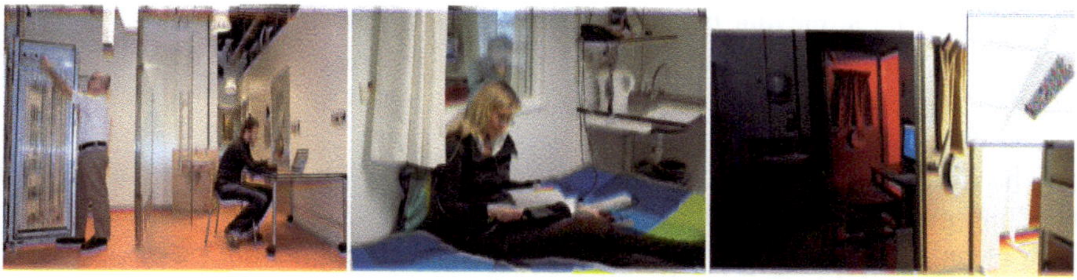

Fig. 7.6 MRUM respiration chambers: overview (*left*), view into one of the chambers with optical connection to the adjacent chamber (*middle*), and light control (*right*) (all photos by MRUM)

close to the ceiling, or through displacement ventilation with a laminar flow provided by a full height mesh. Confined spaces in the chamber were avoided because of the adverse effect on the air-mixing process.

7.2.7 *Institute for Energy Efficient Buildings and Indoor Climate, E.ON Energy Research Center (E.ON ERC), RWTH Aachen University, Germany*

(Möhlenkamp et al. 2015; Müller et al. 2015; Fütterer et al. 2013)

The range of test facilities at E.ON Energy Research Center (ERC), located on the RWTH Aachen University campus, reaches from climate chambers over generic constructions of indoor environments to a living lab in a real building (see Fig. 7.7). The Aachen Comfort Cube (ACCu) is a highly modular climate chamber and enables the evaluation of thermal comfort under very precise boundary conditions. The comfort cube has a floor area of 2 m × 2 m and a height of 2.5 m. Three of the surrounding side walls are divided into four surface segments. The first and the third segment are 0.4 m, the second is 0.5 m, and the highest segment is 1 m high. A test subject can be sitting or standing in the comfort cube. The segmentation provides a heated segment at head height of each test subject. The chamber provides no view contact to the outside.

Each surface segment can be set to a temperature between 8 and 45 °C, including the ceiling, floor, and door segments. As a result, 16 surface temperatures can be controlled independently in the comfort cube. By using a variable air distribution system, displacement and mixed ventilation can be applied. For both setups the inlet temperature is adjustable in a range from 15 to 40 °C. The system allows for a maximum airflow rate of 900 m^3/h (90 ACH). With this installation the cube can simulate a variety of different indoor situations. No specific means are foreseen for subjects to interact with the indoor climate.

For thermal comfort analysis in vehicles the institute operates a real vehicle test facility in which cabin climate conditions can be tested under simulated outdoor conditions with test persons. The communication with the vehicle and control of the air conditioning's relevant parameters is realized via the CAN bus of the vehicle. Besides the possibility of temperature control of the experimental hall for the simulation of summer or winter outdoor conditions, a movable solar simulator with a total output of 12 kW is available.

The institute's acoustic and indoor air quality lab is equipped with its own air handling unit (AHU) including activated carbon filters. The AHU supplies the lab with clean and conditioned air. Emitting odors of e.g., building products can be evaluated with trained test subjects and a reference scale that provides six different acetone-air mixtures. All investigations can be performed at a constant relative humidity (50%) and temperature (23 °C). These values are fixed in a range of ±2 °C for the temperature and ±5% for relative humidity. The surrounding wall structure of the lab is sound absorbing and enables good conditions to measure and optimize the acoustic properties of HVAC components.

For field tests the institute uses the E.ON ERC main building at RWTH Aachen University as a living lab. The building is the home for five institutes in total. The ground floor area is 58.43 m by 34.72 m leading to a useable area of 7500 m^2 over four storeys. Workshops and service areas are located in the basement laboratories. The ground floor hosts the main entrance, the foyer, and administration space, as well as seminar and meeting rooms. The office rooms of the five institutes and computer rooms are in the upper two floors. The roof of the building is used for technical services units.

The energy concept is based on geothermal energy and heat displacement in connection with a heat pump process. Heat and cold base loads are distributed by concrete core activation. A sorption supported air conditioning unit provides fresh air to conference rooms and computer pools. Offices are equipped with façade ventilation units, covering peak loads by supplying cool air during summer and warm air during winter. In order to collect local and global energy data on a highly detailed level, the building is equipped with an extensive monitoring system. The energetic flows of every energy source, energy conversion unit and energy distribution circuit are monitored. Thermal conditions—temperature, relative humidity, as well as CO_2 and VOC concentration—are recorded for every single room. Further sensors are installed in ten so-called "reference rooms", recording energy flows supplied to and extracted from the room. A common building automation system provides control and access to all network data points. The data is centrally stored in an SQL database and accessible from there.

Located at the same University, the Institute for Energy Efficient Building (e3D) runs a test facility for experiments on comfort in vehicles (Schmidt et al. 2015). It incorporates three equally equipped mock-ups with the dimension of a middle-class vehicle (1.55 m height × 0.8 m width × 2.17 m length), constructed on the base of a wooden framing (Fig. 7.8). All enclosing elements are made of chipboard. The cabin has different radiant heating elements, built as sandwich panels which include electrical heating foils (200 W/m^2) and black cloth surfaces towards the interior of the cabin. Each cabin has independent heating devices: radiant panels at the doors on the right and left of the driver, on the middle console as well as in the leg space; further, there is a heated steering wheel and a heated seat.

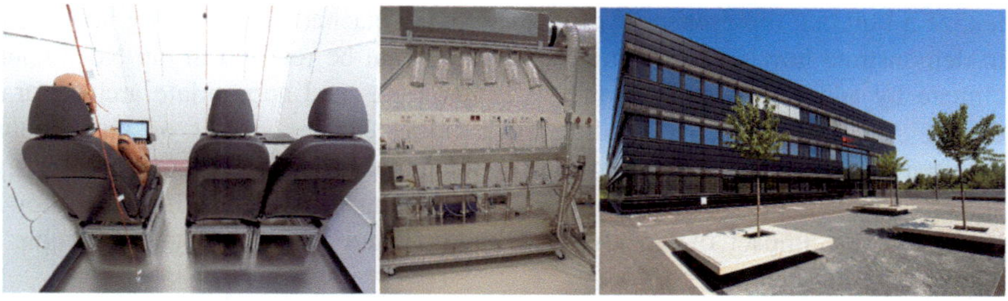

Fig. 7.7 The Comfort Cube ACCu (*left*), the set-up of the reference scale in the Indoor Air Quality lab (*middle*), and the "Living Lab" institute building at E.ON ERC (*right*) (all photos by E. ON ERC)

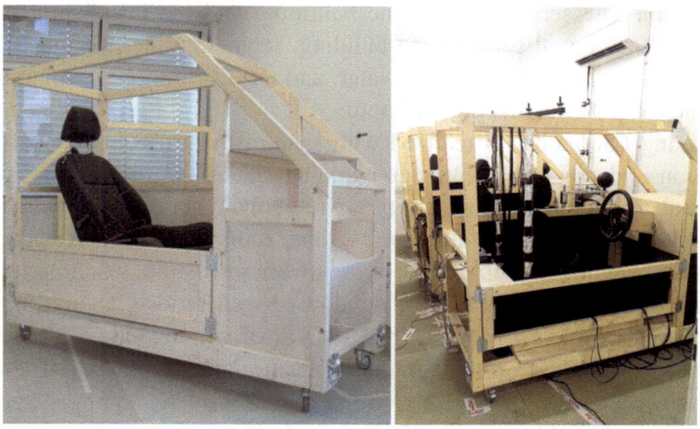

Fig. 7.8 Vehicle test facility at the Institute for Energy Efficient Building (e3D) (all photos by Institute e3D)

The mock-ups are in an air-conditioned container, which has a thermally separated space for the researchers and their control equipment. A window provides visual contact between the subjects and the researchers. The cooling and heating capacity for the container, provided by a split air-conditioning unit, is 6 and 7 kW, respectively.

7.2.8 The ZEB Living Laboratory at the Norwegian University of Science and Technology (NTNU) and SINTEF, Norway

(Goia et al. 2015)

The ZEB Living Laboratory (see Fig. 7.9) is a multipurpose experimental facility located on the main campus of NTNU in Trondheim. One important aim was to realize a building representative of a Norwegian detached single-family house and to demonstrate how a CO_2-neutral construction can be realized in the Norwegian climate by adoption of state-of-the art technologies and appropriate architectural design. As an experimental facility, the building enables studies on interactions between occupants and technology, and on the impact of people's actions on energy and environmental performance of the building. Conversely, the facility also allows studies on the impact of an advanced building and its technology on people's everyday living habits.

The building, with a gross volume of approximately 500 m^3 and a heated floor area of approximately 100 m^2, comprises two main zones: a south-facing living area which integrates a sitting room and a kitchen, and a working/sleeping area towards the north with two bedrooms, a second sitting/working room, and the bathroom. The entrance is situated in the south-west corner, near the technical services room. The services room, the bathroom and the installation wall of the kitchen form the central spine of the building, separating the two main zones, and hosting infrastructure for electricity, water, and air distribution. A small mezzanine is placed above the west-facing bedroom.

The Living Laboratory is realized with state-of-the-art technologies for energy conservation and renewable energies exploitation. On a yearly basis, solar gains from solar thermal collectors and from a photovoltaic generator are larger than the building energy demand. Walls, floors, and roofs are built as wooden-frame structures and employ a double layer of rock wool insulation, resulting in a U-value of approximately 0.11 W/m^2K. The window-to-wall ratio is approximately 20%, with the largest window facing south; U-values of windows lie between 0.65 and 0.97 W/m^2K, depending on the ventilation feature and orientation. There are four more windows in the roof with a U-value of 1.0 W/m^2K, equipped with electric drivers. The double skin window facing south and the large north-facing, single skin window are also equipped with electric drivers for automated opening. Further, approximately 90 m^2 of commercially available boards embodying phase change material (PCM) are installed in the lightweight roof construction to minimize overheating.

The building is equipped with a ground source heat pump connected to a surface collector field, which provides thermal energy to cover heating, ventilation, and domestic hot water demand. A combined tank serves as a buffer for the heating circuits and for domestic hot water, with two auxiliary electric coils for backup purpose. A mechanical system allows balanced ventilation with a nominal airflow of 120 m^3/h and a maximum of 360 m^3/h. Air diffusers supply the living room, the studio and the two bedrooms, while exhaust air is extracted primarily from the bathroom and to a lesser extent from the kitchen. Fresh air is supplied through a compact air handling unit with a rotary heat recovery wheel (nominal efficiency approximately 85%). An electric and a water coil alternatively serve for preheating the air.

Heating is realized either through floor heating panels installed over the entire indoor floor surface, or by just one radiator (with a nominal capacity of 2 kW) in

Fig. 7.9 Exterior view (*left*, photo by Katrine Peck Sze Lim, NTNU) and main living room (*right*, photo by Anne Jørgensen Bruland, NTNU) of the ZEB Living Lab

the living room, which heats up the main areas of the building in combination with floor heating in the bathroom. Alternatively, the ventilation system can be deployed to cover the heating demand in combination with fresh air supply by using the abovementioned electric/hydronic coils. In the latter case, floor heating in the bathroom is combined with the preheated air supply.

Monitored data include all relevant indoor and outdoor environmental quantities, energy flows, occupancy of rooms, operation of windows/shading systems, use and control of appliances, and lighting system. LabView® enables full management of the facility according to virtually any desired specification, including predictive control algorithms and real-time simulation. Moreover, the facility can be entirely operated without occupants by activating different functions which replicate occupancy of the building and related interior/energy load profiles. Different user interfaces are and will be made available to test a wide range of control approaches —from conventional switches to a touch screen installed in the entrance, from mobile app to full automation without user control—according to different experimental settings.

7.2.9 Indoor Environmental Laboratories at the Fraunhofer Institute for Building Physics (IBP), Germany

(Fraunhofer 2017a, b, c)

The Fraunhofer Institute for Building Physics (IBP) features special laboratories for indoor environmental quality tests, the so-called HiPIE-Lab (High Performance Indoor Environment) and the IATC (Indoor Air Test Center), (see Fig. 7.10). The HiPIE-Lab has a dimension of 6.6 m × 6.6 m × 2.8 m (123 m^3). It allows to selectively modify key parameters of buildings physics, such as acoustics, lighting, and indoor climate, so that their impact on human beings can be investigated. For instance, a built-in 3D sound system is used to model real sound fields, e.g., various office scenarios, which are physically reconstructed from original sound fields by a technique called "wave field synthesis". For this realistic presentation, a 64-channel sound field system comprising 412 loudspeakers is installed. It also enables the virtual integration of acoustic effects in order to quickly and efficiently evaluate their effect on human beings. In this way it is possible to simulate the specific acoustic properties of any working environment. To perform tests on specific materials and building components, even the space-enclosing surfaces can be modified. The "daylight wall" located in front of the windows of the lab allows the simulation of solar radiation through windows, taking into consideration the changing lighting conditions during the course of a day. The luminance can be controlled up to10,000 cd/m^2 and the color temperature between 3000 and 6500 K. In combination with the free configuration of artificial light scenarios by means of a DALI bus system, e.g., for adaptive light management, all relevant lighting situations can be simulated and the building components assessed according to their effect on human beings. For the simulation of summer and winter scenarios, the room temperature can be variably controlled in the laboratory between 18 and 30 ° C with varied ventilation rates up to 1800 m^3/h.

The IATC is a space of approximately 175 m^3 (6 m × 7.5 m × 3.9 m), in which different investigations of air quality, air flow, and temperature distribution, as well as effectiveness of active and passive air purification systems can be performed with and without test persons. Using various high-precision dosage systems the indoor air can be loaded with typical indoor pollutants like chemicals, spores, and particles to establish different and well-defined pollution situations. The inner surface of the test room consists of surface elements which can be individually heated and cooled to create different temperature profiles on the walls between 8 and 40 °C. The test room also offers space for a whole car to perform VOC emission measurements outside and inside the vehicle. The ventilation system allows for a maximum airflow rate of 1800 m^3/h. The indoor air can be controlled up to 80 °C and 95% relative humidity.

Innovative façade systems and their effects on indoor environments and energy consumption can be investigated under natural weathering conditions at the Modular Test Facility for Energy and Indoor Environments (VERU). For testing purposes, façade components or shading systems can be attached to predefined fixtures at the east, south, and west fronts of the multi-storey building. Due to the special construction (partially removable intermediate floors) and modular design it is also possible to investigate multi-storey rooms or halls. The interior space(s) can be fully conditioned and different usage scenarios can be simulated.

Fig. 7.10 View into the HiPIE-Lab (*left*), the "daylight wall" of the HiPIE-Lab (*middle*) and the IATC at the Fraunhofer IBP (all photos by Fraunhofer IBP)

7.2.10 Flight Test Facilities, Fraunhofer Institute for Building Physics (IBP) and Institute for Energy Efficient Buildings and Indoor Climate, RWTH Aachen University; Germany

(Fraunhofer 2017d; E.ON ERC 2017)

The Fraunhofer Institute for Building Physics hosts a flight test facility on its Holzkirchen site (Fig. 7.11). The front section of an Airbus A310-200® with a length of approximately 15 m and space for up to 80 test subjects has been placed into a low-pressure chamber (150 hPa). Typical indoor climates with a range of supply temperatures from −10 to +30 °C and relative humidity from 5 to 65% at 20 °C can be provided, and the ventilation system allows for a maximum airflow rate of 3700 m^3/h (40 ACH). Additionally, realistic simulation of the flight sound and seat vibration, as well as the cooling of fuselage shell from −30 to +60 °C enable realistic typical flight conditions on the ground. Besides using the facility to study cabin air quality and thermal comfort, the aircraft can also be studied as an overall system, e.g., in terms of energy aspects and usage requirements of different areas of a plane (cockpit, passenger cabin, avionics, cargo bay). Further, thermodynamic correlations and the appearance of condensation on aircraft components are investigated.

Another aircraft cabin mock-up is situated at the Institute for Energy Efficient Buildings and Indoor Climate at RWTH Aachen University. It comprises comprehensive airflow measurement equipment and offers a flexible design able to host 36 subjects. Indoor climates with a range of temperatures from 5 to 40 °C and relative humidity from 20% to 80% can be provided, and the ventilation system allows for a maximum airflow rate of 100–3500 m^3/h. The ventilation system is modular so that the location of the air supply, the supply air volume, as well as temperature and humidity can be varied. Further, different parts of the cabin, like air diffusers, can be changed easily in order to investigate and optimize HVAC components for aircraft cabin applications (Fig. 7.11).

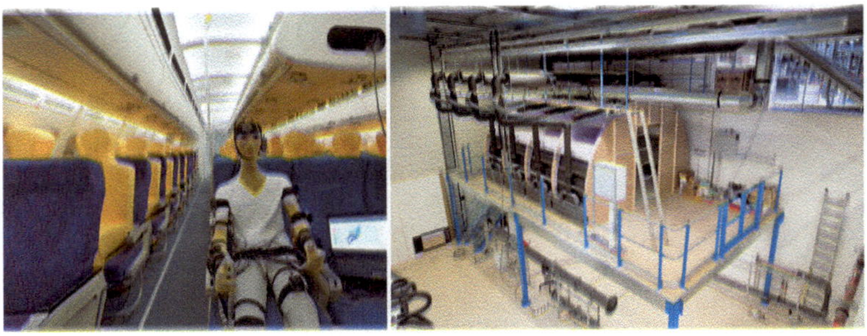

Fig. 7.11 Aircraft cabin mock-ups at the Fraunhofer IBP (*left*) and the EON ERC (*right*) (photos by Fraunhofer IBP and EON ERC EBC)

7.3 Indoor Climate in Labs—Technical Services, Control, Sensors

This section discusses the technical equipment used in different labs in order to condition and control the indoor environments. It also highlights some special features for solar input, (day)lighting, and acoustics, as well as aspects of sensing approaches. The section only refers to test facilities /climate chambers in the traditional meaning—i.e., it excludes the special environments shown above for comfort and air quality in vehicles and aircraft cabins.

7.3.1 Conditioning of Labs

In most of the labs the control of indoor temperature is mainly realized by air conditioning. In the field chambers of ICIEE (see Sect. 7.2.1) additional radiators or convectors are used underneath the windows for heating. The ICIEE classroom is equipped with floor and ceiling radiant heating and cooling systems, as well as convectors (only for heating) underneath the windows. One of the chambers of the IEQ Lab (see Sect. 7.2.3) features passive and active chilled beam systems providing the possibility of cooling the space on the basis of convection of indoor air around the cold surface (Fig. 7.12). Due to a separate water circuit, the SinBerBEST test facility (see Sect. 7.2.5) also offers the possibility to use chilled beams or other water-based components for cooling, but they have to be mounted for this purpose. Only the LOBSTER (see Sect. 7.2.4) relies on a water-based radiant surface heating and cooling system alone. The additional ventilation system preheats or pre-cools nothing but the inlet air of the ventilation system; it does not have enough capacity to condition the whole space. Also the E.ON ERC building, as a living lab (see Sect. 7.2.7), is solely equipped with concrete core activation in the ceilings for radiant heating and cooling.

Fig. 7.12 Passive and active chilled beam systems in chamber 2 of the IEQ Lab (photo by Richard de Dear, University of Sydney)

Air conditioning allows for a uniform spatial temperature distribution with mixed flow ventilation. By applying displacement ventilation together with internal heat sources (subjects, artificial heaters), temperature stratifications can be achieved, as well. Temperature asymmetries could be tested only with special air ducting and outlets directly at/behind the surfaces, being fed by individual air circuits providing temperatures different from the general space supply. This has not been done in any of the labs introduced in Sect. 7.2; in the chamber at Waseda University, Japan, however, conditioned air at a high flow rate was circulated in a 5 cm plenum behind the room-facing surfaces for this purpose (Kimura and Tanabe 1985). They also used sheet-type electric heaters to heat up single surfaces independently. In the ICIEE removable water-based radiant panels are used for this purpose. The individual adjustment of airflow rates for different outlets further allows to induce different air velocities for studying draft perception by the subjects. Another advantage of air conditioning is that temperature and humidity in a space can principally be controlled by one system. But as seen in Sect. 7.2 and the referring literature, often different subsystems and control devices are used to control temperature, humidity, and air change rate separately.

Space conditioning by radiant surface heating and cooling also allows for a uniform spatial temperature distribution if temperature settings of all surfaces are equal and inlet air temperature of the ventilation system does not differ. By varying the surface temperatures stratifications or different asymmetries can be effectively achieved. Consideration has to be given to eventual local disturbances due to the temperature of incoming air, which has to be adjusted likewise (depending on the ventilation system and the corresponding air outlets); outlets should be far enough from subjects to avoid negative influences on their perception. Humidity has to be controlled through the ventilation system or by additional (portable) devices if needed.

In the LOBSTER the surface temperature of the façade cannot be controlled, as the possibility for an easy exchange of the windows or even the whole façade system was an important constraint during design. To reduce the influence of the surface temperature as much as possible a very low U-value for all components was realized (see Sect. 7.2.4).

Apart from flow patterns for general space conditioning and ventilation—and here some of the labs already offer various options—a large variety of individual air flow pattern /temperature level scenarios for the subjects can be studied with different ventilation systems which have to be installed additionally (see Fig. 7.2 as an example). If the lab has operable windows different natural ventilation situations (one-sided, cross flow) can be studied, depending on the temperature difference between indoors and outdoors, wind velocity, and wind direction.

Labs which are mainly designed for air quality tests or for experiments with precise measurements of gases (respiration chambers) require a high air tightness, as well as ventilation equipment and interior materials which do not pollute the air on the way through the system or in the space. Again, a large variety of experiments can be performed in terms of optimum fresh air supply at the workplace with additional (personalized) ventilation devices installed for the specific purpose (e.g., in the ICIEE chambers, see Sect. 7.2.1). Some chambers also allow for altitude simulation by reducing the oxygen level of the incoming air; at the test facility of Loughborough University subjects can be exposed to environments which resemble locations between 0 and 7000 m in altitude (Loughborough 2016).

7.3.2 *Solar Incident, Daylight and Acoustics in Test Chambers*

An important aspect with regard to individual comfort parameters of persons is solar radiation, which can cause thermal disturbances locally on the body (e.g., one side heated up by direct solar incident) or thermal radiation from absorption by glazings and shadings. Another issue is glare due to high luminance levels. This can only be approached realistically with labs which have windows facing to the outside. The azimuth angle of the window façade then determines the time periods for experiments including solar radiation—but this also means that efficient shading devices have to be foreseen in order to eliminate the influence of the sun if required by the experiment (see also Sect. 7.5.2). The LOBSTER (see Sect. 7.2.4) can be rotated in order to adjust the position of the window façade relative to the sun's position. The IEQ Lab (see Sect. 7.2.3) uses a solar simulator in the so-called "environmental corridor". It consists of lamps providing the sun's visible spectrum and electric resistance heating panels on the spandrel panels below the windows for

Fig. 7.13 The daylight emulator at the SinBerBEST test facility (photo by BEARS)

infrared effects of solar heating. The same is true for the chamber at Waseda University where radiation from lamps enters through a window of the chamber (Kimura and Tanabe 1985). Removable panels on the external windows of the environmental corridor of the IEQ Lab permit the introduction of natural (glare-free) daylight. The SinBerBEST test facility (see Sect. 7.2.5), which has no windows to the outside, provides a daylight emulator together with a flexible façade mock-up for visual comfort studies (Fig. 7.13).

The option of rotating a test chamber relative to the sun's position is also a great benefit for experiments on visual comfort with regard to daylight. These experiments also have stricter requirements on the optical quality of indoor surfaces in order to provide an environment without irritations for the eyes (low contrasts, no reflections, etc.). Further, the possibility of rearranging the façade (different windows, shading and blind systems) is of great advantage for these kinds of investigations. Besides various day-lighting possibilities, studies on visual comfort usually require various possibilities of artificial lighting in the test chamber as well. This includes typical direct/indirect lighting or other light distributions in the room to achieve different luminance levels of the ceiling, the walls, and desks/tables. Additionally, measuring devices like illuminance sensors and luminance cameras are needed.

Aural comfort experiments benefit from adjustable reverberation times of a space by altering the acoustical surface properties of the enclosure. The IEQ and the HiPIE Labs (see Sect. 7.2.3 and 7.2.9) also have a system of speakers and sound generating devices to simulate a great variety of internally- and externally-sourced noises at different levels.

7.3.3 *Temperature Control in the Experimental Environment*

Generally, climate chambers aim to provide a well-defined indoor environment in contrast to field studies. This can be e.g., a constant thermal climate without any changes over a certain period of time, as applied in many experiments on thermal comfort. This requires a control system with high preciseness and only small hysteresis for each of the climate quantities. For example, the minimum and maximum steady state temperatures in the CBE chamber (see Sect. 7.2.2) can be maintained in their range (13–35 °C) with a stability of ±0.2 °C, and the relative humidity (10 to 90%) with an accuracy of ±2%.

However, for studying adaptation and behavioral actions, a changing indoor environment is favorable. The way of changing has to be sought out during the experimental design—for example, in the case of thermal comfort studies, typical temperature profiles are ramps with a constant slope over a day (see Fig. 7.14) or step functions with rapid changes of different extents. In order to simulate realistic scenarios, temperature profiles from real buildings which were monitored beforehand, are often applied. During the studies presented in Schweiker and Wagner (2016), the set point temperature was increased linearly depending on the subjects' behavior, as summarized in Table 7.2; due to the chosen states of windows and venetian blinds by the occupants, the temperature ramp gradient was reduced or increased to simulate changes in indoor temperature as they would occur in a real setting. In the IEQ Lab (see Sect. 7.2.3) the slope of temperature ramps depends on the chosen ventilation mode (5 min per K in VAV mode and 15 min per K for UFAD mode). Adequate measures apply to all other environmental parameters respectively.

Labs with windows connecting to the outdoor climate may allow—and also are subject to—influences from outside, like temperature changes with opening or closing the windows, or solar and daylight input. Further, different opportunities for the subjects to interfere with their environment can change the respective environmental conditions. Again, a precise control system is required to allow, compensate, or amplify occupants' interventions if desired by the experiment

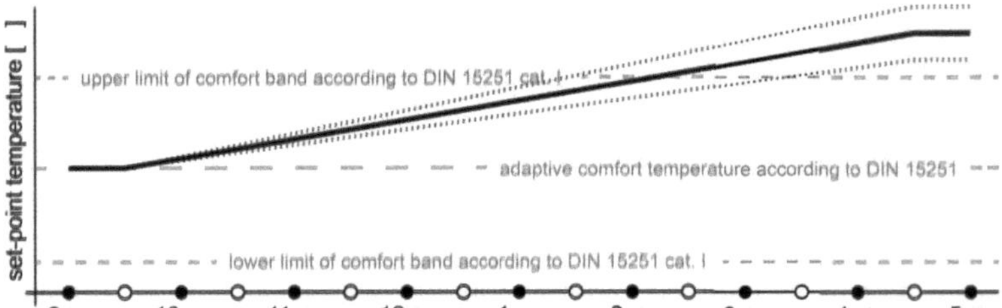

Fig. 7.14 Typical temperature ramp applied at the LOBSTER during a one-day experiment (Schweiker and Wagner 2016)

Table 7.2 Exemplary temperature ramp gradient in [K/hour] applied at the LOBSTER during a one-day experiment depending on the subjects' behavior (T_{in} = indoor air temperature, T_{out} = outdoor temperature) (Schweiker and Wagner 2016)

			Venetian blinds	
			Opened	Closed
Window	Opened	If $T_{in} > T_{out}$	0.5	0.5
		If $T_{in} < T_{out}$	1.1	0.8
	Closed		0.8	0.5

Fig. 7.15 Capillary tube systems for radiant heating and cooling with low thermal inertia; installed in the LOBSTER (photo by Marcel Schweiker, KIT)

(see Table 7.2 as an example). In the IEQ Lab, for example, a humidity sensor can override the signals from temperature sensors which trigger a chilled water valve or activate an electric duct heater for temperature control in a zone in order to guarantee the pre-set zone humidity (Nathwani et al. 2012). For fast reactions on changes in the settings or interactions it is preferred that the construction system of the chamber has a low thermal mass so that the temperature can change rapidly. This is particularly true if the space conditioning is done with a water-based system (and restricted heating/cooling capacity). In the LOBSTER (see Sect. 7.2.4), a capillary system has therefore been chosen (see Fig. 7.15) which has a rather low thermal capacity compared to other tube systems and allows temperature change rates of individual surfaces of up to 20 K/h and a temperature change rate of the operative temperature without asymmetries between surface and air temperatures of up to 6 K/h.

7.3.4 Data Acquisition and Sensing Indoor Environmental Quantities

In order to have greatest flexibility for indoor environmental control, all relevant quantities are normally controlled through the data acquisition system of the lab. This requires tailored solutions which connect the data acquisition system to the building management system (BMS, see Fig. 7.16 as an example) or to the different BUS systems of the facility. Most labs feature pre-programmed scenarios according to the experimental design. This is rarely achievable in field situations due to liability reasons with regard to building operation. In contrast to in situ experiments, the lab offers a higher degree of freedom for sensor installations. Sensing technologies are discussed in detail in Chap. 4; here, only lab-specific aspects will be tackled. In the case of thermal comfort this may reach from measuring spatial distributions of the relevant indoor climate parameters or surface temperatures of the space enclosure to monitoring comfort parameters directly at the subjects' positions. For other environmental parameters, this applies accordingly. Although a high number of sensors in a room may give very precise information about specific indoor environmental parameters under the influence of subjects in the room (e.g., air flow patterns and temperatures close to a person, luminance in the view field of a subject), it is important to consider to which extent these sensors might disturb the subjects and influence their perception of and satisfaction with the environment. This is particularly important for experiments with regard to behavior when subjects should have the feeling of being and acting in a "normal" environment (e.g., usual office space), rather than in a lab environment.

Fig. 7.16 HVAC control system of the IEQ Lab at the University of Sydney (photo by Richard de Dear, University of Sydney)

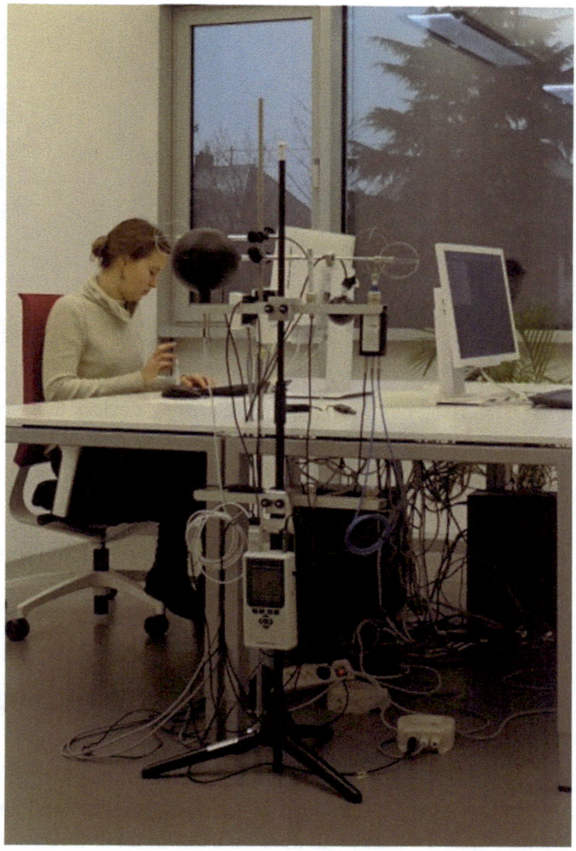

Fig. 7.17 Thermal comfort measurements at a height of 1.1 m (subjects' head level) close to the two workplaces in one of the test rooms in the LOBSTER (photo by Marcel Schweiker, KIT)

Taking into account the lab appearance, all relevant quantities for addressing comfort, adaptive, and non-adaptive behavior can still be measured very precisely in a lab (see Fig. 7.17 as an example) according to the applying standards. This includes the positioning of sensors as well as their quality. In contrast to field experiments where the sensors of the BMS are very often used, the quality of the sensors can be specified explicitly for each data point (see also Fig. 7.18). Table 5.2 in Chap. 5 gives an overview on parameters to be measured in labs in comparison to in situ studies. They also include states of openings (windows, doors) or other devices which the indoor climate/environment can be modified with (blinds, fans, thermostats, artificial light, etc.). A continuous monitoring of these quantities is necessary, and is achieved by connecting the sensors and actors to the BMS of the lab. Movement or any action of the subjects can be monitored by cameras, but this has to be accepted by the subjects beforehand and approved by the ethics commission. Devices with non-imaging measuring processes like infrared (IR) or luminance cameras might be an alternative if subjects would not like to be pictured

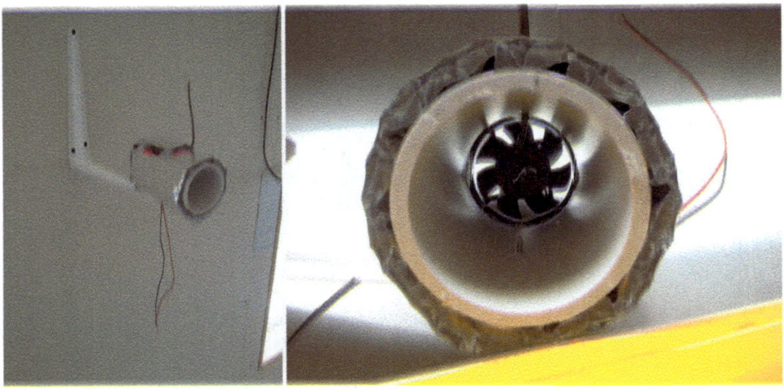

Fig. 7.18 Hand-built air temperature sensing devices used in a lab at Carleton University. The device sucks air through a reflective pipe using a fan to prevent measuring stagnant air or being heated by the sun. The thermocouples are at the centre of the pipe (photo by Sarah Brown, Carleton University)

directly; but in a lab environment with a restricted number of subjects this still does not preserve anonymity (see also Chap. 11).

7.4 Sensors for Personalized Monitoring

In Chap. 4 the most relevant sensing technologies for field studies are presented and evaluated, with a focus on occupant presence and interactions with built environment. Many of them can also be used in a lab environment, but as the subjects in lab experiments can be addressed personally the usage of enhanced sensors is possible, which allows to collect physiological and other data directly related to an individual person. This gives further insight into drivers of behavioral actions due to the indoor environment and particularly the indoor climate (see also Sect. 7.6). Some of the sensors are also applicable in the field, but others are restricted to be used in the lab only due to the processing of samples, costs, etc. Again, disturbance of the subjects might influence the experiments and particularly the application of on-body equipment might cause a negative perception which could lead to biased results. Transparency and strong efforts to ensure highest possible comfort while affixing and wearing special measurement gear are a prerequisite for valuable experiments.

For experiments related to adaptive thermal comfort heart rate, skin temperature (at different points of the body), body core temperature, as well as the moisture level of the skin are of great interest to understand physiological adaptations or to find trigger levels for behavioral actions. Heart rate is measured with a sensor which is attached to the body by a chest strap and which delivers high-resolution data to either an on-body (wrist watch) or a remote receiver. There are devices on the market which take the signal at the wrist (wristband) or at the ear (ear clip or in-ear sensor), but their accuracy is lower.

7.4.1 Measuring Skin and Body Temperature

Skin temperature can be measured using thermocouples or thermistors. Thermocouples are very small and therefore very accurate, as they only cover the skin to a small extent (Fig. 7.19). However, they need to be calibrated often and are connected with wires to the measurement device. Nowadays, wireless iButton® temperature sensors with thermistors are often used (Fig. 7.19). They can be fixed to the skin with tape (van Marken Lichtenbelt et al. 2006) and constrain the subject at a minimum due to their low weight. Their disadvantages are that they are relatively large (diameter of 16 mm) and thereby cover a larger part of the skin, and that their thermal mass retards response time under transient conditions. Measurement positions depend on the planned experiment, and the coinciding number of sensors also varies. The relevant standard is (ISO 9886:2004 2004); an overview of possible positions and related formulas to calculate mean T_{skin} can be found in Parsons (2014). For warm environments, the following four positions were used in experiments in the LOBSTER: hand, neck, shoulder, and shin. Body core temperature can be measured precisely by sensors which the subjects swallow like pills (length ca. 22 mm and diameter ca. 9 mm, see Fig. 7.19) (Zhang 2003), but this might be disagreeable for some persons and also increases costs for large sample sizes.

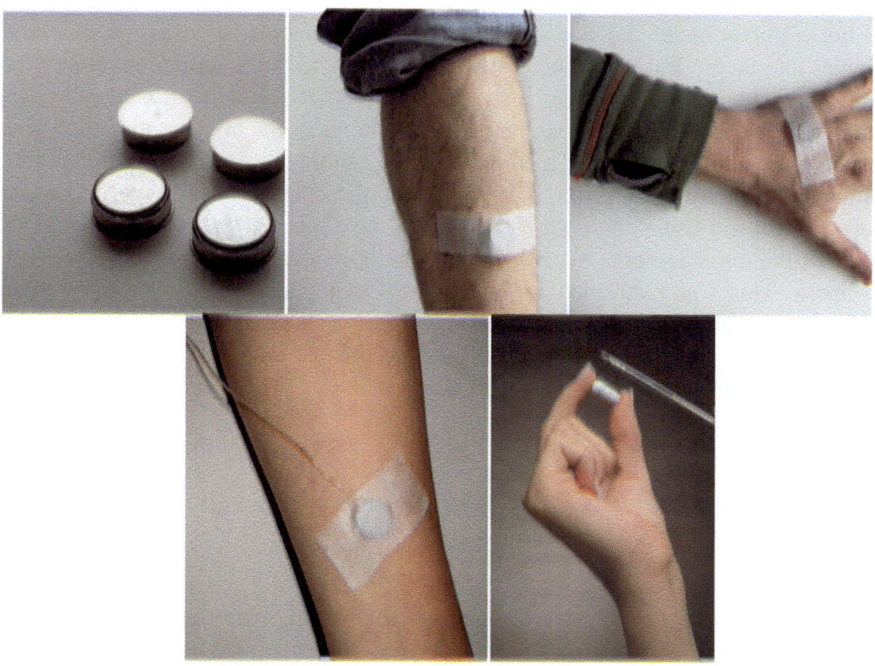

Fig. 7.19 Different sensors for temperature measurements on the body. *Top* iButton® temperature sensors (*photos* by Michael Kleber (KIT); *bottom* thermocouple (*left*) and encapsulated CorTemp® ingestible core body temperature sensor which can be swallowed (*left* photo by Hui Zhang (CBE), *right* photo by HQ Inc.)

Alternatively, the in-ear, forehead, or armpit temperature can be measured periodically during an experiment with a clinical thermometer; however, this is not precise and reliable, as it depends very much on the shape of the ear canal and on sweating, which leads to false measurements. More precise would be a thermocouple placed against the tympanic membrane, which is rather uncomfortable. Using an infrared camera, which is focused onto the subject's forehead (used at airports for screening passengers with probable diseases), is also rather inaccurate and cannot be recommended for scientific purposes.

7.4.2 Measuring the Skin Wettedness

Skin Wettedness is a measure for the (restrained) evaporative heat dissipation of the human body over the skin. There are different approaches for measuring skin wittedness, but they all show shortcomings either in terms of accuracy or practicability. In the past, skin wettedness was assessed using humidity sensors at the skin surface together with air temperature, humidity level, and barometric pressure of the environment (Storaas and Bakkevig 1996). Another measuring principle is to use the capacitance of the skin; with a Corneometer® (Clarys et al. 2011) periodic measurements are taken, as the probes cannot be fixed to the body in a way to allow continuous monitoring. In experiments in the LOBSTER, however, measurements using a corneometer did not show a reliable correlation between the measured values and theoretical values of skin wettedness. In addition, individual differences were high. Another device, an Electro-dermal activity (EDA) meter displays the change of electrical conductance between two points of the body continuously over time.[1] This active measuring method involves sending a small amount of current through the body (Wikipedia 2016). Due to (the change in) skin perfusion and sweat produced in the skin which might not have reached the skin surface at the time of the measurement, this method may be not valid to measure the actual skin wetness. This was confirmed by experiments in the LOBSTER that showed no reliable relationship with skin wettedness. More accurate is the use of ventilated capsules (e.g., used in experiments in the MRUM, see Sect. 7.2.6), but this makes the measurement procedure more complicated.

7.4.3 View Tracking, Measuring Hormone Levels and Monitoring Movements

In terms of visual comfort, devices are available for tracking the view of subjects. An EyeSeeCam® worn by a subject performs binocular video oculography and

[1]EDA meters may also be used to indentify stress symptoms of persons, e.g., in situations where a decision has to be made.

records real-time head-centered and gaze-centered moves (Khanie et al. 2011). Together with a high dynamic range (HDR) camera this allows to determine luminance levels at the subject's eyes and therefore a more precise assessment of glare situations. Other devices, e.g. "Luxblick®" (Vandahl et al. 2010), are mounted on spectacles and used to measure the vertical illuminance at the eyes, as well as the circadian/melatonin suppressing irradiance at eye level.

Further measurements without sensors at the body include e.g., the investigation of changes in hormone levels which might be related to indoor environment. This requires samples of blood, urine, or saliva. All three methods need further medical or biological labs for analysis and, as a result, are rather expensive. Blood sampling also requires medical staff at the lab. Saliva tests were used in experiments on the biological effects of light on the human body, particularly melatonin production (Chellappa et al. 2011). Also, glucose levels can be measured using ambulant devices.

Movements of subjects in a lab space can be recorded by accelerometers applied to the subjects at a variety of locations, and need to be validated to actual whole body movement. Often wrist sensors are used, such as Actiwatch®; others can be mounted on the waist belt. Rooms can also be equipped with an ultrasonic system for the registration of the physical activity (Schoffelen et al. 1984).

7.5 Lab Studies on Occupant Behavior and Considerations with Regard to Lab Design and Equipment

A literature review shows that experimental studies were carried out in the context of thermal comfort and individual adjustments long before 2010, e.g., (Olesen and Fanger 1973; Fanger et al. 1980). In Wyon et al. (1975), Olesen et al. (1979), subjects were e.g., allowed to adjust the indoor temperature (indirectly by a dial voting apparatus) towards their individual comfort temperature, which was then taken as a basis for further experimental steps. In other experiments, subjects could make choices about environmental parameters and their settings with regard to personalized control, see e.g., (Clausen and Wyon 2008; Veselý and Zeiler 2014). These studies did not, however, investigate behavioral actions to adapt to a changing indoor climate as a topic itself. There also have been few lab studies with regard to light control and glare prevention, e.g., (Newsham et al. 2004), but again, no behavioral studies. The area of lab studies on air quality and aural comfort has not been reviewed, but it is assumed that the situation is similar.

Since 2010, an increasing number of laboratory studies on occupant behavior itself can be seen, including the areas of visual comfort, e.g., (Meerbeek et al. 2016), and thermal comfort, e.g., (Schweiker and Wagner 2015a, b, c, 2016; Schweiker et al. 2012, 2013a, b, 2014; Boerstra et al. 2015; Zhai et al. 2013).

In these experiments, subjects are basically allowed or explicitly asked to react on given constant or changing indoor environmental conditions, with constraints defined by the specific experimental design.

7.5.1 Options for Experimental Settings

With regard to thermal comfort, experimental settings allow for different adaptations (clothing, drinking, changing posture, metabolic rate) and for interactions with different technical devices (changing the thermostat setting for heating/cooling, switching the ventilation on/off, using fans, operating blinds, opening windows, switching lights, etc.). Moving away from sources of discomfort towards places with a higher level of comfort, e.g., air outlets or places with direct solar incident, would be another option. Considering visual comfort, occupants' reaction on illuminance and luminance levels is of interest: allowing the operation of blinds or change of position or view direction relative to the light source, or the adjustment of the illuminance level by dimming or switching the lights. As both thermal and visual comfort strongly depend on outdoor climate (in non air-conditioned spaces), experiments have to be planned accordingly, considering also repetitions to include a variety of different conditions.

The latter is also true if experiments with regard to air quality are carried out in naturally ventilated test rooms. Adaptive reactions of occupants to be investigated in the context of air quality may include the opening of windows to allow for more fresh air, the adjustment of mechanical ventilation devices (switching ventilation on/off, increasing air change rate), or the changing of the position relative to a source of bad odor. Behavioral actions in terms of aural comfort are the operation of windows with regard to external noise, raising the voice against high interior (background) noise, or the use of headphones/ear protection.

It is obvious that decisions on certain actions, e.g., opening/closing a window or operating a blind, can have different backgrounds, including all fields of comfort (see also the definition of adaptive and non-adaptive triggers in Chap. 2). This has to be considered in the experimental design, particularly for decisions involving sensor equipment and questionnaire design. Labs offer a much better opportunity to research for combined effects in occupant behavior. However, other factors, like the number of subjects in a room, also influence decisions and have to be considered, as well (see Sect. 7.6).

Presence detection or movements have been of minor interest in comfort-related lab experiments so far, as they are normally designed for a pre-defined number of subjects, while movements in and out of test rooms (e.g., for toilet visits or lunch) are registered by a study assistant. Other movements within experiments, like working out on a stepping platform, were for the purpose of the experiment, mostly to vary the metabolic rate of subjects.

7.5.2 Recommendations for Lab Design

From the overview on different lab facilities and related studies on comfort and occupant behavior the following recommendations can be given for the design of climate chambers to be used for behavioral studies.

Labs for occupant behavior studies should be designed with real windows to the exterior (better yet, with a façade system which allows the mounting of different window types). Windows connect to the most healthy and appreciated source of light, help for orientation, and offer a view to the outside. These factors have been proven to be very important for well-being indoors, which is particularly necessary for day-long (or longer) experiments. Further, windows can connect indoor and outdoor climate and are therefore an undisputable prerequisite for studying thermal comfort adaptation by natural ventilation and cooling. Finally, operating windows is an important behavioral action with regard to different personal needs or preferences (indoor climate, air quality, acoustics) and must not be neglected in research on occupant behavior. Windows have to be equipped with effective shading devices to prevent the space from overheating (see Sect. 7.2.4 for example); this again might interfere with other comfort areas—in this case mostly with visual comfort by blocking the view to the outside and cutting off daylight which can influence behavioral decisions of subjects.

On the other hand, the existence of real windows might significantly influence the sample size if the experimental design focuses on parameters which are unrelated to windows (and their operability). Real windows can also influence the experimental periods, as the effects of daytime or season cannot be controlled. Further, optical properties of the (coated) glazings have to be considered when selecting the fenestration for a test room. Finally, windows require sufficient and fast responding heating and cooling capacity in the lab due to thermal losses and solar gains. A number of test facilities are equipped with artificial/virtual windows or windows to the interior of the building which hosts the test chamber. Further, artificial light sources are used for "sunshine" or day-lighting. Depending on the given wavelength spectrum of the light source this might have an influence on the occupants' subjective perception.

In order to simulate a most "non-artificial" and familiar environment for subjects, test rooms should provide all necessary furniture and means for the purpose which the experiment is designed for. However, sensor equipment must not be compromised (e.g., by pictures or cupboards on or in front of a wall). The requirements increase for long-running experiments (more than one day), which require a bathroom and a bed in the close vicinity of the test chamber (or even inside, see MRUM, Sect. 7.2.6). Flexible room partitioning and furniture help for experiments with different numbers of occupants by not letting subjects feel alone in a large space or constricted in a room with larger samples. Basically, a high standard of furnishing should be achieved in order to avoid side effects on well-being. In particular, care has to be taken with acoustics (appropriate reverberation times) and shading (adjustable blinds). For example, the IEQ Lab has been approved for a grade-A commercial office space, also in terms of acoustic quality.

With regard to providing a "normal" environment, care has to be taken in terms of placing sensors in a lab. As already mentioned in Sects. 7.3.4 and 7.4, subjects should not be disturbed or even inhibited by any sensor to perform intended (comfort-related) actions in a lab. A good balance has to be found between an ideal situation in which the subjects are not permanently reminded that they participate in a study, and the scientific reality which requires the sensors as invaluable data sources. A thorough experimental design should avoid an over-instrumentation of a lab and favor solutions with a maximum of hidden sensors.

As lab facilities for occupant behavior studies have to allow user reactions on indoor environmental conditions on all levels—i.e., besides personal adaptation the interaction with all technical services devices (HVAC and lighting) and the façade (windows, blinds)—easy access with appropriate interfaces has to be provided for the subjects. This can be realized with the monitors of computers or by extra (remote) devices to be operated by the subjects (Fig. 7.20). The experimental design then regulates by activation/deactivation within the control system to which extent and under which circumstances occupants are allowed to take advantage of the different options.

Technical services and controls have to be designed in a way that all desired indoor climate profiles and combinations of different environmental parameters can be simulated without major restrictions. Aspects like ample heating and cooling capacity (also for compensating for effects of occupants' interactions), as well as

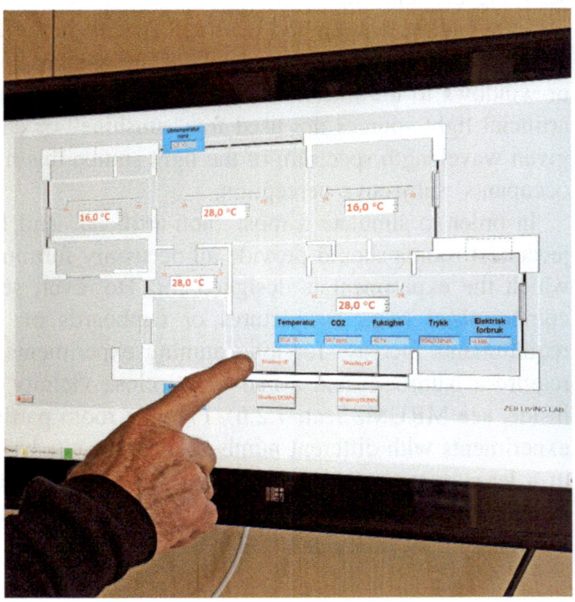

Fig. 7.20 Touch screen for indoor environmental control by the occupants, realized in the entrance area of the ZEB Living Lab at NTNU and SINTEF [photo by Ole Tolstad (NTNU)]

different options for conditioning, e.g., mixed-mode, have to be considered carefully during design. Further, the type of medium and heat transfer (air/water, air-conditioning/radiant surfaces), as well as independent medium flows for controlling temperature, humidity, and air change rate have to be taken into account. Low thermal inertia of the chamber construction and the heating/cooling system is beneficial for fast reacting systems (Fig. 7.15).

Apart from design aspects which directly refer to the lab itself the following issues are probably worthwhile being mentioned to support a smooth experimental procedure: good noise protection between experimental rooms and the researchers' area/operation room; ample space for the subjects to change clothes and to put on personalized sensors; pleasant area for acclimatization before the experiment; and sanitary facilities in close vicinity and reachable without leaving conditioned areas.

7.6 Influencing Factors Driving Occupant's Behavior and Their Impact on Experimental Design

As for any study, an indispensible basis for lab experiments is a robust research design; Chap. 3 introduces into this topic at length. Further, Chap. 6 gives valuable information about the preparation of experiments in general, as well as documentation and publication. More specifically, for designing lab experiments on occupant behavior it is necessary to have basic knowledge of the different influencing factors driving occupants' behavior (see Sect. 2.3 for an overview). In the context of this book, these factors were grouped into adaptive triggers, non-adaptive triggers, and contextual factors (Chap. 2). Within these groups, further distinctions are made between physical environmental triggers and factors, physiological triggers and factors, psychological factors, social factors, and non-adaptive triggers such as time of day and scheduled activity.

The adaptive triggers (physical environmental triggers such as indoor air temperature and physiological triggers such as skin temperature) can be controlled to a large extent and precisely in a test facility (depending on the chamber design and the technical services system), which enables to expose subjects to a great variety of well-defined indoor environmental settings. This includes typical real-life situations (e.g., temperature ramps) which can be repeated as often as needed to vary other parameters, or extreme situations which would not—or only very seldom—occur in situ. A typical research question to address the variability in personal behavior could be: How large is the deviation in reaction of a person who is exposed to the same conditions repeatedly? For example, will a person open the window at the same indoor temperature (or skin temperature) if exposed to the same ramp several times? This also tackles the number of (physical) parameters changing in an experiment; if the air quality is not kept constant during the abovementioned experiment and the outside temperature appears to be different during the series of experiments, there are already three driving factors influencing the probable action.

The question would then be how different subjects balance perception and preferences between several IEQ parameters. Generally, subjects do not know in the beginning how the test chamber reacts to their interventions (in contrast to their usual environment), and so a "learning effect" has to be considered. It is also important to consider whether subjects show a different tolerance for unfavorable thermal conditions over a short period in a lab experiment (e.g., 8 h), compared to real-life situations—e.g., a heat wave over three or more days.

The non-adaptive triggers (time of the day and scheduled activities) can be included in the design of the experiment by specifying the period of experiments and forced intermediate activities, such as changing the room.

Within the contextual factors, some of the physical environmental factors (building quality, view to outside) are generally fixed by the experimental setting in a test facility (type of room, quality of furniture and equipment). They also partly determine the degree of interaction allowed to the subjects by providing the respective devices (e.g., thermostats, remote controls) and permitting access to windows, blinds, and so on. Some labs, however, allow varying the type and degree of controls given to the subjects. A typical research question relating to this area could be whether there are preferences or specific orders for using devices to improve thermal comfort (e.g., window, fan), or which devices would be used under which physical indoor climate conditions. Also, the type of controls given to subjects could be investigated in the context of "effort and habit" (see also psychological factors). Physical environmental factors are important moderators for experiments, as they influence the appearance of the room(s) and therefore the general well-being of subjects.

Addressing psychological factors is probably one of the most difficult tasks, both in labs and in situ. In Hawighorst et al. (2016) it is outlined that personality traits have to be considered in order to understand occupants' expectations, habitual adaptations, or context-related reactions. The construct of thermo-specific self-efficacy was used to analyse differences in the perception of thermal comfort, assumed temperature, perceived control, and physiological parameters. Other psychological factors are awareness (e.g., financial concern, environmental concern, building technology), attitudes (e.g., towards energy saving), cognitive resources (e.g., knowledge), and habits or personal lifestyle (Fabi et al. 2012). In addition, individual differences in considering future consequence, energy saving attitudes, and political orientation could influence people's reactions on tasks with regard to indoor environmental parameters (Xu et al. 2015).

Social factors have to be considered if more than one occupant is in an experiment. Interactions between occupants depend on the number of persons in a space, personality traits, and backgrounds, as well as possible differences in access to devices used to modify indoor environmental parameters. Further, contextual factors such as group norms or organizational culture can influence individuals' behaviors. Lab experiments, on the one hand, can explicitly address these topics by the respective experimental design; for example, it can be examined whether the number of subjects has an influence on perception and acceptance of the prevailing indoor environmental conditions or on variations of changes applied by the subjects

(Schweiker and Wagner 2016). On the other hand, it can be questioned whether subjects do behave the same way in a one-day (or only half-day) experiment, in which they might accept even unfavorable persons, compared to a real-life situation when they have to meet them regularly. Other social factors, such as ownership, can hardly be addressed by the experimental setting.

Finally, physiological factors like age, gender, health state, activity level, or intake of food and beverages can be very well controlled in lab experiments. Therefore, a lot more specific, but also broad research topics can be tackled in this area compared to in situ experiments. On the other hand, care has to be taken to cover a wide range of the single parameters in order to be able to generalize results. This requires a large number of experiments because sample sizes are usually significantly smaller in comparison to in situ studies where the sample mostly represents a larger cross-section over physiological factors. In addition, the measurement of physiological quantities is easier to handle or in some cases only possible in the lab, as described in Sect. 7.4.

In summary, lab experiments may have the advantage of a more tailored selection of subjects according to certain characteristics, a better control of influencing factors, or the potential to perform more specific experiments in this field than in situ. Additionally, qualitative research methodologies like semi-structured interviews could support this approach to investigate the reasons (or barriers) behind why some occupants behave in a certain way. Another probably more common approach to comprehensively assess social-psychological factors relating to occupant behavior is the survey methodology (Chen and Knight 2014) (see also Chap. 8).

7.7 Conclusion

This chapter first introduced a variety of different types of laboratories which have been built at different places in the world over the last almost 50 years. Typical technical equipment and sensor technologies were then presented with regard to the experimental opportunities foreseen in lab environments. It was shown that lab facilities offer a large number of features to simulate a wide range of indoor environmental scenarios under carefully controlled conditions and to measure different physical, physiological, and even psychological quantities in detail. This makes lab studies advantageous against in situ experiments. However, a fundamental question remains: Can occupant behavior be studied within a controlled lab environment, or do occupants behave differently compared to a "normal" environment where they are not observed and instructed? In other words, can results derived from a controlled lab environment be generalized to occupant behaviors in real-life situations? There is certainly not a simple "yes" or "no" to this, and so this last section will point out different aspects related to this question in order to sensitize with regard to experimental design.

While designing experiments for labs, researchers have to be aware that occupants are exposed to an environment and situations in which they likely feel observed, if not personally by the researcher then still indirectly through sensors and control devices. In order to keep the so-called "Hawthorne effect" as small as possible, the real purpose and (scientific) objective of the experiment should not be explained to the subjects before and no feedback should be given during the experiment. Instructions should be given before the experiment in an objective (not encouraging or discouraging) way and information about the study should be kept as general and short as possible. The environment should be as "experientially realistic" or "typical" as possible with a minimum of disturbance by sensing equipment, as discussed above. Participant compensation could be an issue if participants regard this as a financial reward, rather than as compensation for their time spent in the lab.

A point to be raised is whether an environment which is too pleasant might influence the perception and acceptance of the subjects—e.g., placing students in labs equipped like executive suite offices. Further, it has to be considered whether the instructed activities match the subjects' background. For example, who would be a suitable "office worker"? Often students are recruited for experiments, but students normally have varying daily routines and, therefore, might behave differently compared to office workers. Another issue is to which extent certain situations can really be simulated. For example, if an office scenario with its typical behaviors is to be replicated, how can typical real-life stress situations of the occupants be provoked? Or, how can responsibility for energy-conscious behavior be stimulated if the participants of an experiment are not charged for energy consumption (like in their real homes)?

As lab experiments generally offer great possibilities to address very specific questions in terms of behavior—particularly with regard to situations or scenarios which are difficult to find or provoke in situ—the proper design of experiments is of utmost importance (see Chap. 3). It is important to be aware of the number of dependent variables in a specific setting and to eliminate side effects prior to experiments. Regarding the different driving factors discussed above (especially psychological and sociological parameters) and the possible approaches to investigate them in lab experiments, the overall objective has to be kept in mind: the improvement of models to simulate and predict occupant behavior during design and operation of buildings. Within this context, it is important to consider whether and how very specific results from experiments can be applied in practice at all (beyond pure academic interest). Which of the discussed parameters are really known during the design of a building? These are mainly physical, environmental, and contextual parameters—so maybe models excluding social-psychological factors are sufficient at this stage?

On the other hand, it is probably worthwhile to adopt advanced models, which e.g., include behaviors of different occupant types, for a risk analysis of different designs in terms of investment and operation costs. For optimizing building performance, such advanced models could be useful if the behavior of specific occupants is known through surveys. Another important field of application lies in personalized control of indoor environments. The understanding of occupant behavior can be used for more intelligent and learning control units in (decentralized) technical services systems, as well as for better ergonomics of remote control devices for occupants.

In conclusion, this chapter has shown that laboratories offer the possibility to study occupant behavior in a very detailed manner. Main advantages are that a wide range of indoor environmental scenarios can be simulated under precisely controlled conditions and that subjects can be selected based on pre-defined criteria. The degree of control over experiments is high and a large number of physical, physiological, and psychological quantities can be monitored. However, there are limits in investigating occupant behavior given the “artificial” environment and the fact that subjects always feel observed to some extent. The experimental design determines the results, which then add to findings from field studies and surveys.

References

Bauman F, Arens E (1988) The development of a controlled environment chamber for the physical and subjective assessment of human comfort in office environments. In: Kroner W (ed) Proceedings of international symposium on advanced comfort systems for the work environment. Troy, NY

Boerstra AC, te Kulve M, Toftum J et al (2015) Comfort and performance impact of personal control over thermal environment in summer: results from a laboratory study. Build Environ 87:315–326

Chellappa SL, Steiner R, Blattner P et al (2011) Non-visual effects of light on melatonin, alertness and cognitive performance: can blue-enriched light keep us alert? PLoS One 6(1):e16429. doi:10.1371/journal.pone.0016429

Chen C-F, Knight K (2014) Energy at work: Social psychological factors affecting energy conservation intentions within Chinese electric power companies. Energ Res Soc Sci 4:23–31

Clarys P, Clijsen R, Taeymans J et al (2011) Hydration measurements of the stratum corneum: comparison between the capacitance method (digital version of the Corneometer CM 825®) and the impedance method (Skicon-200EX®). Skin Res Technol 18(3): 316–323. doi:10.1111/j.1600-0846.2011.00573.x

Clausen G, Wyon DP (2008) The combined effects of many different indoor environmental factors on acceptability and office work performance. HVAC&R Res 14(1):103–113. doi:10.1080/10789669.2008.10390996

de Dear R, Nathwani A, Cândido Ch et al (2013) The next generation of experimentally realistic lab-based research: the University of Sydney’s Indoor Environment Quality Laboratory. Architect Sci Rev 56(1):83–92. doi:10.1080/00038628.2012.745807

E.ON ERC (2017) https://www.ebc.eonerc.rwth-aachen.de/go/id/fdql/lidx/1. Accessed 21 Apr 2017

Fabi V, Andersen RV, Corgnati S et al (2012) Occupants’ window opening behaviour: A literature review of factors influencing occupant behaviour and models. Build Environ 58:188–198

Fanger PO, Bánhidi L, Olesen BW et al (1980) Comfort limits for heated ceilings. ASHRAE Transactions, vol 86, Part 2

Fraunhofer (2017a) https://www.ibp.fraunhofer.de/en/Expertise/Acoustics/Projects/hipie-test-room.html. Assessed 6 Mar 2017

Fraunhofer (2017b) https://www.ibp.fraunhofer.de/content/dam/ibp/de/documents/Informationsmaterial/Abteilungen/BBH/Produktblaetter/IBP_PB_IndoorAirTestCenter.pdf. Assessed 6 Mar 2017

Fraunhofer (2017c) https://www.ibp.fraunhofer.de/content/dam/ibp/en/documents/VERU_Flyer_e_tcm1021-85279.pdf. Assessed 6 Mar 2017

Fraunhofer (2017d) https://www.ibp.fraunhofer.de/en/Expertise/energy-efficiency-and-indoor-climate/VehicleClimateControlSystems.html. Accessed 21 Apr 2017

Fütterer J, Constantin A, Schmidt M et al (2013) A multifunctional demonstration bench for advanced control research in buildings—monitoring, control, and interface system. In: Proceedings of the 39th Annual Conference of the IEEE Industrial Electronics Society, IECON 2013

Goia F, Finocchiaro L, Gustavsen A (2015) The ZEB Living Laboratory at the Norwegian University of Science and Technology: a zero emission house for engineering and social science experiments. In: Proceedings of the Conference "7. Passivhus Norden | Sustainable Cities and Buildings", Copenhagen, 2015

Hawighorst M, Schweiker M, Wagner A (2016) Thermo-specific self-efficacy (specSE) in relation to perceived comfort and control. Build Environ 102:193–206

ICIEE (2017) http://www.iciee.byg.dtu.dk/Facilities/Climate-Chambers. Accessed 21 Apr 2017

ISO 9886:2004 (2004) Ergonomics—evaluation of thermal strain by physiological measurements. ISO/TC 159/SC 5 Ergonomics of the physical environment

Khanie S, Andersen M, 't Hart B M et al (2011) Integration of eye-tracking methods in visual comfort assessments. Presented at CISBAT 11: cleantech for sustainable buildings—from Nano to Urban Scale, Lausanne, Switzerland, September 14–15, 2011. http://infoscience.epfl.ch/record/166209/files/Integration%20of%20Eye-Tracking%20Methods%20in%20Visual%20Comfort%20Assessments.pdf. Accessed 10 Mar 2016

Kimura K, Tanabe S (1985) Design of Waseda University environmental test chamber and its performance characteristics. In: Fanger PO (ed) Proceedings of CLIMA 2000, Vol. 4, Indoor Climate: 121–126, Naples

Loughborough (2016) http://www.lboro.ac.uk/departments/design-school/research/environmental-ergonomics/. Accessed 21 April 2016

Meerbeek BW, de Bakker C, de Kort YAW et al (2016) Automated blinds with light feedback to increase occupant satisfaction and energy saving. Build Environ 103:70–85

Möhlenkamp M, Schmidt M, Wick A et al (2015) Thermal comfort of displacement ventilation in environments with different temperature gradients. In: Proceedings of the Healthy Buildings Europe Conference, Eindhoven

MRUM (2017) https://www.maastrichtuniversity.nl/research/institutes/nutrim/facilities/metabolic-research-unit-maastricht-mrum. Accessed 21 April 2017

Müller D, Hegemann D, Braunstein V et al (2015) Measurements of the perceived air quality in shopping centers. In: Proceedings of the healthy buildings europe conference, Eindhoven

Nathwani A, de Dear R, Cândido C (2012) The missing link—next generation IEQ LAB with ultimate flexibility. In: Proceedings of 7th Windsor Conference: the Changing Contest of Comfort in an Unpredictable World, London

Newsham G, Veitch J, Arsenault C et al (2004) Effect of dimming control on office worker satisfaction and performance. In: Proceedings of IESNA Annual Conference: 19–41, Tampa

Olesen BW, Fanger PO (1973) The skin temperature distribution for resting man in comfort. Arch Sci Physiol 27(A):A385–A393

Olesen BW, Schøler M, Fanger PO (1979) Discomfort caused by vertical air temperature differences. In: Fanger PO, Valbjørn O (eds) Indoor climate. Danish Building Research Institute, Copenhagen, p 561

Parsons KC (2014) Human thermal environments, 3rd edn. CRC Press, London

Pasut W, Zhang H, Arens E et al (2015) Energy-efficient comfort with a heated/cooled chair: results from human subject tests. Build Environ 10:10–21. doi:10.1016/j.buildenv.2014.10.026, https://escholarship.org/uc/item/2tq3z4cw

Schmidt C, Veselá S, Nabi Bidhendi M et al (2015) Zusammenhang zwischen lokalem und globalem Behaglichkeitsempfinden: Untersuchung des Kombinationseffektes von Sitzheizung und Strahlungswärmeübertragung zur energieeffizienten Fahrzeugklimatisierung. FAT Schriftenreihe 272:2015

Schoffelen PFM, Westerterp KR, Saris WHM et al (1997) A dual-respiration chamber system with automated calibration. J Appl Physiol 83:2064–2072

Schoffelen PFM, Saris WHM, Westerterp KR et al (1984) Evaluation of an automatic indirect calorimeter for measurement of energy balance in man. In: Van Es AJH (ed) Human energy metabolism: physical activity and energy expenditure measurements in epidemiological research based upon direct and indirect calorimetry. Koninklijke Bibliotheek, The Hague, pp 51–54

Schweiker M, Wagner A (2015a) On the effect of the number of persons in one office room on occupants physiological and subjective responses, and performance under summer conditions. In: Proceedings of the Healthy Buildings Europe Conference, Eindhoven

Schweiker M, Wagner A (2015b) The effect of thermal inertia on office workers subjective and physiological responses; and performance under summer conditions. In: Proceedings of the 6th International Building Physics Conference, Torino

Schweiker M, Wagner A (2015c) A framework for an adaptive thermal heat balance model (athb). Build Environ 94:252–262. http://www.sciencedirect.com/-science/-article/-pii/-S0360132315300998

Schweiker M, Wagner A (2016) The effect of occupancy on perceived control, neutral temperature, and behavioral patterns. Energy Build 117:246–259. doi:10.1016/j.enbuild.2015.10.051

Schweiker M, Brasche S, Bischof W et al (2012) Development and validation of a methodology to challenge the adaptive comfort model. Build Environ 49:336 – 347. http://www.sciencedirect.com/-science/-article/-pii/-S0360132311002459

Schweiker M, Brasche S, Bischof W et al (2013a) Explaining the individual processes leading to adaptive comfort: Exploring physiological, behavioural and psychological reactions to thermal stimuli. J Build Phys 36(4):438–463. http://jen.sagepub.com/-content/-36/-4/-438.short

Schweiker M, Hawighorst M, Wagner A (2013b) Quantifying individual adaptive processes: first experiences with an experimental design dedicated to reveal further insights to thermal adaptation. Architectural Sci Rev 56(1):93–98. http://www.tandfonline.com/-doi/-abs/-10.1080/-00038628.2012.744297

Schweiker M, Brasche S, Hawighorst M et al (2014). Presenting LOBSTER, an innovative climate chamber, and the analysis of the effect of a ceiling fan on the thermal sensation and performance under summer conditions in an office-like setting. In: Proceedings of 8th Windsor Conference: Counting the Cost of Comfort in a Changing World, London

Storaas G, Bakkevig MK (1996) Correlation between measured skin wettedness and subjective sensations of skin wetness. In: Proceedings of the 7th International Conference on Environmental Ergonomics (ICEE), Jerusalem

Vandahl C, Wolf S, Schierz C (2010) Messung von Beleuchtungsstärken am Auge mit dem mobilen Messgerät LuxBlick. In: Tagungsband zur Konferenz Licht, Wien, Oktober

van Marken Lichtenbelt W D, Daanen HAM, Wouters L et al (2006) Evaluation of wireless determination of skin temperature using iButtons. Physiol Behav 88:489–497

Veselý M, Zeiler W (2014) Personalized conditioning and its impact on thermal comfort and energy performance—a review. Renew Sustain Energ Rev 34:401–408

Wikipedia (2016) https://en.wikipedia.org/wiki/Electrodermal_activity. Accessed 10 Mar 2016

Wyon DP, Fanger PO, Olesen BW et al (1975) The mental performance of subjects clothed for comfort at two different air temperatures. Ergonomics 18(4):359–374

Xu X, Arpan L, Chen C-F (2015) The moderating role of individual differences in responses to benefit and temporal framing of messages promoting residential energy saving. J Environ Psychol 44:95–108

Zhang H (2003) Human thermal sensation and comfort in transient and non-uniform thermal environments. Center for the Built Environment. http://escholarship.org/uc/item/11m0n1wt#

Zhai Y, Zhang H, Zhang Y et al (2013) Comfort under personally controlled air movement in warm and humid environments. Build Environ 65:109–117

Zhang H, Arens E, Kim DE et al (2009) Comfort, perceived air quality, and work performance in a low-power task-ambient conditioning system. Build Environ 45(1):29–39

Chapter 8
Survey and Interview Approaches to Studying Occupants

Julia K. Day

Abstract This chapter provides guidance for survey development related to building occupant research. Many researchers studying occupant behavior have used survey methods to collect self-reported data of occupant behaviors in buildings, either exclusively or in tandem with data gathered in field or laboratory studies. The chapter also serves as a how-to guide for issues such as: (a) how should survey questions be conceptualized, (b) are the questions measuring what was intended, (c) how should questions be written so that participants understand the intent, (d) how can the validity be increased for the survey itself, (e) how does one select the appropriate sample for a survey, and (f) how should one select the appropriate survey tool for data collection? Real examples of occupant behavior survey research and case studies offer lessons learned and precedent for future research efforts. Finally, the last section of the chapter presents a brief discussion of interview methods.

8.1 Introduction

Many studies on occupant behavior have used surveys to gain a better understanding of occupants (Nicol 2001; Haldi and Robinson 2008a, b; Konis 2013; Day and Gunderson 2014). Surveys can provide a cost-effective solution for obtaining both a large sample size and useful information related to occupant behaviors, perceptions, and preferences. In some studies, surveys have been used to develop models—for example, Haldi and Robinson (2008a, b) collected data from surveys and observations to better understand occupant behaviors. They applied advanced statistical analysis methods to the results of the surveys to form stochastic models to predict building occupants' behaviors related to window and shading openings (see Humphreys and Nicol 2002; Yun and Steemers 2008, 2009; Haldi 2013; Herkel

J.K. Day (✉)
School of Design and Construction, Washington State University, PO Box 645815, Pullman, WA 99164, USA
e-mail: julia_day@wsu.edu

A. Wagner et al. (eds.), *Exploring Occupant Behavior in Buildings*,
DOI 10.1007/978-3-319-61464-9_8

et al. 2008; Haldi and Robinson 2008a, b for additional examples). While surveys are not an appropriate tool in all instances or contexts, a well-designed survey can offer useful and rich data for better understanding occupant behaviors in buildings.

Some researchers choose not to use survey methods because they are unsure how to create and deliver surveys, or they are simply unaware of how surveys can be used to better understand self-reported occupant behaviors. Chap. 5 provides a discussion on which research methods are most suitable for understanding the occupant phenomena of interest. In many instances, surveys can be used in tandem with physical data collection, observations, through laboratory studies and in situ studies, to holistically understand occupant behaviors and perceptions in any given building.

Regardless of the data collection method(s) used, there are differing requirements and levels of rigor required for both implementation and analysis. Surveys rely on self-reporting of personal behaviors rather than observations and measured site data; some researchers may fear untruthful or inaccurate responses and avoid surveys altogether, especially because of established issues such as the Hawthorne effect, Pygmalion, social desirability, participant survey fatigue, etc. However, a well-designed survey can help to alleviate these concerns and provide rich data if proper steps are taken to maximize the chances that occupants are reporting both valid and required/desired information. Alternatively, if surveys are poorly designed or poorly worded, participants may not understand the questions and misread the intent behind questions and respond accordingly.

It should be briefly mentioned that, in most cases, there is a slight difference between purely post-occupancy evaluation (POE) studies and surveys aimed at understanding behavior. POE surveys are typical in the both the architectural and engineering fields, and their primary objective is to learn if building systems are working properly. Based on responses, designers can utilize feedback and implement these data into their next project. Meanwhile, building operators can also use POE results to improve building operations. A POE may include questions related to occupant behavior, but this is not typical. Surveys developed for occupant behavior research are typically implemented to better understand a specific research question or hypothesis related to building systems and occupant behaviors. They may also be implemented because indoor environmental quality (IEQ) complaints have been made and the owner/designer/researcher is attempting to better understand underlying factors or building issues related to occupant behaviors. The focus of this chapter is on occupant behavior surveys, as opposed to POEs. Some of the guidelines may apply to both types of surveys, but discretion should be used.

In general, the purpose of survey research is to collect quantitative responses to generalize results from a sample to a population (Fowler 2009). However, surveys may also include open-ended questions that are qualitative in nature; at times, other qualitative methods such as interviews and focus groups may also be appropriate for understanding the causes of specific occupant behaviors (Day and Gunderson 2014). In this case, responses would be best used to better understand a specific building population, as opposed to using results to generalize to a broader population. In either scenario—whether the results are generalizable or specific to a building—both survey and interview methods can be used to better understand occupant behaviors. This chapter aims to explore the advantages and disadvantages of using surveys in place of, or in combination with, physical data collection (e.g., see Sect. 8.8.1). Throughout the chapter, examples of surveys from occupant behavior research offer lessons learned and precedent for future research efforts. The chapter also serves as a how-to guide for issues such as: (a) how should survey questions be conceptualized, (b) are the questions measuring what was intended, (c) how should questions be written so that participants understand the intent, (d) how can the validity be increased for the survey itself, (e) how does one select the appropriate sample for a survey, and (f) how should one select the appropriate survey tool for data collection? Toward the end of the chapter, a brief discussion of interview methods are also discussed, which is followed by the chapter conclusion.

8.2 Constructing the Survey

A survey can provide "a quantitative or numeric description of trends, attitudes, or opinions of a population by studying a sample of that population" (Creswell 2014, p. 155). Surveys can also elicit qualitative data, if desired and if questions are formatted appropriately. To ensure a survey instrument is delivering reliable and valid data, it is important to recognize and thoughtfully consider all elements of a survey. A good survey design should be based on theories, research questions, hypotheses, and well-defined variables and measurements (scales). There are several forms of data collection (also called "survey modes") for conducting a survey, including: mail/email, telephone, the Internet, video conferencing, personal interviews, and focus groups (Fink 2012).

8.2.1 Before Designing the Survey

According to Creswell (2014), a general list of pertinent questions should be considered before designing the survey (pp. 155–156):

1. What is the nature of the survey design (e.g., cross-sectional vs. longitudinal, see Dillman, 2000)?
2. Is the population defined and its size mentioned?
3. Will the population be stratified (e.g., by climate type, office location, building orientation)? If so, how?
4. How many people will be in the sample? On what basis was this size chosen?
5. What will be the procedure for sampling these individuals (random or non-random)?

6. What instrument will be used? Who developed the instrument?
7. What are the content areas addressed in the survey? The scales used?
8. What procedure will be used to pilot or field-test the survey?
9. What is the timeline for administering the survey? Will reminders be sent?
10. What are the variables in the study?
11. How do these variables cross-reference with the research questions and items on the survey?
12. What is the data analysis procedure? How can these data be interpreted?

Additional questions, related specifically to building/occupant research include:

1. Will the survey responses be compared with findings from other previously conducted surveys, and if so, are the same questions being asked? If so, the researcher would want to ensure that the measures used are consistent (for instance, if the original survey used a 7-point Likert scale, then the survey developed should also use a 7-point scale, as opposed to a 5-point scale).
2. Are the survey data being compared to any measured data (from in situ, laboratory or observation studies)? If so, surveys should be deployed during in situ data collection.
3. Should the survey(s) be deployed at a specific time of year? For example, if collecting information about the effectiveness of shading systems and occupant behaviors related to glare during both the summer and winter equinox, then the timing of the survey deployment should be planned accordingly (i.e., near both June 21 and December 21).
4. Has a survey or POE already been conducted in the study building(s)? If so, are those data available to help inform additional survey questions?

In addition, researchers should consider how the data will be analyzed and presented in advance (and use the appropriate analysis tools).

8.3 Developing Questions and Constructing the Survey Tool

When developing a survey, it is crucial to think carefully about how questions are written, composed, and measured. In terms of measurement and the overall length of a questionnaire, many researchers may want to reduce the risk of irritating respondents by either asking a very minimal number of questions or by using a more succinct measurement scale. However, using only a few questions, or even a scale that is too brief, may be detrimental to the research and conclusions that can be drawn from the results (DeVellis 2012, p. 15). In addition, using any measure that does not truly reflect the researchers' assumptions or theories can lead to unreliable outcomes. For example, if the questionnaire is seeking responses related to thermal comfort and occupant behaviors, then it may not be wise to add large sections of questions about acoustics or other unrelated building characteristics.

In occupant behavior research, although it may lead to a longer survey, it may be best to use multi-item measures to best understand psychological attributes. A single-item measure, such as overall indoor environmental quality (IEQ) satisfaction, is less desirable than breaking down IEQ into several, more nuanced measures (thermal, visual, acoustics, privacy, lighting, etc.). In general, when trying to understand occupant behaviors—and associated reasons for certain behaviors—being able to break down responses in a more granular fashion may yield the best results. When measuring any psychological or behavioral attribute, it is likewise important to implement multi-item measures. Nunnally and Bernstein (1994, pp. 66–67) provide the following reasons for using multi-item measures: (a) individual items generally correlate poorly with the particular attribute in question; (b) each item tends to relate to multiple attributes in addition to the one being measured; (c) each item has a degree of specificity in the sense of not correlating with any general attribute or factor (i.e., factor analysis); (d) individual items have considerable random measurement error, i.e., are unreliable; and (e) an item can categorize people into only a relative small number of groups. A dichotomously scored item (yes or no) can distinguish between only two levels of attributes.

When developing measurements and scales for survey questions, it is crucial to understand what type of data are being collected since measures relate specifically to how the measures can be compared and analyzed. When analyzing survey questions, statistical analyses must be selected to suit the types of measures—if they are not, then the results are unreliable. Different measures and scales generally fall into four types: nominal, ordinal, interval, and ratio (Nunnally and Bernstein 1994). Nominal scales determine if two objects (or more) are equivalent to one another (e.g., male, female, or other). It should be noted that a numerical value may be used as a proxy for a nominal scale item, but it is an arbitrary code, i.e., with no numerical value—for example, gender may be assigned a numerical value (male=1, female=2, other=3). Ordinal scales determine whether (a) one object is greater or lesser than the other, or (b) if variables can be ranked in some meaningful way. For example, ordinal variables in occupant behavior research may relate to no control, some control, and full control. These responses could be ranked in order from least to most control, but there is no true absolute value that separates the measures. Interval scales contain three criteria: (1) the rank ordering is known, (2) the distances among objects on the attribute are also known, but (3) the absolute magnitudes are unknown (Nunnally and Bernstein 1994, p. 14). Ratio scales are interval scales with a true zero—for example, zero height or weight. In occupant behavior research, temperature can serve as a prime example for both interval and ratio data types. For instance, temperature readings in Fahrenheit (°F) and/or Celsius (°C) are considered interval scales, whereas Kelvin (K) is considered as a ratio variable since it has an absolute zero value. Additional discussion on measurement, scales, and general survey constructs are provided in Chap. 3.

8.3.1 Writing Survey Questions

A well-developed survey question will encourage respondents to interpret questions in the same way, respond to questions accurately, and increase the participant's willingness to complete the questionnaire (Dillman 2000).

Choosing the most appropriate words and sentence structure is the first step to writing good survey questions. Guidelines for writing survey questions have been extensively published (see Dillman 2000, pp. 51–78 for a comprehensive list of guidelines). In general, questions should be simple, clear, and easily understandable to participants. Questions are typically more understandable when complete sentences are used and when technical jargon is avoided. Using straightforward questions that respondents can reasonably answer, as well as keeping a survey as short as possible, will often result in a higher sample size and quality responses. A successful survey seeks to reduce survey errors including: coverage, sampling, measurement, and nonresponse, which are further discussed later in this chapter.

8.3.2 Types of Questions

There are essentially three types of question structures that can be used in surveys: (1) open-ended; (2) closed-ended with ordered categories; and (3) closed-ended with unordered categories (Dillman 2000). A question could also be partially closed-ended, which would take the form of a closed-ended question with an additional open-ended option at the end of the answer options—usually "other" or "please specify".

An open-ended question allows the participant to answer the question openly. For instance, "Please tell us what features you like in your building," where the respondent can fill in a blank answer field. A closed-ended question with ordered categories would provide specific ordered response options. For instance: "Please answer the following: 'I am cold at my office desk'; Answer options: All of the time; Most of the time; Some of the time; Seldom; Never." In a closed-ended with unordered categories question, the provided answers would have no particular order (i.e., red, green, blue, yellow). Unordered responses can be more difficult to analyze than ordered responses. It is always best to *not* mix the two together for one question. For instance, it would be impossible to analyze a question accurately that has the following response options: once per day, once per week, blinds, curtains, other—they simply do not relate.

8.3.3 Criteria for Examining Each Survey Question

To ensure questions are effectively written, there are several criteria for examining each survey question.

1. Does the question *require* an answer? Will each respondent be given the opportunity to answer every question? It is important to think carefully about which responses must be answered, and to what extent skip logic is used throughout the survey. For instance, if an occupant does not have local control of a thermostat, then it would be senseless to ask them how often they control the thermostat. In this case, it would be best to ask first, "Do you have access to temperature/thermostat control in your office space?" If the occupant says "yes", they can then be directed to further questions regarding their reported behaviors. If an occupant reports "no", they would be directed to the next set of survey questions.
2. Consider whether people can accurately recall and report past behaviors. For example, if an occupant's behaviors are habitual (e.g., flipping on a light switch every day upon arrival to their office), then they may recall this action less than a more purposeful action (e.g., flipping on a light switch because the lights turned off automatically while they were already working). To mitigate this type of issue, it is important to be familiar with the building. In the case of the example provided previously, does the building have manual light switches, or is it controlled by occupant sensors?
3. Will the participant be willing to share personal information (e.g., age, salary, reported productivity)? If questions are sensitive in nature, then it is helpful to give ranges as response options, rather than absolutes. Also, these sensitive questions should not be required, so that respondents can move forward in the survey if they choose not to answer.
4. Is the survey information being collected by more than one survey mode (email, phone call, etc.)? If more than one survey mode is used, then it will be important to maintain validity by ensuring that questions are the same in all modes. It is also important to recognize that occupants may provide different answers based on the survey mode used. For instance, when doing a phone survey with a researcher, the occupant may be more shy or more vocal than in an online survey form.

8.3.4 Questionnaire Structure

In writing a survey, the researcher's two main objectives are to maximize the response rate and to reduce measurement errors (Dillman 2000). An effectively constructed questionnaire will use both verbal and visual cues to help maintain the interest of respondents, while encouraging them to provide accurate answers to all questions. According to Dillman, there are three primary steps for designing surveys (see Dillman 2000, pp. 96–133 for further details):

- "Step 1: Define a desired navigational path on each page.
- Step 2: Create visual navigational guides that will assist respondents in adhering to the prescribed navigational path and correctly interpret the information.
- Step 3: Develop additional visual navigational guides, to interrupt established navigation behavior and redirect respondents, for example, through skip patterns." (p. 96)

8.3.5 Ordering Questions

It is important to have logical flow when ordering questions. A questionnaire is a way to communicate with respondents (Schwarz 1996), and it is more likely that the researcher will receive more accurate responses if questions are ordered in a logical way. The following provides guidance for ordering questions (see Dillman 2000, pp. 87–88):

1. Question topics and corresponding questions are generally grouped from most significant to least significant to the respondent.
    a. This same idea is also beneficial from a research perspective. If the most important questions are asked first, then, if someone decides to not complete the survey, the researcher will still have data for the early and important questions asked.
    b. In addition, it may be best to place questions about demographics or sensitive topics near the end of the survey, after the respondent becomes interested in the questionnaire.
2. Consider using descriptive questions before requesting evaluations of the experiences (e.g., ask if occupants *can* control a given building feature before asking *how* they control that same feature).
3. It may also be beneficial to group questions together that have similar answer categories.

8.4 Survey Instrument Assessment

After developing well-written questions and responses, as well as an overall survey logic, the researcher should also consider additional issues of survey instrument assessment, including reliability, validity, and survey error. The next few sections provide guidance and examples for understanding these concepts.

8.4.1 Reliability & Validity

As discussed in both Chaps. 3 and 9, researchers should thoughtfully consider validity and reliability when designing the survey tool. In survey research, validity is achieved when the measurement tool (i.e., the survey) provides accurate measurements of the concepts being studied. In general, reliability is the ability of a survey tool to provide consistent measurements over a population and a period of time with other groups of respondents. Researchers should be cognizant of both validity and reliability throughout the survey process—from the process of conceptualization of survey variables, to survey construction, and throughout the process of collecting and analyzing data.

Measuring validity and reliability can differ greatly in quantitative and qualitative research, and a survey can measure both quantitative and qualitative variables. For instance, a survey will likely include variables that are easily measurable and statistically analyzable (quantitative), as well as open-ended responses, which are best analyzed through qualitative methods. As previously noted, in quantitative research, validity is achieved when the measurement tool provides accurate measurements of the intended concepts and constructs of study (Creswell and Plano Clark 2011). Qualitative reliability means that both the researcher's methods (Creswell 2014) and the participants' responses are "consistent and stable over time" (Creswell and Plano Clark 2011, p. 211). In any qualitative research, as well as occupant behavior research, reliability essentially means that the study, or the instrument used to gather qualitative data, are replicable. Although a researcher may prove reliability, it does not necessarily mean that that the study is also valid (Golafshani 2003). There are many types of validity, but those that most closely align with survey research are mentioned below.

Face validity has been defined as "the degree that respondents or users judge that the items of an assessment instrument are appropriate to the targeted construct and assessment objectives" (Hardesty and Bearden 2004, p. 99). This is the simplest type of validity, which involves a subjective evaluation of the survey items to determine whether they appear logical and appropriate for the concepts being studied. For example, in occupant behavioral question development, if a researcher is trying to determine whether a participant can control their blinds for glare and lighting preferences, then they would want to be specific in how they ask this question. Simply asking: "Do you have controls in your office?" would not be effective for measuring the specific concept of daylight/blind control. A better question would be: "Do you have access to your blind controls in your office?" Follow up questions would then be used to further understand how and why the occupants control the blinds.

Construct validity is "the extent to which a measurement instrument (i.e., a question or set of questions) measures the intended construct and produces an observation distinct from that produced by a measure of a different construct" (De Leeuw et al. 2008, pp. 16–17). In other words, this type of validity is obtained when the variables of interest being measured within a research instrument are logically connected with one another and align with the theoretical model being utilized by the researcher (Babbie 2004). For example, a set of questions may be used to assess the construct of thermal comfort and how that might relate to occupant behaviors and energy bill responsibility (See Example 9.3 below). If the questions selected do indeed probe at issues of thermal comfort and potential occupant behaviors relating to their perceived thermal comfort, then the measurement instrument (for those questions) would likely have strong construct validity if the questions also demonstrate evidence of both convergent validity and discriminant validity. Convergent validity is accomplished when data shows that variables that should be related are in fact related, which can be ascertained by using related measures of the same construct. For example, survey measures surrounding thermal comfort and actual physical data measurements of temperature readings would help to show that the measures are indeed related. Conversely, when data reveal that unrelated variables do not have a relationship, the researcher has demonstrated discriminant validity (Corral-Verdugo and Figueredo 1999). See Chap. 3 for additional information on convergent and discriminant validity as they pertain to occupant research.

8.4.2 Types of Survey Errors

Due to the complexity and variability related to any research, it is not always possible to completely eliminate error in the research process. To effectively enhance a survey's validity and reliability, the researcher must be mindful of various sources of error frequently encountered during the research process. What follows is a summary of the most common types of research errors in survey research and examples in the context of occupant behavior research for each type of error.

Coverage error. Coverage error occurs when the sample does not adequately represent the population. Under-coverage is the result of omitting individuals who should have been included in the study population. Wrongful inclusion of individuals will result in over-coverage. In either case, the researcher will be drawing a sample from a group that does not accurately represent the population of interest (Dillman 2000). One example of coverage error in terms of occupant behavior research might relate to commercial building-wide listservs. In some buildings, a master email list may exist that would allow the survey to be sent to all employees. However, in some cases building employees may work from home or telecommute, and only a portion of the employees work in the study building. In this case, the researcher(s) would want to only send the survey to people that work in the building. For this reason, in one study, the researchers (i.e., Day and Gunderson 2014) specifically asked survey participants (a) what building they worked in, (b) how often they worked from the office building, and (c) if they were in their office space at the time the survey is being completed.

Nonresponse error. Survey research frequently encounters this type of error, which is the result of missing data that renders the collected data insufficient for statistical analysis.

Unit nonresponse describes a low response rate to the survey by the sample population (a higher response rate is ideal, and sending up to two friendly reminders during the survey period is one strategy for increasing the response rate). Incentives (e.g., money or gift cards) can also be offered to increase the response rate. In addition, different types of survey modes, the layout, the length of the survey, and even the font size and typeface selected can all affect the response rate.

Item nonresponse is when research participants do not provide responses to specific items within the survey. One way to avoid this type of error is to require participants to answer questions before they move on the next question (most survey programs have this feature), but in some cases this might not be desirable and the researcher may want to leave a certain question response optional.

Measurement error. This type of error is encountered when responses to survey items are inaccurate or cannot be effectively compared to the responses of other survey participants. Measurement error is most frequently associated with poorly constructed survey instruments, poorly worded questions, and/or confusing or incomplete instructions given to survey respondents (Dillman et al. 2009). This type of error is part of what Babbie (2004) refers to as non-sampling error. Social desirability response (SDR) bias (i.e., the tendency for occupants to present a positive image of themselves on questionnaires) is one form of measurement error, along with the Hawthorne effect, etc. Another example of measurement error is cited as a lessoned learned below (see survey story in Sect. 8.8.5).

Processing error. This type of error is the result of mistakes made to data during collection and analysis. Common mistakes include recording inaccurate responses, coding mistakes, errors in inputting data into analysis software, and mistakes in statistical analysis. This is the second type of non-sampling error (Babbie 2004). One example of this might relate to a survey that uses Likert measurement scales in two separate sections (e.g., IEQ satisfaction and assessment of occupant productivity). If one scale uses the positive appraisal as a (7) and the other scale uses the positive appraisal as a (1), this can present errors in the analyses if these are not coded on input correctly into the software analysis. This type of issue would also be a problem in terms of confusing occupants. If positive appraisals (e.g., "strongly agree") typically appear on the right side, and then they appear on the left for a different set of questions, that could alter the accuracy and validity of the survey responses received.

Sampling error. This type of error is the result of selecting a sample that is not accurately representative of the entire study population due to either an insufficient number of participants from the sampling frame, or due to the use of another method of sampling that does not ensure a truly random sample (e.g., convenience sampling). This type of error is similar to coverage error; however, the key difference is that coverage error is the result of erroneously including (or not including) individuals that should have been part of the population and sampling frame, whereas sampling error occurs because a random selection was not achieved from the intended sampling frame. Sampling error prevents the generalization of data to the entire study population (Babbie 2004). This type of error is also referred to as "selection error". In occupant behavior-related research it is often very difficult, if not impossible to avoid this type of error. A thorough discussion of this topic is covered below in both 6.2 Sample size and Chap. 3.

In a perfect world and a perfect research design, all types of reliability and validity would be met and all types of potential errors would be minimized or eliminated. However, it is often the case in building research that not all of these conditions can be met. For example, in many instances, it will be impossible to avoid sampling error or achieve a true sample of a population when studying occupant behavior in buildings—this is not ideal, but does mean that results should be completely discarded. There is a fine line between striving for perfection in reliability and validity of measures and becoming consumed in the process. Best practices should be followed if possible, but when building studies or study populations do not allow for the preferred methods, then these instances should simply be reported as study limitations and not hidden away during dissemination of results.

8.4.3 Pre-testing and Pilot Studies

"Even after years of experience, no expert can write a perfect questionnaire… If you do not have the resources to pilot-test your questionnaire, don't do the study" (Sudman and Bradburn 1982, p. 283). To ensure reliability, validity, and error factors have been considered, and that they are working as intended (i.e., that questions are measuring what they are supposed to measure, questions are written well, participants understand what is being asked of them), it is necessary to deploy a pilot survey before the final survey is deployed. For in-depth guidance on preparing a pilot study, see Chap. 10 in De Leeuw et al. (2008), an open-source guidebook for survey methodology.

The primary purpose of the pilot study is to ensure that the survey instrument is working as intended; as such, the results of the pilot study should typically not be included in the results of the full survey deployment, especially if the sample selection differed between the two phases (i.e., convenience sampling for the pilot study and random sampling for the full survey implementation). However, if the sampling method is the same and no changes to the survey instrument were made after the pilot study was deployed, then some researchers may choose to include their pilot study results with the full survey to increase the overall sample size.

8.5 Participant Selection and Sample Design

In general, sampling of populations of interest is necessary because it would be impractical/expensive/impossible to survey every single person in a population; therefore, many methods and sampling strategies have been used in research. However, when conducting survey research, probability sampling is the most widely accepted method in most fields: "Probability sampling methods give a known probability of selection for all possible samples from the sampling frame. They thus provide protection against selection bias and give a means of quantifying sampling error" (De Leeuw et al. 2008, p. 112). There are several types of probability sampling (i.e., simple random sampling, stratified random sampling, cluster sampling). Random sampling is widely used, and it refers to a sample in which all participants have an equal chance of being selected (Gliner et al. 2011). Oftentimes, researchers do not understand the true meaning of random sampling. For example, randomly calling people for a phone survey is not random sampling, and neither is knocking on every 5th door on a street between 5PM and 7PM in the evening, etc. True random sampling follows specific guidelines based on inferential statistics, and as such, it assumes that every person in each population has an equal chance of being selected (Creswell 2014).

The sampling method selected will depend on many factors such as budget, timeline, projects goals, and the type of population being surveyed. Additional key sampling terms and concepts are identified in the glossary.

8.5.1 Alternative Sampling Strategies in Building Research

Ultimately, the above sampling strategies (i.e., simple random sampling, stratified random sampling, cluster sampling) are best practices in survey research for maintaining validity and reducing error. If the intention is to generalize results to the broader population, then it is imperative to utilize a combination of probability sampling and a sufficient sample size. A more robust discussion on sampling strategies, as they relate to our field, is presented in Chap. 3.

However, sometimes in building research it can be very challenging to employ some of these probability sampling strategies due to the nature of occupant behavior research (e.g., it may be difficult to obtain a truly representative sample, some buildings may have differing design strategies of different floor, or buildings and associated samples may have to be purposefully selected, rather than randomly selected, to address a phenomenon). In these cases, an alternative sampling strategy may be used, such as purposeful sampling or convenience sampling. For example, in the field of occupant behavior, it is often the case that we want to know information about a specific building, or even a specific floor or section of a building. Obviously, in this example, the sample for the survey would not be random, but would represent respondents from a population of a given building. These types of constraints are certainly present and should be mentioned and considered in the limitations section of any write up, yet they are often unavoidable in this field.

In other cases, research projects may begin with the intention of drawing a random sample, but factors outside of the researcher's control may exist. For instance, it may be extremely difficult obtaining permissions from building owners for surveys, locating or recruiting buildings that perfectly align with the researcher's specific hypotheses or research questions, or identifying the population and defining a suitable sample frame. For these reasons, a purposeful sampling or convenience sampling method that includes very specific buildings may be required. The technique of purposeful sampling (also "purposive sampling") is frequently used when a researcher decides to select units or cases of study rather than sample randomly (Teddlie and Tashakkori 2003); in addition, purposeful samples are usually selected by the expert judgment of the researcher (Teddlie and Tashakkori 2009). While this type of sampling strategy is a valid research method, it does limit the generalizability of the results. In these cases, the individuals who take the survey are not randomly selected, which should be reported as a study limitation.

8.5.2 Sample Size

A commonly asked research question in survey research is, what sample size do I need? Formulas are available to determine what sample size is needed based upon the overall population (see Dillman et al. 2009 for a detailed explanation of a sample size formula and related variables). There are also several software programs available for calculating sample size if needed. In some cases, it may also be appropriate to refer to previous studies and their respective sample sizes. To be clear, the studied building may dictate the available sample. If a building has very few occupants, then a lower response rate is expected, which is typically lower than what would be required in a traditional sampling method. In this type of research, these issues are typically unavoidable, and while the findings will likely not be generalizable to a broader context, responses will still be valuable for better understanding the study building and should not be dismissed. See Chap. 3 for a more comprehensive explanation on sampling.

8.6 Available Tools for Survey Delivery

There are many survey tools available for survey delivery; this chapter does not advocate for one tool over the other, but simply provides options for survey development and delivery. The selected tool (or method) should ultimately align with the project' goals and budget. There are many free survey tools, such as limesurvey and Google Forms, as well as paid tools, which vary in price (e.g., Surveymonkey, SurveyGizmo, Qualtrics, Amazon M-Turk tool). Some of these firms will assist researchers in implementing the survey for an additional fee. However, researchers should be aware that ethics restrictions may be present if the server is hosted elsewhere. For example, because of the Freedom of Information Act, Canadian researchers are discouraged from using USA-based servers (e.g., Google Forms). See Chap. 11 for additional ethical considerations.

Once again, as previously mentioned, it is wise to conduct a pilot survey before the final survey is released. This will help ensure that any potential inconsistencies, errors, or software/survey platform glitches are caught before the official survey deployment.

8.7 Interviews

Oftentimes, when conducting on-site data collection or surveys, it is valuable to also conduct individual face-to-face interviews with building occupants. Collecting qualitative data in addition to quantitative data via surveys and field studies can help provide valuable insights into *how* and *why* occupants may be acting a certain way. For example, in Day et al. (2012) the researchers found that nearly half of building occupants surveyed were not interacting with the shading controls simply because they could not reach the control for the blinds. This behavior was not evident through the online survey or site visits. However, open-ended interviews with building occupants allowed the research team to discover this design mistake. Occupants wanted more daylight and to use less electric light, but they were unable to in this case. Interviews (and open-ended survey questions) are a great way to uncover complicated behavioral problems like the one mentioned above.

The primary focus of this chapter is the use of surveys in occupant behavior research. However, surveys and interviews often occur sequentially or concurrently in building research, and so a few basic guidelines for interviewing occupants will also be briefly mentioned.

8.7.1 Interview Formats

There are several formats for conducting interviews: individual face-to-face, telephone, video conferencing, email, and focus groups. Individual face-to-face interviews are ideal if the researcher will be on-site collecting data already. In addition, one-on-one phone interviews can be conducted if the researcher cannot visit the actual research site, or if there was not enough time to conduct in person. It is also possible to conduct interviews via email. One of the cons of this method is that the responses received may be much shorter; however, one of the pros is that the researcher does not need to spend additional time transcribing the interview responses. Email interviews are an effective option if there are time or budgetary constraints and/or if the researcher(s) have structured questions to ask (discussed in Sect. 8.7.2).

Finally, focus groups can be conducted with several people at a building site at once. This has many advantages, such as: multiple perspectives can be obtained at one time, thus saving both time and resources; the researcher may discover issues that may not have arisen in a structured survey; and responses can build upon one another. For example, one person might complain about something in a building, and then the other participants can include their opinions and perceptions, as well. In this format, it is important to try to elicit responses from all participants, as some types of personalities may dominate the conversation (see Liamputtong (2011) for additional focus group methods).

8.7.2 *Types of Interviews*

Different types of interview question formats include open-ended, semi-structured, and fully-structured. In open-ended formats, a general subject is selected to discuss, but very few pre-planned questions may be available; in this context, the interviewer is allowing the participant to openly respond to questions. Valuable data may emerge from this format; however, it may be difficult to compare answers across different interview responses. In semi-structured interviews, some of the questions are prepared beforehand, but there is still flexibility in adding additional questions as new topics emerge during the interview. In fully-structured interviews, all interview questions are predetermined and the same questions are asked to all participants. This type of format is easier to compare across responses and is the best method for email interviews. However, this type of format may not provide a deeper understanding of the research problem, which may be gleaned from open-ended responses and occupant stories.

8.7.3 *Conducting the Interview*

When conducting interviews, the interviewer should be clear, objective, and avoid leading questions that may influence the respondent's answers. It is good practice to prepare questions (depending on the format and type of interview) ahead of time. The interviewer should ask clear, singular questions and be a good listener. In addition, if applicable, the interview should be recorded (with permission) and the interviewer should take thorough notes. If the interviews are recorded, then ideally the recordings should be transcribed. Ultimately, the format and type of interview method will largely depend on the goals of research and any participant, budget, or research constraints. More specific interview protocols and examples are available elsewhere (i.e., Berry 1999; Creswell 1998; Doyle 2004; Teddlie and Tashakkori 2003, 2009).

Analyzing results from interview collection is beyond the scope of this chapter, but there are many resources and books that discuss best practices and rigorous methods for analyzing interviews and other qualitative data (e.g., authors Creswell, Plano-Clark, Teddlie and Tashakkori). It is important to note that open-ended responses and interviews do not result in large-scale generalizable results across a population. However, studies in occupant behavior research (e.g., Langevin et al. 2013) have found that interviews offer a method for better understanding little known areas of study. Research topics can be explored in a way that allows the researchers to discover key themes while also providing a rich qualitative context: "This information can then serve as a rich basis for larger-sample, structured surveys that will be necessary to generate more broadly conclusive findings" (Langevin et al. 2013, p. 1368).

Ultimately, it is critical to report survey and interview data in a way that both designers and owners can understand it. In other words, be aware of the audience; findings should be presented in ways that are understandable by academics, designers, owners, and the public.

8.8 Survey Stories and Lessons Learned in Occupant Behavioral Research

This final section of this chapter includes brief survey stories and lessons learned from researchers who have used surveys in occupant behavior related research. These stories were written by the researchers, specifically for inclusion in this chapter. Survey story authors were asked to provide a short story ($\sim$500 words), which included the following information:

1. author name and affiliation
2. name of study
3. research explanation
4. types of survey data collected
5. key findings
6. lessons learned.

Each of the stories below have been included to help readers understand the unique challenges that may emerge with this type of research. These stories were written from the perspective of the researchers.

8.8.1 Field Study of Thermal Comfort and Occupant Satisfaction in Canadian Condominiums

Isis E. Bennet and William O'Brien

Department of Civil and Environmental Engineering, Carleton University, Ottawa, Canada

The research intent of this project was to explore comfort implications of living in modern highly glazed condominiums, in addition to corresponding occupant behaviors. This was completed by performing a field study of 20 condominium units: 10 with large window areas and 10 with medium-sized window areas.

Occupant comfort was assessed in summer and winter, through two open-ended surveys, infrared thermography, and recorded interior temperature and humidity. The elegance of this approach is that it allowed the researchers to survey the occupants during the extreme seasons, while also leaving equipment in the condos for about six months. The surveys covered a range of subjective topics including condo preference, daylighting, view, privacy, acoustics, indoor air quality, thermal comfort, adaptive behaviors, and energy efficiency. Participants were asked for additional feedback during informal interviews.

The infrared thermography was used in winter to look for thermal bridging in the building enclosures. Construction differences between buildings could account for some thermal comfort issues when cross referenced with survey responses. For a follow-up study on thermal comfort, the researchers may choose to eliminate construction quality as a factor by analyzing one building that has both large and medium window areas.

Assessing the effect of thermal comfort implications from large windows can be a challenge, as each condominium uses its space differently, with varying time spent near windows. If the study were to focus more heavily on thermal comfort, the researchers would ask the occupant to sit near the window for an extended period (e.g., during the interview or survey), and would also ask the participant to rate their thermal comfort both before and after sitting near the windows. To improve this study, heating and cooling energy use should be collected for each condominium unit. This would allow for conclusions on energy use associated with large glazing.

Overall, this study accomplished its goals of identifying relevant comfort issues in highly-glazed Canadian condominiums as well as giving context and direction for future research. The survey responses showed the greatest difference between the studied groups, compared with the other data sources. The results of the survey determined that occupants with large window areas reported uneven interior temperature distributions more frequently, and windows were a source of discomfort in winter.

8.8.2 Energy Consumption in Residential Buildings and Occupants' Behavior. An Investigation in Mediterranean Climatic Conditions

Marilena De Simone[1], Cristina Carpino[1], Dafni Mora[2]

[1]Department of Mechanical, Energy and Management Engineering DIMEG—University of Calabria —Italy

[2]Centro de Investigaciones Hidráulicas e Hidrotécnicas (CIHH) Universidad Tecnológica de Panamá Avenida Domingo Diaz, Panama City, Panama

The purpose of the study was to evaluate which factors affect the energy performance of a housing stock located in Mediterranean climatic conditions. The residential sector was used to test the relative roles of socioeconomic and behavioral aspects of occupants, as compared with climatic and physical building characteristics. Data collection started in 2012, and the participants were the families of engineering students of the University of Calabria. The investigated area was the Calabria Region located in Southern Italy. The sources of information were questionnaires, interviews, energy bills, and statistical data available at the regional and national level.

Several variables were considered to best understand occupant behaviors and related energy consumption. The survey consisted of 63 questions divided into six groups of parameters: general information, household energy consumption, conditioning and domestic hot water, domestic appliances, occupants' behavior, and renewable energy systems.

Based on this study, the questionnaire presented a few issues. In the future, the research team will take care with initial survey development, verification, question wording, and answer selections (see Sect. 8.3.1) to alleviate some of the low response rate issues.

The experience was fruitful and important information was collected regarding the technical characteristics of buildings and energy consumption. On the other hand, many interviewees did not answer questions about income because they did not want to reveal personal information. Also, some questions were without coherent answers as revealed by data processing. For these reasons, it is necessary to improve the questionnaire by means of more specific questions, while also adding details about the occupants' behavior with another type of question (e.g., closed response format, Likert scale, and forced-choice response formats instead of open ended questions). Also, it would have been useful to have a larger sample, and in the future, the researchers would prefer to use electronic surveys instead of the paper forms.

The surveys should be adapted to each region, and should consider lifestyles, attitudes, personal background, and types of heating/cooling systems that people use. Another important aspect to keep in mind is the language. It is necessary to propose clear questions and translate the questionnaire to native speech. Better understanding the territorial context could also be useful to gather information for later analysis.

8.8.3 *On the Behavioral Effects of Residential Electricity Submetering in a Heating Season*

H. Burak Gunay[1], William O'Brien[1], Ian Beausoleil-Morrison[2], Isis Bennet[1]

[1]Department of Civil and Environmental Engineering, Carleton University, Ottawa, Canada

[2]Department of Mechanical and Aerospace Engineering, Carleton University, Ottawa, Canada

A quarter of Canadian households do not directly pay for their energy bills. Instead, the landlord, property manager, or the condominium corporation are responsible for energy bills. As such, there is no monetary incentive for tenants to make energy-saving retrofits or modify their behaviors to save energy. In this study, the researchers conducted a survey with 20 participants who were responsible for their energy bills and 20 participants who were not responsible for their energy bills. Two previous studies (Becker et al. 1981; Seligman et al. 1979) formed the basis for the questionnaire used in this study. A few new questionnaire items were added after consulting a population scientist from University of Ottawa (Dr. Andrea Perna). In addition to the survey, temperature loggers were placed in the living rooms and bedrooms of these participants' apartments during the heating season in Ottawa, Ontario.

Observations revealed that the occupants who were responsible for their energy bills were more diligent and active in controlling indoor air temperature. These occupants chose to heat different areas of their units at different times. On the contrary, the occupants, who were not responsible for their energy bills rarely adjusted their thermostat settings or setback their thermostats. The occupants who were not responsible for their energy bills maintained their apartments about 2 °C higher than their counterparts who were responsible for their energy bills.

Furthermore, the participants living in submetered apartments reported being more sensitive to engaging in personal adaptive behaviors (e.g., wearing heavier clothes) and were more sensitive about their thermal comfort (e.g., reducing their thermostat settings). Participants living in submetered apartments were found to be more influenced by health concerns related to indoor temperature compared to the participants living in the bulkmetered apartments.

If the study were repeated, the researchers would monitor the occupants' thermostat keypress behavior as well as the indoor temperature. In addition, the researchers would log the window states (open or closed), as well, because a considerable fraction of the occupants reported keeping their windows open for airing purposes during the heating season. Occupants' window use behavior in residential buildings in cold climates such as Eastern Canada has not been studied. The researchers would also choose to collect the suite level electricity use data (from the breaker panels in the bulkmetered units). In Gunay et al. (2014), the survey results and temperature data were divided, and then analyzed independently. In the future, as a method for recruiting a higher number of participants to increase the overall sample size, the team would reduce the number of items in the survey. This would permit the research team to conduct a comprehensive statistical analysis between the survey responses and the field measurements.

In the study, the researchers intended to demonstrate that the utilities included in rent or condominium fees led to a remarkable extravagance in heating and cooling equipment use. The research team provided some preliminary evidence of this phenomenon and raised awareness of the importance of utility submetering in occupants' behavior patterns. The ultimate objective was to provide convincing statistical evidence to policy makers to encourage submetering in apartment and condominium buildings to become mandatory. See Gunay et al. (2014) for additional details.

8.8.4 A Norwegian Survey Story: The Use of Qualitative Methods

Thomas Berker, Department of Interdisciplinary Studies, Centre for Technology and Society, Norwegian University of Science and Technology, Trondheim, Norway

The Department of Interdisciplinary Studies of Culture has been involved in qualitative research on building use(rs) since 1995. Researchers mostly conduct semi-structured interviews, where some questions are pre-determined but other topics are left open for discussion. Another useful tool is focus group discussions, where 5–10 individuals are selected according to given criteria (e.g., where they are based in the building) and encouraged to discuss certain topics. Surveys are typically used for quick checks of indoor environmental qualities to register if there are symptoms for problems (e.g., headaches, dry air, concentration problems).

In this team's experience, the strength of qualitative data is that it gives the researcher access to uncensored occupant opinions and provides additional insights into incorrect uses of buildings (such as taping over the light sensor, or bringing a space heater to work). Qualitative data also enable researchers to understand why occupants act in certain ways. Interviews tend to reveal a certain rationality in the occupants' actions.

Alternatively, qualitative research's weakness is its inability to produce data that then can be used directly to produce models of occupant behavior. However, it can be used to make sense of data produced in other ways (sensors, surveys). Often anomalies in energy consumption (and other data) are easily explained if the occupants are asked directly ("that spike in energy consumption during the whole night - that was when we had a party at the office").

Current and future research efforts include experimenting with a building, which is equipped with many types of sensors, and another where the research team is looking for better ways to triangulate quantitative and qualitative data (i.e., use multiple data sources to best understand the research problem at hand).

8.8.5 Occupants' Behavior Patterns for Air-Conditioning, Windows, and Lighting

Xiaohang Feng, Tsinghua University, Beijing, China

This study used questionnaire surveys to better understand occupants' behavior patterns on air-conditioning, windows, and lighting. Participants were asked to fill in a form to describe air-conditioning use behaviors. For this study, instruments were installed in several households and offices to record the indoor thermal conditions and the air-conditioning state. Based on the collected data, occupant behavior models were built. The questionnaire survey was used to discover and understand the stimuli for certain behavior in the studied buildings.

The diversity of occupant behavior was reflected in this study. The stimuli for occupant behavior was different from person to person. Another finding was that the questionnaire was difficult for people to understand (based on feedback), or their understanding was unexpected. The stimuli from the questionnaire were used in the model. Another goal of the large-scale questionnaire survey was to identify the distribution of occupant behavior, or some kinds of typical occupant behaviors, as a simplified representation of the occupant behavior diversity. The prerequisite for this study was to understand the quantitative parameters in the model from the questionnaire, which is rather difficult. There are still some defects in the design of the questionnaire used, such as the lengthiness, the ambiguity of the question or the options, and the lack of validity from a sociological view (lack of theory implementation). In the future, the research team would elect to re-design the questionnaire together with experts from psychology and sociology before redistributing the survey again.

8.9 Additional Considerations for Occupant Behavior-Related Surveys

Cross-sectional surveying of occupants' behavior can also support sustainable building designers and consultants to learn more about the human side of their projects. Especially in the case of low and net-zero energy buildings, more and more designers are interested in post-occupancy surveying—for example, to get feedback from occupants on the real-world performance of their energy-efficient designs. This type of data acquisition is essential to fine-tune building operation and to give feedback and lessons to designers to make optimized and energy-efficient and cost-effective developments. There are established POE surveying frameworks for this purpose, such as the well-known CBE Occupant IEQ Survey (CBE 2017); also see Hauge et al. 2011). However, these surveys primarily measure comfort and general satisfaction of building occupants in current practice. In the future, it will be important to include occupant behavior-related questions into existing survey databases, or develop a set of well-developed and well-vetted questions on this topic for use by researchers. By including such questions, researchers can learn more about occupant behavior patterns, drivers, and use of control options in general, have larger sample sizes worldwide for research purposes, and empower designers, facility managers and owners to understand how and why people behave in a certain way in their buildings.

8.9.1 Informed Consent for Online Surveys

One additional factor to consider when conducting surveys is to ensure ethical guidelines are followed. When sending surveys to building occupants, it is of the utmost importance to include language for informed consent. Oftentimes, the first page of an online survey, for example, will include informed consent, and if the participant selects "they agree," then the survey will advance to the questions; if they select "do not agree" in reference to the informed consent language, then the survey will not appear. Alternatively, a university's ethics board will typically accept clicking into a survey link in an e-mail as informed consent, provided this is explained in the e-mail message. These are simple ways to ensure that informed consent information is provided to each participant. Additional ethical considerations are thoroughly covered in Chap. 11. A brief example of informed consent language for an online survey is provided below.

8.9.2 Example of Informed Consent Statement for Survey Study (Day 2014)

Introduction

You have been invited to participate in this survey based on ______. The research team would greatly appreciate your participation in a short (_____) minute survey about __________.

General Information

The information on this page is intended to help you understand exactly what we are asking of you so that you can decide whether or not you would like to participate in this study. Please read this consent form carefully before you decide to proceed with the survey. If you decide to not participate, it will not be held against you in any way. You may exit out of the survey at any time.

Privacy and confidentiality

Your participation in this survey is completely voluntary, and your responses will not be shared with your employers. Your answers will be kept confidential and your identity protected. All data will be transmitted by a secure, encrypted internet connection and stored in a password protected file. The ***insert university/institution here** Office of Research Assurances has determined that this study satisfies the criteria for Exempt Research at 45 CFR 46.101(b)(2) [***this statement will change based on different studies/universities/Ethic Review Board protocols.***]

Potential harms/benefits

There are no known harms associated with your participation in this research (if there are potential harms, state them here). **State how they will or will not benefit from participation.**

If you agree to the terms listed above, please proceed to the survey (***if they click onto the next page, that serves as their informed consent***). Thank you in advance for your time and cooperation. Please be honest with your answers. Your responses are extremely valuable to our research! If you have any questions, please do not hesitate to ask.

Thank you,

Contact information here

***Note: each university may have different requirements for what this consent page should look like in a survey. Obtain approvals based on individual requirements as necessary.*

8.10 Conclusion

Ultimately, surveys can be used to ascertain a great deal of information from building occupants. Since these behaviors are self-reported, it is helpful to triangulate results with physical measurements, interviews, and/or observations when possible (see Chap. 5). Both surveys and interviews can be performed either in situ or remotely. In situ studies are advantageous because they allow the researcher to further investigate observations or issues that arise from survey or interview responses on site. Alternatively, online-based surveys can also be tremendously beneficial, especially if there are budget, time, or travel constraints; this method may also aid in increasing the overall sample size. Each type of survey has advantages and disadvantages, which should be carefully weighed and considered while designing the study.

Overall, survey research should be conducted with rigor and thoughtfulness. It is important to carefully craft survey questions, order the questions logically, and ensure that the questions are truly providing desired answers relevant to the research question, constructs, and theories. Questions can be both closed-ended or open-ended, and each type can yield different types of results. Consistency among measures and scales is also important. As discussed previously, different types of survey modes, the layout, the length of the survey, and even the font size and typeface selected can all affect the response rate.

This chapter provided guidance for survey and interview development in occupant behavior research; however, for many of the methods mentioned in this chapter (surveys, interviews, focus groups, etc.), it is necessary to seek out additional resources for more detailed explanations of appropriate and valid data collection methods and associated analyses. In addition, respected survey methods and resources should still be appropriately sought out for the following topics: longitudinal survey methods, additional available tools for survey delivery, incentives for increasing survey response rate, valid methods of evaluating surveys, proper triangulation of survey responses with additional, and more.

References

Babbie E (2004) The practice of social research, 10th edn. Wadsworth/Thomson Learning, Belmont, CA

Becker LJ, Seligman C, Fazio RH, Darley JM (1981) Relating attitudes to residential energy use. Environ Behav 13:590–609

Berry, RSY (1999) Collecting data by in-depth interviewing. Presented at the British educational research association annual conference, University of Sussex at Brighton

Center for the Built Environment (2017) Occupancy indoor environmental quality (IEQ) survey. http://www.cbe.berkeley.edu/research/survey.htm

Corral-Verdugo V, Figueredo AJ (1999) Convergent and divergent validity of three measures of conservation behavior: The multitrait-multimethod approach. Environ Behav 31(6):805–820

Creswell JW (1998) Qualitative inquiry and research design: choosing among five traditions. Sage Publications, Thousand Oaks, CA

Creswell JW (2014) Research design: qualitative, quantitative, and mixed methods approaches, 4th edn. Sage Publications, Thousand Oaks, CA

Creswell J, Plano Clark V (2011) Designing and conducting mixed methods research, 2nd edn. SAGE Publications, Inc., Thousand Oaks, CA

Day JK, Gunderson DE (2014) Understanding high performance buildings: the link between occupant knowledge of passive design systems, corresponding behaviors, occupant comfort and environmental satisfaction. Build Environ 84:114–124. doi:10.1016/j.buildenv.2014.11.003

Day JK (2014) Occupant training in high performance buildings: an assessment of environmental satisfaction, learning, and behaviors in high performance buildings. Dissertation, Washington State University.

Day J, Theodorson J, Van Den Wymelenberg KG (2012) Understanding controls, behaviors, and satisfaction in the daylit perimeter office: a daylight design case study. J Int Des 31(1):17–34

De Leeuw ED, Hox JJ, Dillman DA (2008) International handbook of survey methodology. European Association of Methodology/Lawrence Erlbaum Associates, New York

DeVellis RF (2012) Guidelines in scale development. In: DeVellis RF (ed.) Scale development: theory and applications, 3rd edn. Sage, Thousand Oaks, CA

Dillman DA (2000) Mail and internet surveys: the tailored design method, 2nd edn. John Wiley & Sons, Inc., New York

Dillman DA, Smyth JD, Christian LM (2009) Internet, mail and mixed-mode surveys: the tailored design method, 3rd edn. Wiley, Hoboken, NJ

Doyle JK (2004) Introduction to interviewing techniques. In: Doyle JK (ed) Handbook for IQP advisors and students. Worcester Polytechnic Institute, Worcester, MA, pp 26–27

Fink A (2012) How to conduct surveys, 2nd edn. Sage, Thousand Oaks, CA

Fowler FJ (2009) Survey research methods, 4th edn. Sage, Thousand Oaks, CA

Gliner JA, Morgan GA, Leech NL (2011). Research methods in applied settings. Routledge, Abingdon, Oxon

Golafshani N (2003) Understanding reliability and validity in qualitative research. Qual Rep 8(4): 597–606. http://nsuworks.nova.edu/tqr/vol8/iss4/

Gunay HB, O'Brien W, Beausoleil-Morrison I, Perna A (2014) On the behavioral effects of residential electricity submetering in a heating season. Build Environ 30(81):396–403

Haldi F (2013) A probabilistic model to predict building occupants' diversity towards their interactions with the building envelope. Paper presented at the BS2013 13th conference of international building performance simulation association, Chambéry, France, 26–28 Aug 2013

Haldi F, Robinson D (2008a) On the behavior and adaption of office occupants. Sol Energy Build Phys Lab 43:2163–2177

Haldi F, Robinson D (2008b) On the behavior and adaptation of office occupants. Build Environ 43(12):2163–2177. doi:10.1016/j.buildenv.2008.01.003

Hardesty DM, Bearden WO (2004) The use of expert judges in scale development: Implications for improving face validity of measures of unobservable constructs. J Bus Res 57:98–107

Hauge Å, Thomsen J, Berker T (2011) User evaluations of energy efficient buildings: Literature review and further research. Adv Build Energy Res 5:109–127. doi:10.1080/17512549.2011.582350

Herkel S, Knapp U, Pfafferott J (2008) Towards a model of user behaviour regarding the manual control of windows in office buildings. Build Environ 43:588–600

Humphreys MA, Nicol JF (2002) Adaptive thermal comfort and sustainable thermal standards for buildings. Energ Build 34:563–572

Konis K (2013) Evaluating daylighting effectiveness and occupant visual comfort in a side-lit open-plan office building in San Francisco, California. Build Environ 59:662–677. doi:10.1016/j.buildenv.2012.09.017

Langevin J, Gurian P, Wen J (2013) Reducing energy consumption in low income public housing: Interviewing residents about energy behaviors. Appl Energy 102:1358–1370. doi:10.1016/j.apenergy.2012.07.003

Liamputtong P (2011) Focus group methodology: principles and practice. Sage, London

Nicol JF (2001) Characterising occupant behaviour in buildings: towards a stochastic model of occupant use of windows, lights, blinds, heaters and fans. Paper presented at the Building Simulation, Rio de Janeiro, 13–15 Aug 2001

Nunnally J, Bernstein IH (1994). Psychometric theory, 3rd edn. McGraw-Hill, New York, NY

Schwarz N (1996) Cognition and communication: judgmental biases, research methods, and the logic of conversation. Erlbuam, Mahwah, NJ

Seligman C, Kriss M, Darley JM, Fazio RH, Becker LJ, Pryor JB (1979) Predicting summer energy consumption from homeowners' attitudes. J Appl Soc Psychol 9:70–90

Sudman S, Bradburn NM (1982) Asking questions: a practical guide to questionnaire design. Jossey Bass, San Francisco

Teddlie C, Tashakkori A (2003) Major issues and controversies in the use of mixed methods in the social and behavioral sciences. Handbook on mixed methods in the behavioral and social sciences. Sage Publications Inc, Thousand Oaks, CA, pp 3–50

Teddlie C, Tashakkori A (2009) Foundations of mixed methods research: integrating quantitative and qualitative techniques in the social and behavioral sciences. Sage Publications Inc, Thousand Oaks, CA

Yun JY, Steemers K (2008) Time-dependent occupant behavior models of window control in summer. Build Environ 43:1471–1482 The Martin Centre for Architectural and urban Studies

Yun JY, Steemers K (2009) Corrigendum to thermal performance of a naturally ventilated building using a combined algorithm of probabilistic occupant behaviour and deterministic heat and mass balance models. Energ Build 41:489–499

Chapter 9
Validation and Ground Truths

Da Yan, Chuang Wang, Xiaohang Feng and Bing Dong

Abstract It is essential to ensure the validation of measurements and the reliability of the collected data. This chapter discusses several topics related to measurement validation and ground truth in occupancy and occupant behavior observations. It introduces the basic concept of measurement quality and calls for attention to the measurement of occupancy and occupant actions. It provides general guidelines for verifying and validating the reliability of collected data. It also offers suggestions for how to construct ground truth data. In this chapter, questions about measurement validation and ground truth are raised and the particularities of occupancy and occupant behavior observations are discussed.

9.1 Introduction

In previous chapters, the readers were introduced to possible measurands, measurement means, and scenarios in occupancy and occupant behavior observations. Measurands that were discussed included: (1) occupancy, (2) occupant actions, (3) indoor and outdoor environmental conditions, and (4) occupants' perception of the environment. Measurement means included instrument monitoring

D. Yan (✉) · C. Wang · X. Feng
Building Energy Research Center, Tsinghua University, Tsinghua Park #1,
100084 Beijing, China
e-mail: yanda@tsinghua.edu.cn

C. Wang
e-mail: wangchuang@tsinghua.edu.cn

X. Feng
e-mail: fengxh12@mails.tsinghua.edu.cn

B. Dong
Built Environment Science and Technology (BEST) Lab,
Department of Mechanical Engineering, The University of Texas at San Antonio,
One UTSA Circle, San Antonio 78249, USA
e-mail: bing.dong@utsa.edu

A. Wagner et al. (eds.), *Exploring Occupant Behavior in Buildings*,
DOI 10.1007/978-3-319-61464-9_9

(e.g., sensor-based) and questionnaire surveys or interviews. Measurement scenarios included in situ and laboratory.

This chapter maintains that regardless of what is measured and how, there are basic concerns to address. Are the measurement procedures and collected data reliable? How can these be validated? How much uncertainty is there in the measured values? To begin to answer these questions, this chapter discusses several topics related to measurement validation and ground truth in occupancy and occupant behavior observations. Although there are general concepts, methods, and procedures for measurement validation and collecting ground truth data, they have not been fully discussed in the context of occupant research. It is important to do so in order to ensure the quality of occupancy and occupant behavior observations.

Generally speaking, there are three basic questions for occupancy and occupant behavior measurements:

1. How do we know that the collected data is reliable/truthful?
2. What steps should researchers take to ensure data reliability?
3. How and for which purpose can we construct ground truth datasets for occupant behavior?

To answer these questions, the following key topics will be discussed in this chapter:

- Concepts of measurement validation and ground truth;
- Measurement for occupancy and occupant behavior;
- Verification and validation of measurement procedures; and
- Constructing a ground truth dataset.

Questions and issues particular to occupancy and occupant behavior observations are also raised throughout this chapter. Finally, suggestions are made regarding how to validate occupancy and occupant behavior measurements, as well as how to construct ground truth data.

9.2 Basic Concepts of Measurement Quality

The sections that follow introduce the basic concepts, terms, and methods of measurement validation and ground truth; these are then discussed in the context of occupant behavior research.

9.2.1 Basic Terms of Measurement Quality Performance

In the context of any measurement system, terms such as "trueness", "precision", "accuracy", etc., have been used to describe the quality of measurements. Figure 9.1 shows the interrelations between different error types (systematic, total,

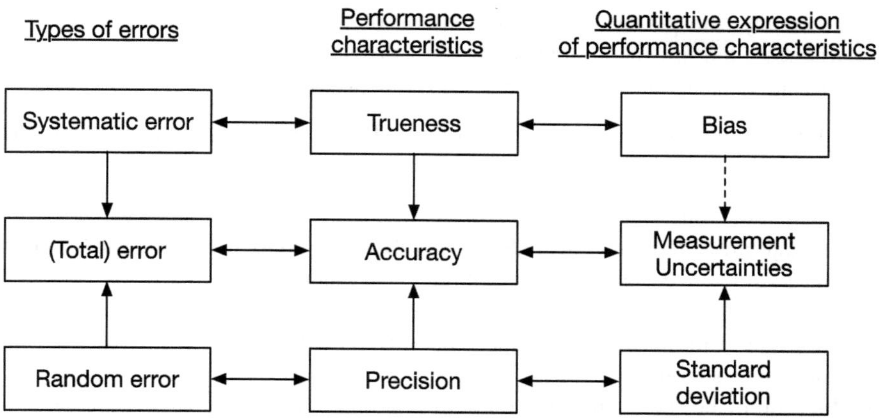

Fig. 9.1 Measurement system variance and uncertainties (adopted from Menditto et al. 2006)

random), their corresponding performance characteristics (trueness, precision, accuracy), and the parameters for quantitatively expressing these performance characteristics (standard deviation, bias, measurement uncertainty). These terms and relationships are unpacked throughout this section.

To understand the concept of measurement error, two terms must first be introduced: true value and measured value. True value cannot be found by experimental means and is defined as the average of measured values derived from a sequence of repeated measurements. In contrast, measured value is a single measurement of an object that is intended to be as accurate as possible. The difference between these two measurement concepts is referred to as the error, or total error. This error can be divided into two types: random error, which refers to having a different magnitude and sign in the case of repeated measurements; and systematic error, which refers to having the same or systematically changing magnitude and sign in the case of repeated measurements. Errors cannot be known exactly, but can be estimated.

The International Vocabulary of Metrology (BIPM 2012) defines three performance characteristics as follows:

- Trueness: the performance characteristic of the systematic error, i.e., how closely aligned the average value obtained from a number of measurement results and the true value are. For determining trueness, a reference value needs to be know, (sometimes called a "conventional true value"). The reference value usually has a small uncertainty compared to the true value.
- Precision: the performance characteristic of the random error, i.e., the closeness of agreement between the results obtained from multiple repeated measurements. For determining precision, usually multiple repeated measurements must be taken on the same or similar objects under specified conditions. Precision can also refer to repeatability, intermediate precision, and reproducibility, depending on the conditions under which precision is determined.

- Accuracy: the performance characteristic of the total error. It embraces both trueness and precision, indicating the closeness of agreement between the measured value and the true value of the measurand.

The International Vocabulary of Metrology (BIPM 2012) further describes how performance characteristics can be quantified:

- Bias: the numerical expression of trueness, i.e., the difference between the average measured value of multiple repeated measurements on the same sample and its reference value (or conventional true value).
- Standard deviation: the numerical expression of precision, i.e., the amount of variation in the results from repeated measurements on the same sample using the same method.
- Measurement uncertainty: the numerical expression of accuracy, characterizing "the dispersion of the quantity values being attributed to a measurand". It is a combination of bias and standard error.

These terms are used to assess measurement instruments' performance and use (i.e., methods or procedures). High quality measurements require high accuracy and small uncertainty; however, due to the limits of feasibility and cost, investigators should define suitable requirements of measurement quality for each specific research project.

Looking at the terms, it should be noted that accuracy and precision must not be used interchangeably. As defined above, precision refers to how close together and how repeatable the measured quantity values are, while accuracy refers to how close a measured quantity value of a measurand is to the true quantity value. Refer to Chaps. 3 and 8 for more details on these terms.

9.2.2 *Validation and Verification of Measurement Methods*

Validation and verification are procedures that are used to determine realistic expectations and confidence. According to the International Vocabulary of Metrology (BIPM 2012), verification is "provision of objective evidence that a given item fulfills specified requirements", whereas validation is "verification, where the specified requirements are adequate for the intended use".

Method validation is a particular type of validation described as "the process of defining an analytical requirement, and confirming that the method under consideration has performance capabilities consistent with what the application requires" (De Biévre et al. 1998). The validation of a measurement method should establish the performance characteristics (e.g., trueness, precision, accuracy) and the limitations (e.g., scope, working range) of the method and determine whether it fulfills the particular requirements for its intended use (ISO 2005a, b; Theodorsson 2012). Method validation should be performed to differing extents depending upon a method's intended use (Theodorsson 2012).

Method verification includes "procedures to test to what extent the performance data obtained by manufactures during method validation can be reproduced with the environments of end-users" (Theodorsson 2012). If the measurement method (including sensors, instruments, and procedures) is established by a reliable source that has carried out appropriate method validation and provides detailed validation results, a method validation is then not required and method verification is sufficient (Theodorsson 2012).

Validation and verification of measurement methods are the primary guarantee of the quality of measurement results and need to be performed before performing a field study. If these processes are carried out *after* field study and it is found that the measurement system needs improvement, the earlier field study results are not valid. The field study should be performed again after improving the measurement system.

9.2.3 Ground Truth in Measurements

The terms used to describe the concept of ground truth vary from field to field. Generally, constructing ground truth refers to the method of measurement that, regardless of cost, most accurately and precisely measures that which one is trying to measure. Thus, ground truth is typically collected by direct observation. The term was coined in geological/earth sciences to describe the validation of data by going out in the field and checking "on the ground" (Gupta 2013). It has been adopted in other fields, such as biometrics and computer vision, to refer to the underlying absolute state of information or express the notion of data that is "known" to be correct (Anonymous 2011; Krig 2014).

In theory, ground truth data perfectly represent the thing being measured with no error or uncertainty; in practice, this is simply not possible (BIPM 2012). It would be more accurate to say that constructing ground truth refers to the process of gathering data leading to a set of measurements known to be the *most accurate using the current instruments.* It is important to be aware that all measurements are, to some extent, an approximation of the underlying concept being measured. For this reason, the term "ground truth error" is also in wide use, illustrating the fact that what is "known" is not always entirely correct. This is why some fields prefer the term "best available measure".

Due to its overall—if slightly imperfect—accuracy, precision, and reliability, ground truth data are regarded as the "gold standard" or a solid baseline to check the validity of a theoretical model, algorithm, or technical means of measurement. In some areas, such as computer vision, there have been a number of public ground truth datasets contributed by researchers or institutions; these have been established for various purposes and have served as fundamental materials for the development, calibration, training, and testing of new models or algorithms (Krig 2014).

In the context of occupant behavior observations (e.g., monitoring and surveys), there is still considerable work to be done in terms of agreeing what constitutes ground truth data for different measurands. There are few examples in the literature that explicitly touch on this gap (Haldi and Robinson 2008, 2009). On the one hand, the credibility and reliability of data used in some literature on occupant behavior model development and validation are not sufficiently described. For example, a common approach to determine a room or dwelling's occupancy status (presence/absence, number of occupants, etc.) is by the monitored CO_2 concentration. Although this method is preferential to disregarding occupancy completely, it is known to result in notable uncertainties (Andersen et al. 2013). On the other hand, the sharing of open occupant behavior datasets is rare in this research community, which has limited the growth of occupant behavior research. One probable reason for this situation is the complexity and particularity of occupant behavior observations, as well as data protection issues, for example as described in Chap. 11.

9.3 Measurement of Occupancy and Occupant Behavior

The target of occupant behavior research is to understand and quantify occupant behaviors, including occupancy and occupant actions. To accomplish this goal, related variables need to be obtained through the most accurate approaches. As mentioned in Sect. 9.1 and as shown in Fig. 9.2, the two general approaches for observing occupant behaviors are (1) instrument monitoring, and (2) questionnaire surveys or interviews. These approaches can be utilized in both in situ and laboratory scenarios. Readers can refer to Chaps. 6, 7, and 8 for further information on these scenarios.

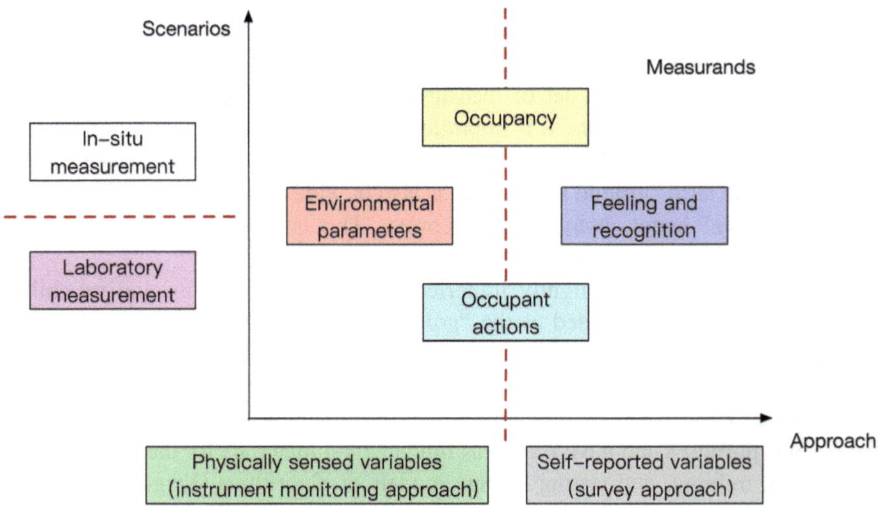

Fig. 9.2 Categories of measurands for occupancy and occupant behavior

From the viewpoint of verification and validation, regardless of the measurement approach or scenario, there are two main types of information that can be collected: (1) physically sensed variables, which can be measured and monitored via instruments; and (2) reported variables, which can be derived from occupants through interviews, surveys, and manual observation (Cozby 2001).

Physically sensed variables include:

- Occupancy—for example, presence/absence, number of occupants, occupant location, occupant identity;
- Occupant actions—for example, turning on/off lights, turning on/off air conditioners, turning on/off computers, opening/closing window, adjusting curtains/blinds; and
- Indoor and outdoor environment—for example, indoor temperature, humidity, illuminance, CO_2 concentration, and outdoor temperature, humidity, noise, solar radiation intensity, wind speed, wind direction.

Reported variables include:

- Occupant perception and mood—for example, thermal comfort, satisfaction degree, attitude, self-reported habits; and
- Occupancy and occupant actions. Some questionnaires are designed to ask the pattern information about occupant behaviors (Ren et al. 2014, Wang et al. 2016)—for example, first arrival time, last departure time, when or on what condition the occupant would turn on/off air conditioners. For these variables, a questionnaire survey approach and an instrument monitoring approach can validate each other.

In the following sections, physically sensed variables and reported variables which are possibly involved in occupant behavior observations are further discussed alongside issues that should be paid attention to in measuring both variable types.

9.3.1 Physically Sensed Variables

To measure physically sensed variables like occupancy, occupant actions, and environmental parameters, various instruments or sensors can be deployed for observation and monitoring. The reader can refer to Chaps. 4 and 6 for the principles of sensing techniques and their typical applications to in situ monitoring studies.

Before beginning a measurement, it should be determined which variables should be measured and what sensors or instruments should be used. Based on existing literature, Table 9.1 summarizes some of the possible monitoring variables in an occupant behavior experiment and the applicable instruments or sensors for these measurands. It should be noted that not all variables will be measured in one experiment; rather, researchers should choose the most suitable variables for their purposes—for example, to observe presence/absence status (Dong and Andrews

Table 9.1 Physically sensed parameters and instruments

	Parameters	Instrument
Occupancy	Presence/absence Number of occupants Location Identity (i.e. who is present)	Infrared motion sensor Infrared counting sensor RFID sensor Video monitoring system
Occupant actions	Air conditioner (AC) on/off Light on/off Window open/closed Curtain open/closed Degree of window/curtain opening	Power meter Temperature meter Magnetic sensor Angle/distance meter
Environment	Temperature Humidity CO_2 concentration Illuminance Solar radiation Wind speed Wind direction	Temperature sensor Humidity sensor CO_2 sensor Illuminometer Radiometer Wind speed meter Wind direction vane

2009; Chang and Hong 2013, Timilehin et al. 2015); to observe occupancy status, window open/closed state, and related environment parameters for window opening behavior (Nicol and Humphreys 2004; Nicol et al. 2006, Yun et al. 2008; Haldi and Robinson 2008, 2009; Fabi et al. 2012; Andersen et al 2013); or to observe similar items for curtain/blinds adjusting behavior (Haldi and Robinson 2010; Correia da Silva et al. 2013), lighting behavior (Hunt 1979; Reinhart and Voss 2003), and air conditioning behavior (Tanimoto and Hagishima 2005; Schweiker and Shukuya 2009; Ren et al. 2014).

To obtain accurate results, instruments or sensors with high trueness and precision, as well as stability and linearity must be used. Linearity refers to consistency over the measurement range. Stability is the measurement variation over time. A sensor or instrument should be calibrated regularly to address drift and ensure its stability. Stability and linearity errors are to be addressed during calibration of the instrument. Additionally, the layout and installation of instruments or sensors should be carefully carried out as Chaps. 4 and 6 discuss.

For the measurement of environmental parameters, many reliable sensors which are easy to deploy and use can be obtained on the commercial market. For example, the precision of a conventional temperature sensor can be less than 0.5 K; more precise (e.g., ± 0.1 K) temperature sensors are also available. Environmental instruments should be calibrated regularly; their inaccuracy should be checked (and, if necessary, corrected) and it should be ensured that they are in good working condition, especially before an experiment. Calibration is usually carried out by scientific-grade instruments or in advanced facilities. In this context, the results of scientific-grade instruments would constitute ground truth data, i.e., be the best available measure for that measurand.

That said, even if the instrument itself is in good condition and has a very high accuracy, the reliability of the values obtained depends very much on where the instrument is placed. For example, it could be quite different to measure illuminance at a placement close to window, below the ceiling, at floor level, etc. This issue should be carefully considered in the research design. The layout and installation of instruments will depend on the specification of the measurand. For example, if the measurand is the indoor temperature of a room, what exactly is to be measured should be carefully defined.

To illustrate, there are several different types of temperature (ambient, radiant, operative, etc.) and temperatures vary over space and time within a room. If the measurand is the room's mean ambient temperature, then an individual temperature sensor logging half-hourly data will almost certainly be inaccurate spatially and imprecise temporally. While logging every second with a single sensor will decrease temporal imprecision, it will not reduce spatial inaccuracy. Ground truth data in this context would be a dense spatial array of highly accurate (i.e., well calibrated) and precise (say, $\pm$ 0.1 K) sensors sampling at a temporal frequency relevant to the problem under investigation. In this example, it should be noted that the accuracy and precision of the ground truth data are not only related to the precision of the sensors, but also determined by the design of the sensor array; thus, the instrument in this context is the sensor array, not the sensor. In practice, obtaining a good approximation of the mean ambient temperature requires the careful placement of one or more sensors in the area. In many fields, such as thermal comfort research, the locations (including heights) of the sensors are stipulated in multiple standards (see Sect. 9.4).

The measurement of occupancy and occupant actions is far less mature than traditional environmental measurement and thus faces more challenges. Indeed, in many areas of occupant behavior research no measurement standards exist. Accuracy of sensors in real environments tends to be poorer than for laboratory tests because of constraints on positioning, the possibility of obstruction, unquantified sources of noise (e.g., heat gains), etc. For example, existing occupancy sensors are not yet sufficiently accurate or reliable to substitute for direct observation (manually or by video monitoring), but video systems usually pose ethical concerns in real houses or offices due to privacy. In addition, it is very time-consuming to manually view and code this kind of data.

Of all the existing measures to collect data on occupancy and occupant actions, a video monitoring system is usually regarded as the best measure for constructing ground truth, as it can record all activities of investigated occupants. However, this approach is usually high cost and involves issues of privacy—indeed, it is sometimes entirely infeasible due to these and other practical constraints. In those cases, there are alternative approaches for some measurands. For instance, if there is a lack of occupancy sensors, some researchers use the indoor CO_2 concentration monitored by a CO_2 sensor to estimate the number of occupants in a room (Andersen et al. 2013). In some instances, due to the difficulty of deploying power meters, some researchers have used the temperature as a proxy for air conditioner state (Ren et al. 2014). These and other alternative approaches are practical, but they do not constitute gathering ground truth data and should be carefully validated by direct observation. All measurement methods, including sensors, procedures, and instruments should be tested for accuracy and precision against best available measures as well.

Although there is much technological development and standardization work to be done with respect to observing occupancy and occupant actions, the best practice for measuring occupants' physically sensed variables should be to use instruments with validated accuracy and reliability, and to collect in situ measurements whenever possible.

9.3.2 Reported Variables

To study occupants, reported variables related to occupant behaviors, social science research methods (e.g., questionnaire surveys, time use surveys, self-record surveys, diaries, face-to-face interviews, etc.) can be employed. The reader can refer to Chaps. 8 and 11 for more information on the principles of survey techniques and their typical applications to occupant behavior studies. In short, investigators ask questions and get answers about what they want to know from people in the environment of research interest. Table 9.2 summarizes some of the usual parameters and means of occupant behavior data collection.

The survey approach is a traditional and popular approach adopted in the social, psychological, and behavioral sciences. Surveys help to observe the internal processes of occupancy and occupant actions which cannot be captured by sensors. This approach is much less costly than in situ monitoring approaches and is thus suitable for large-scale research. In an in situ monitoring study, surveys can act as a supplementary measure whereby participants declare their behaviors and clarify the cause of their occupancy or actions. Survey results can thus help to illustrate and explain monitoring results; in return, monitoring results can validate the credibility of the survey results, resulting in a more reliable analysis.

Overall, the survey approach is not as widely used as in situ monitoring in occupant behavior research. One known example of a large-scale survey is Nicol and Humphreys (2004), a study about adaptive thermal comfort and window-opening behaviors in naturally ventilated buildings. The survey included questions about comfort responses, the clothing and activity of the subjects, the

Table 9.2 Self-reported parameters and measurement means

	Parameters	Means
Occupant perception and mood	Thermal comfort (TSV) Degree of satisfaction Narration of habits Driving factors Attitude or expectations Self-record of activities Presence/absence AC/window states	Questionnaire Survey Time use survey Self-record survey Diary (to report daily activity) Face-to-face interview

indoor and outdoor temperature and humidity, and the use of various controls (windows, doors, lighting, fans, heaters, and, where applicable, AC). The final database comprised 5000 full records from the UK and 7000 from Pakistan.

There are two main drawbacks of Nicol and Humphreys (2004)'s survey results and regression models: first, the results are not generalizable. In other words, they

> cannot be divorced from the population from which they are derived and applied to other contexts—at least not without losing much of their statistical credence. A diversity of factors (e.g., climate, cultural issues, building type and functions, organizational specifics, building systems peculiarities, space orientation, interior design features) may influence occupants' behavioral tendencies and their dependencies on hypothesized independent variables. (Mahdavi 2011)

Second, regression models predict, based on a set of given environmental conditions, "the probability for the outcome state variable to take the value 1, rather than predict the transition of this variable between states. Therefore, they do not describe the real dynamic processes of the system to be modelled" (Haldi and Robinson 2008). These two drawbacks limit the use of the survey approach in the establishment and validation of occupant behavior models.

To obtain reliable and valid results, a survey approach should carefully consider the questionnaire design, sampling methods, sample size, etc., to control for errors in the data. Any questionnaire should be piloted before being carried out on a larger scale, and data cleansing is required to eliminate invalid survey responses. As introduced in Chap. 8, the researcher should also consider issues of survey instrument assessment, such as reliability, validity, and survey error. Reliability is the extent to which the results of a survey tool (e.g., questionnaire) are consistent and stable. Statistical confidences are used to evaluate the reliability of survey data. Validity is how effectively a survey tool measures what it is intended to measure (Fowler 1995). A reliable and valid survey tool ensures the quality of survey data, i.e., that data reflects reality as much as possible. The ground truth data to validate survey data can be from interviews, manual observations, or in situ monitoring.

There are several types of reliability. Phelan and Wren (2005) define four main types as follows:

- Test-retest reliability: "a measure of reliability obtained by administering the same test twice over a period of time to a group of individuals";
- Parallel forms reliability: "a measure of reliability obtained by administering different versions of an assessment tool (both versions must contain items that probe the same construct, skill, knowledge base, etc.) to the same group of individuals";
- Inter-rater reliability: "a measure of reliability used to assess the degree to which different judges or raters agree in their assessment decisions"; and
- Internal consistency reliability: "a measure of reliability used to evaluate the degree to which different test items that probe the same construct produce similar results". Average inter-item correlation and split-half reliability are two subtypes of internal consistency reliability.

Refer to Fowler (1995) and Phelan and Wren (2005) for detailed explanations and examples.

There are also several types of validity that have been introduced in Chap. 8, including face validity, content validity, internal validity, to name a few. The reader can refer to Chaps. 3 and 8 for more details on how a survey or interview is carried out and how issues such as reliability and validity should be addressed.

9.4 Verification and Validation of Measurement Methods

To obtain ground truth data or establish best practice in occupant behavior measurements, there are some general rules and steps that investigators should follow. First, before a measurement is carried out, the investigator should consider: What do they expect to obtain from the measurement? What objects and behaviors will be observed? What types of measurement instruments will be used? What type of survey will be deployed? A verification and validation plan should be developed according to ISO 17025 (ISO 2005a, b), ISO 15189 (ISO 2012), or other similar quality systems.

To realize the aim of the experiment, the investigator should select measurement schemes or procedures which have been validated. Preferred schemes are "those that have been published in authoritative textbooks, peer-reviewed texts, or journals, in international consensus standards or guidelines, or in national or regional regulations" (ISO 2012). There are international standards for the measurement of local values of air temperature, humidity, illuminance, CO_2 concentration, thermal comfort, etc. The investigator is suggested to follow these standards for best practice. These standards include: ISO standard 7726 and 7730 (ISO 1998, 2005a, b), ASHRAE standard 55, 62 and 113 (ASHRAE 2013a, b, 2016), and Europe Standard EN 13182 (CEN 2002).

In case there is no such scheme or standards available, the investigator will have to establish their own scheme. The new scheme should be verified and validated before a real experiment. Of course, the measurement of environmental variables involved in the new scheme should follow the existing standards.

9.4.1 *Verification of Measurement Methods*

In an ideal case, the investigator can directly apply validated measurement scheme without modification for the intended use. Before being routinely used, however, these validated procedures still need to undergo independent verification by the investigator. Information from the instrument manufacturer or the measurement method developer can be useful starting point for determining the performance characteristics of a procedure (e.g., the accuracy and precision of occupancy meters or the validity of questionnaire sampling methods).

The investigator should provide independent verification that the performance claims have been fulfilled using performance characteristics, a type of objective evidence. For example, a thermocouple unit can be used to check for—and correct where appropriate—a lack of accuracy or precision in the temperature meters. Or, a video monitoring system could be used to confirm the accuracy and precision of infrared-based occupancy meters. The measurement procedure performance claims yielded during the process of verification should be related to the intended purpose of the measurement results.

In this process of verification, the investigator should document the procedures used and keep a record of the obtained results. Preferably a person with appropriate authority and experience (e.g., supervisor or lab director) would carry out a review of the verification results and keep a record of the review. Specifically, they should review: (1) whether the measurement procedures satisfy the intended use; (2) whether the investigator took each step properly; (3) whether the performance claims for the measurement procedure are confirmed by ground truth data. This review would improve the reliability of the verification procedure.

9.4.2 Validation of Measurement Methods

If there are no existing validated measurement schemes or procedures which can be applied without any modification for the intended use, the investigator probably needs to develop a "new" scheme. This type of measurement scheme may be derived from the following sources (ISO 2012):

- Non-standard methods;
- Laboratory designed or developed methods;
- Standard methods used beyond their initial intended scope; or
- Validated methods subsequently modified.

It is essential for the investigator to validate the new measurement scheme. There are four general steps for the validation:

(1) Set a baseline scheme as ground truth;
(2) Carry out an experiment to gather both ground truth data and ordinary measurement data;
(3) Do any necessary cleaning or post-processing on the obtained data, and compare the two results; and
(4) According to the errors between the two results and the performance requirements, decide how much is due to the uncertainty of the new measurement scheme and whether or not it can be adopted. The metrics to be used for performance evaluation are introduced in Sect. 9.4.3.

Several prerequisites should be satisfied prior to method validation (Theodorsson 2012):

- The method should be fully developed;
- The standard operation procedure for the method should be available;
- The measurement instruments to be used in method validation should be regularly calibrated and well maintained;
- The persons carrying out the measurements should have sufficient training and experience for the task; and
- The needs of the end user (on the basis of the fit-for-purpose principle) should be known.

By providing objective evidence about performance characteristics, the validation should confirm the fulfillment of particular requirements for the measurement scheme (ISO 2005a, b). The performance characteristics of a measurement scheme include consideration of (ISO 2005a, b, 2012): measurement trueness, precision, accuracy, sensitivity (detection limit and quantitation limit), ruggedness, and so on.

As for verification, the validation procedure should be documented by the investigator and a record kept of the results obtained. Validation results should be reviewed by a person with relevant experience and authority (e.g., project leader or supervisor) and a record kept of the review.

For any alterations made to a validated measurement procedure, the influence of such alterations should be tested and documented (ISO 2012). Where appropriate a further validation process should be undergone.

9.4.3 Measurement Uncertainties

Regardless of the instruments or methods used, there are always errors or uncertainties in measurement procedures, whether measuring physically sensed variables or using reported variables of occupant behaviors. In all cases, it is important for the investigator to quantify any measurement uncertainties in the observations.

In order to understand uncertainty analysis, data types must first be introduced (see also Chaps. 8 and 11). Briefly, the quantitative measurement results of instrument monitoring or questionnaire surveys can be categorized as either (1) continuous variables—for example, environmental parameters, such as temperature and humidity, or window opening ratio or angle; or (2) categorical variables—for example, occupancy state (presence/absence), occupant actions (on/off state of lights, AC units, windows, etc.). This chapter focuses mainly on error in instrument measurements of occupancy and occupant actions, but the reader can refer to Chap. 8 for an overview of uncertainty analysis of survey errors.

Results obtained by instrument monitoring are usually logged as time series of continuous or discrete values. Measurement errors include systematic errors and random errors, as introduced in Sect. 9.2.1. Gross errors can also be considered.

Gross errors are errors that cause the measured values to be very far off from the known/accepted values. These errors are usually the result of experimenter carelessness or instrument failure. It may happen when an instrument for in situ monitoring was moved or dropped from the installation place and touched by something else. These extreme values or “outliers” are so far from the true value of a measurand that they are usually discarded when processing, or “cleaning”, data. The Q-test based on 3σ and Grubbs’ criterions is a frequently used method of determining whether a data point should be discarded (Lü 2001).

After cleaning the extreme values, the total error e, systematic error e_s, and random error e_r in the measurement system can be expressed as follows, where x, x_0 are the measured and true value, and t is time.

$$e(t) = x(t) - x_0(t)$$

$$e(t) = e_s(t) + e_r(t)$$

Usually the systematic error e_s can be expressed as a linear function of true values:

$$e_s = f(x_0) = ax_0 + b$$

If this relation is known, it is possible to eliminate or correct the system error from raw data and leave random error only.

If the investigator is verifying or validating a measurement method, ground truth data is necessary to determine the overall performance of the method. The performance characteristics of the method can be evaluated by the following equations (Fei 2010). Hereby, M is defined as the method to be verified or validated, T is defined as the reference method to gain ground truth data, and n is the total number of temporal points for the measurement comparison.

For the continuous variables, the total error of M (i.e., the difference between the results of M and T) can be expressed by:

$$e(t) = x_M(t) - x_T(t)$$

The centerline (i.e., average of the differences), $\bar{e}$, can be expressed by:

$$\bar{e} = \frac{\sum e_t}{n}$$

The standard deviation of the errors, σ_e, can be expressed by:

$$\sigma_e = \sqrt{\frac{\sum (\bar{e} - e_t)^2}{n-1}}$$

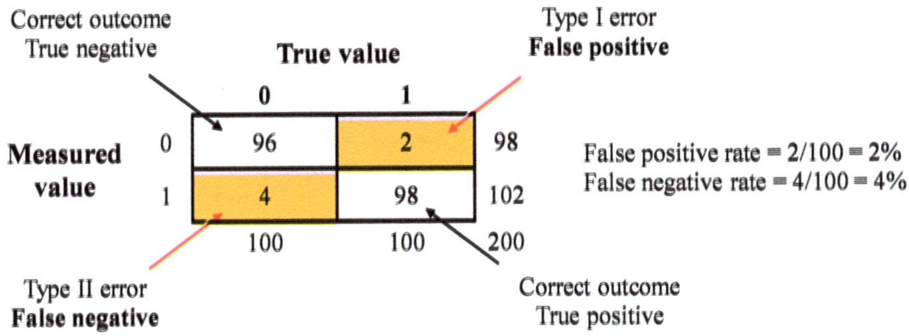

Fig. 9.3 An example of error analysis for categorical variables

The upper and lower limits of the measured value can be expressed by:

$$\bar{e} \pm 3\sigma_e$$

For categorical variables, the processes of error analysis and uncertainty calculation should be somewhat different. Take a binary variable for example: the general format for the analysis is a 2 × 2 table. The critical statistics are the false positive rate (type I) and the false negative rate (type II), as in Fig. 9.3.

Confidence intervals can be calculated and reported for false results. Typically, this calculation requires the exact binomial confidence limits, since the false positive and negative rates may be fairly low and the normal approximation is not appropriate (Agresti 1996).

There are several general guidelines to follow for an uncertainty analysis. The first is to determine the uncertainty associated with each measurand in the whole experiment used to report quantity values on occupants' samples. Performance requirements for measurement uncertainty should be determined by the investigator in addition to regular reviews of the estimates of measurement uncertainty.

Quantity values obtained by the measurements under intermediate precision conditions can be used to calculate measurement uncertainties. These include routine alterations to the standard operation procedure, e.g., changes of instruments, different operators, scheduled instrument maintenance (ISO 2012). The total measurement uncertainty can be calculated by adding the contribution of each components of variation in the measurement procedure. When interpreting measured quantity values, the investigator should consider measurement uncertainty and make the estimations of measurement uncertainty available to other users when requested (ISO 2012). If a measurement procedure includes a measurement step that does not offer a measured quantity value, the investigator should estimate its uncertainty where the measurement step has an influence on the final reported results or has an influence on evaluating the reliability of the measurement procedure.

One example of the calculation and analysis of measurement uncertainties is from Xi (2014), where an infrared counting system was designed and experimented. The system consisted of two pairs of infrared sensors deployed alongside the doorway of a room. By recognizing the direction of occupant entering and leaving the room, the system could count the number of occupants in the room (increasing or decreasing by 1). The system was tested in two typical working days and the measurement results were compared to the ground truth data from a video camera. The accuracy of the infrared system was 87% on average, defined by the number of detected events by the infrared system divided by the total number of entrance and exit events detected by the camera. The author concluded that the infrared system works well for most situations, but still faces an under- or over-reporting risk.

The reader should refer to Chaps. 3, 4, 6, 7, and 8 for more operation details and instructions on how to perform different occupant behavior observations in current technical conditions, including occupant behavior experiment design, in situ monitoring studies, laboratory studies, and survey studies.

9.5 Constructing Ground Truth Datasets

The possible measurands, approaches, and verification and validation steps of occupant behavior observations have been introduced above. This section discusses practical points for constructing ground truth datasets.

9.5.1 Validation of Occupant Behavior Measurement

The purpose of gathering ground truth data is usually to construct occupant behavior measurement methods that can correct for known biases and quantify the uncertainties in ordinary measurement data; ultimately, this allows the collection of less costly (than ground truth) data at scale for use in occupant behavior research. The intended use of a ground truth dataset is for verification and validation of occupant measurement methods, especially for non-matured instrument measurements of occupancy and occupant actions. When verifying or validating a measurement method, ground truth datasets are the referenced true values to determine the overall performance of the method. In the instrumental measurements, several questions should be considered: What is the nature of the ground truth dataset? Which instruments with higher measurement accuracy should be used for ground truth? How big or how long is the sample dataset (i.e., the length of ground truth measurement duration)? What is the time sampling interval of the measurement?

In practice, relatively inexpensive and simple measurement tools are commonly used for long-term or large-scale occupant behavior research—for example, using infrared sensors instead of a video recording system to detect occupant presence or absence, or using questionnaires instead of face-to-face interviews to gather occupant social and psychological variables. To validate and guarantee the reliability of these less expensive measures, ground truth datasets are required to correct the potential biases in ordinary measurement data and quantify the uncertainties in ordinary measurement data. Only if the biases and uncertainties in ordinary measurement data are small enough can the less expensive measures (relative to ground truth measures) and the ordinary measurement data be adopted. The biases and uncertainties in ordinary measurement data should be indicated after the validation.

Ground truth datasets are generally gathered by more accurate, more precise, and more reliable measures that follow a standard operation sequence. The required accuracy of the instruments for ground truth is usually a grade higher than those used for ordinary measurement data (Fei 2010). For example, in situ video monitoring systems can be carried out to validate measurement procedures for building occupancy (presence, number, and identity). The video system should be initialized with an exact startup time and full storage capacity, and it should cover all points of interest to record the moving activities of investigated occupants. In case that a video system is not available, manually observation and recording can work as well.

Ground truth datasets should be typical and cover many representative samples for the experiment scenarios in which the less expensive measures are applied, thus ensuring these measures are fully validated. In the example of using a video system to validate occupancy infrared sensors, how long should the ground truth datasets be? Of course, they should be as long as possible. Realistically, it is essential to record the whole period that occupants repeatedly enter, leave, and stay, including disturbances from sunbeams and pets, if any. A typical sunny working day is a minimum requirement. Also, the time spent for video data post-processing should be considered; although some computer vision algorithms may help, manual recognition, though time-consuming, is preferred in order to get the best ground truths.

To ensure the sampled data are able to represent the real dynamic process of measurands, the time sampling intervals for both ground truth and ordinary measurement data should be less than the theoretical interval given by Shannon's sampling theorem (Marks 1991). The interval of ground truth data should be smaller than that of ordinary measurement data, and can be a half or one fourth of the latter, if possible (Fei 2010).

The more cost-efficient measures to be validated should be deployed and collect measurement data at the same time of gathering ground truth data. Ground truth data are regarded as true values, while ordinary measurement data are regarded as measured values: by comparing the two datasets, the biases and uncertainties of ordinary measurement data can be determined. More details on the analysis of measurement errors or uncertainties can be found in Chap. 3 and in the monograph (Rabinovich 2005).

There are three points to be noted about error analysis. First, since the real occupant living environment is usually not steady, but dynamic, the obtained measurement data contain time-dependent errors and should be dealt with dynamic error estimation methods. Second, the measurement data logged by instruments might have two types: state at an instant in time (e.g., window is open or closed) or state change (e.g., opening window, or closing window). This should be indicated in the document and conversions not suggested before the accuracy evaluation. Third, if the measurement data are 0 or 1 values, type I errors (false positives) and type II errors (false negatives) should be considered separately in the error analysis (Haldi and Robinson 2009). In the example of the infrared occupancy sensor, it is a type I error if the sensor reports absence (tagged as 0) when there is someone, and a type II error if the sensor reports presence (tagged as 1) when there is no one. See Chap. 3 for more discussion on error types.

Based on the validation procedures, any biases found in ordinary measurement data should be corrected and the uncertainties of the corrected measurement data should be claimed. The investigator should judge whether the uncertainties of the tested measures are acceptable or not according to the intended use and the cost involved (time and/or money). There is always a trade-off between accuracy, precision, and cost.

A premise to evaluate the accuracy and precision of less expensive measures by ground truth datasets is that all instruments used should follow good operation and be in good working condition, including having an accurate clock counter (timer), reliable data storage, a full battery, etc.

The ground truth datasets gathered for validation might not be reusable, especially for the measurements of occupancy (movement) and occupant actions (turn on/off light, open/close window, etc.) because the real occupant environment and occupant activities are non-repeatable. This is why the ground truth data and ordinary measurement data should be gathered at the same time: each new validation needs new ground truth datasets. Thus, each ground truth experiment should be carefully carried out and documented for an internal check or external peer review. As more validations are carried out and more ground truth datasets are established, the evaluation of occupant behavior measurement procedures will become more stable and reliable.

9.5.2 Appropriateness, Robustness, and Openness

The following are some final thoughts for preparing ground truth data for the validation of occupant behavior measurement methods.

(1) **Appropriateness**

A principle requirement for a ground truth dataset is that it should be appropriate for the intended application and analysis, where the use cases and application goals should be involved into the ground truth data. Hence, the ground truth dataset should cover all typical scenarios of an occupant living environment. For example, if the focus of an investigation is the seasonal effect of occupant behavior, one year might be a minimum measurement length.

It is necessary to carefully select the variables to be measured or involved in a ground truth dataset. The potential impacts from those contextual variables that are not involved in the ground truth dataset might be considered as well.

(2) **Robustness**

The robustness of a ground truth dataset depends on its intended uses and capabilities. For example, a dataset for testing occupant movement locations can be used for a test of occupant presence measurement, but not vice versa. Likewise, a one-year dataset of lighting behavior has better robustness than a three-month one, since the former dataset can represent the seasonal effect of lighting behavior. Overall, the robustness of a ground truth dataset depends on how much the ground truth data represent real use cases. In general, more coverage of variables and a smaller interval and longer length in the measurement lead to a more robust ground truth dataset.

(3) **Openness**

The ground truth datasets collected can be public or proprietary. Usually proprietary ground truth data is an obstacle to the independent evaluation of occupant behavior measurement methods. It is necessary for different parties to publish their data and share them with the wider occupant behavior research community so that their work can be compared to and externally checked by others and contribute to further research. However, credibility and legal barriers exist for open-sourcing any proprietary ground truth data. The associated privacy and legal concerns are very real and should be dealt with carefully (Krig 2014).

9.6 Conclusion

It is essential to ensure the validation of measurements and the reliability of collected data. In the current contact, validation and verification are procedures for ensuring that collected data meet the required specifications for the purpose at hand. Historically, the various fields of occupant behavior research have lacked a shared theoretical and practical basis, not only due to variations in the tasks undertaken, but also due to differences in terminology and practices (e.g., calibration, validation, quality control) within the field.

In order to validate or verify measurement methods, ground truth data are necessary. The purpose of gathering ground truth data is usually to construct occupant behavior measurement methods that can correct for known biases and quantify uncertainties in ordinary measurement data, thus allowing the collection of less costly data (relative to ground truth data) at scale for use in occupant behavior research, especially for the non-matured instrument measurements of occupancy and occupant actions.

In this chapter, the basic concepts of validation, verification, and ground truth in occupant behavior measurements were introduced. The principles of validating occupant behavior measurements and obtaining ground truths were presented, as were the validation and verification procedures of measurement methods. Finally, it was discussed why and how to construct ground truth datasets. By following these guidelines and best practices, reliable measurements of occupant behavior can be obtained.

That said, the reality is that the practice of occupant behavior measurements and ground truths is has significant need for improvement. Moreover, there is much work to be done on the construction of occupant behavior measurement models and ground truth datasets. These should be collected, published, and, importantly, shared—this will contribute to the advancement of research on occupant behavior and boost the development of the research community as a whole.

References

Agresti A, (1996) An Introduction to categorical data analysis. John Wiley & Sons

Andersen R et al (2013) Window opening behaviour modelled from measurements in Danish dwellings. Build Environ 69:101–113

Anonymous (2011) "Ground-truth data cannot do it alone" Editorial. Nat Methods 8(11): 885

ASHRAE (2013a) ASHRAE Standard 55—Thermal Environmental Conditions for Human Occupancy

ASHRAE (2013b) ANSI/ASHRAE Standard 113-2013—Method of Testing for Room Air Diffusion

ASHRAE (2016) ANSI/ASHRAE Standard 62.1-2016—Ventilation for Acceptable Indoor Air Quality

BIPM (2012) JCGM 200:2012 International vocabulary of metrology—basic and general concepts and associated terms (VIM 3). www.bipm.org/en/publications/guides/vim.html

CEN (2002) EN 13182—Ventilation for buildings—Instrumentation requirements for air velocity measurements in ventilated spaces

Chang W-K, Hong T (2013) Statistical analysis and modeling of occupancy patterns in open-plan offices using measured lighting-switch data. Build Simul 6(1):23–32

Correia da Silva P, Leal V, Andersen M (2013) Occupants interaction with electric lighting and shading systems in real single-occupied offices: results from a monitoring campaign. Build Environ 64:152–168

Cozby PC (2001) Measurement Concepts. Methods in Behavioral Research, 7th edn. Mayfield Publishing Company, California

De Biévre P, Böttger D, Eastwood C et al (1998) The fitness for purpose of analytical methods. a laboratory guide to method validation and related topics. www.eurachem.org/guides/pdf/valid.pdf

Dong B, Andrews B (2009) Sensor-based occupancy behavioral pattern recognition for energy and comfort management in intelligent buildings. In: Proceedings of building simulation

Fabi V et al (2012) Occupants' window opening behaviour: a literature review of factors influencing occupant behaviour and models. Build Environ 58:188–198

Fei YT et al (2010) Error theory and data processing. China Machine Press

Fowler FJ (1995) Improving survey questions: design and evaluation. SAGE Publications, Inc

Gupta RP (2013) Remote sensing geology. Springer Science & Business Media

Haldi F, Robinson D (2008) On the behaviour and adaptation of office occupants. Build Environ 43(12):2163–2177

Haldi F, Robinson D (2009) Interactions with window openings by office occupants. Build Environ 44(12):2378–2395
Haldi F, Robinson D (2010) Adaptive actions on shading devices in response to local visual stimuli. J Build Perform Simul 3(2):135–153
Hunt DRG (1979) The use of artificial lighting in relation to daylight levels and occupancy. Build Environ 14(1):21–33
ISO (1998) ISO 7726:1998 Ergonomics of the thermal environment—Instruments for measuring physical quantities. https://www.iso.org/standard/14562.html
ISO (2005a) ISO/IEC 17025:2005 General requirements for the competence of testing and calibration laboratories. https://www.iso.org/standard/39883.html
ISO (2005b) ISO 7730:2005 Ergonomics of the thermal environment—analytical determination and interpretation of thermal comfort using calculation of the PMV and PPD indices and local thermal comfort criteria. https://www.iso.org/standard/39155.html
ISO (2012) ISO 15189:2012 Medical laboratories—Requirements for quality and competence. https://www.iso.org/standard/56115.html
Krig S (2014) Computer vision metric: survey, taxonomy, and analysis. Apress
Lü CD (2001) Measurement and analysis of thermal parameters. Tsinghua Press
Marks RJ (1991) Introduction to shannon sampling and interpolation theory. Spinger-Verlag
Mahdavi A (2011) The human dimension of building performance simulation. In Proceedings of building simulation 2011: 12th conference of international building performance simulation association. Sydney
Menditto A, Patriarca M, Magnusson B (2006) Understanding the meaning of accuracy, trueness and precision. Accred Qual Assur 12(1):45–47
Nicol JF, Humphreys MA (2004) A stochastic approach to thermal comfort-occupant behavior and energy use in buildings. ASHRAE transactions 110:554–568
Nicol F, Wilson M, Chiancarella C (2006) Using field measurements of desktop illuminance in European offices to investigate its dependence on outdoor conditions and its effect on occupant satisfaction, and the use of lights and blinds. Energy Build 38:802–813
Phelan C, Wren J (2005) Exploring reliability in academic assessment. https://chfasoa.uni.edu/reliabilityandvalidity.htm
Rabinovich SG (2005) Measurement errors and uncertainties: theory and practice. 3rd edn. Springer Science and Media, Inc
Reinhart CF, Voss K (2003) Monitoring manual control of electric lighting and blinds. Lighting Res Technol 35(3):243–260
Ren X, Yan D, Wang C (2014) Air-conditioning usage conditional probability model for residential buildings. Build Environ 81:172–182
Tanimoto J, Hagishima A (2005) State transition probability for the Markov Model dealing with on/off cooling schedule in dwellings. Energy Build 37(3):181–187
Theodorsson E (2012) Validation and verification of measurement methods in clinical chemistry. Bioanalysis 4(3):305–320
Timilehin L et al (2015) Occupancy measurement in commercial office buildings for demand-driven control applications—a survey and detection system evaluation. Energy Build 93:303–314
Schweiker M, Shukuya M (2009) Comparison of theoretical and statistical models of air-conditioning-unit usage behaviour in a residential setting under Japanese climatic conditions. Build Environ 44(10):2137–2149
Wang C et al (2016) A generalized probabilistic formula relating occupant behavior to environmental conditions. Build Environ 95:53–62
Xi X (2014) The infrared-based and RFID-based indoor occupant counting system. B.Sc. Thesis, Tsinghua University, Beijing
Yun GY et al (2008) Time-dependent occupant behaviour models of window control in summer. Build Environ 43(9):1471–1482

Chapter 10
Structured Building Data Management: Ontologies, Queries, and Platforms

Ardeshir Mahdavi, Mahnameh Taheri, Matthias Schuss, Farhang Tahmasebi and Stefan Glawischnig

Abstract Building data monitoring, in general, and occupancy-related data collection in particular have the potential to provide deep performance feedback for: (1) operational optimization of existing facilities and (2) improving future designs. For instance, building monitoring can support energy and performance contracting, preventive building maintenance, smart load balancing, and model-predictive building systems control. Nevertheless, currently this potential is not sufficiently realized. To address a major gap in the current practice, the present chapter first introduces an ontology for the representation and incorporation of various kinds of building monitoring data in a number of applications such as building performance simulation tools and building automation systems. Subsequently, common data processing requirements are addressed and a number of typical queries are exemplified that building monitoring data repositories must support. Finally, data repository specifications and implementations for structured collection, storage, processing, and multi-user exchange of monitored data are described.

A. Mahdavi (✉) · M. Taheri · M. Schuss · F. Tahmasebi · S. Glawischnig
Department of Building Physics and Building Ecology, TU Wien, Karlsplatz 13, 1040 Vienna, Austria
e-mail: amahdavi@tuwien.ac.at

M. Taheri
e-mail: mahnameh.taheri@tuwien.ac.at

M. Schuss
e-mail: matthias.schuss@tuwien.ac.at

F. Tahmasebi
e-mail: farhang.tahmasebi@tuwien.ac.at

S. Glawischnig
e-mail: stefan.glawischnig@tuwien.ac.at

A. Wagner et al. (eds.), *Exploring Occupant Behavior in Buildings*,
DOI 10.1007/978-3-319-61464-9_10

10.1 Introduction

Systematic and continuous scanning of buildings' operational states can offer multiple benefits. It can provide performance feedback for operational optimization of existing facilities and improve future designs. For instance, it can support energy and performance contracting, model-predictive building systems control, smart load balancing, and preventive building maintenance. Likewise, systematically monitored high-resolution data can advance the state of knowledge in a broad range of domains in building science (e.g., building automation, indoor environment, and human factors). This potential is not unknown to the relevant professional community. Accordingly, there are various instances of commercially implemented building monitoring systems, as well as research-oriented data collection campaigns (e.g., Roda and Musulina 2014; Guerra-Santin and Tweed 2015; Böhms and Rieswijk 2015). However, most currently implemented technical infrastructures do not appear to be mature enough. Likewise, the associated hardware resilience and software interoperability could benefit from major improvements.

In addition to the above considerations, the present chapter has a specific motivational grounding as it emerged in a response to requirements formulated within the International Energy Agency Annex 66 (IEA 2016) pertaining to the computational representation of building occupants in view of their presence and actions in buildings. Thereby, it is necessary to address the paucity of comprehensive approaches to the collection, storage, sharing, and analyses of monitored data relevant to the occupants' presence/activities in the buildings and the related impacts. This circumstance requires efforts that go beyond including classes of real and virtual sensors and meters in building information systems. In this chapter, first, an ontology is introduced for the representation and incorporation of multiple data streams in computational applications, such as building performance simulation tools and building automation systems (see Sect. 10.2). Such data streams include, aside from those relevant to the detection of people's presence, movement, and (control-oriented) actions in buildings, external and internal boundary conditions (e.g., weather conditions, indoor environment), as well as relevant states of buildings' devices and systems.

Note that, the ontology presented in this chapter does not cover the vast amount of data that environmental control systems can generate (e.g., information regarding the internal states of various system elements required for real-time control processes). Instead, the focus is on those data streams that pertain either to systems' interface with spaces via device terminals (e.g., radiators, diffusors, luminaires) or capture occupants' interactions with buildings' control components. Then, the common data processing requirements are addressed and a number of typical queries that building monitoring data repositories need to support are exemplified (see Sect. 10.3). Finally, general requirements and prototypical implementations of data repository solutions for the structured collection, storage, processing, and multi-user exchange of monitored data are described (see Sect. 10.4).

10.2 Outline of an Ontology for Building Monitoring

10.2.1 General Categories

In order to construct a well-formed schema for building monitoring, first it is needed to identify a number of fundamental data categories. Based on the prior efforts in this area (Mahdavi and Taheri 2016; Mahdavi et al. 2005, 2016b; Mahdavi 2011a, b; Zach et al. 2012b), the following six data categories are suggested. These categories can provide a coherent framework to classify the multiplicity of empirical information collected via buildings' monitoring systems. These are: (1) occupants (OC), (2) indoor environmental conditions (IC), (3) external environmental conditions (EC), (4) control systems and devices (CS), (5) equipment (EQ), and (6) energy flows (EN).

Figure 10.1 provides an overview of these categories together with associated sub-categories and illustrative examples of corresponding monitored variables (see also Mahdavi and Taheri 2016).

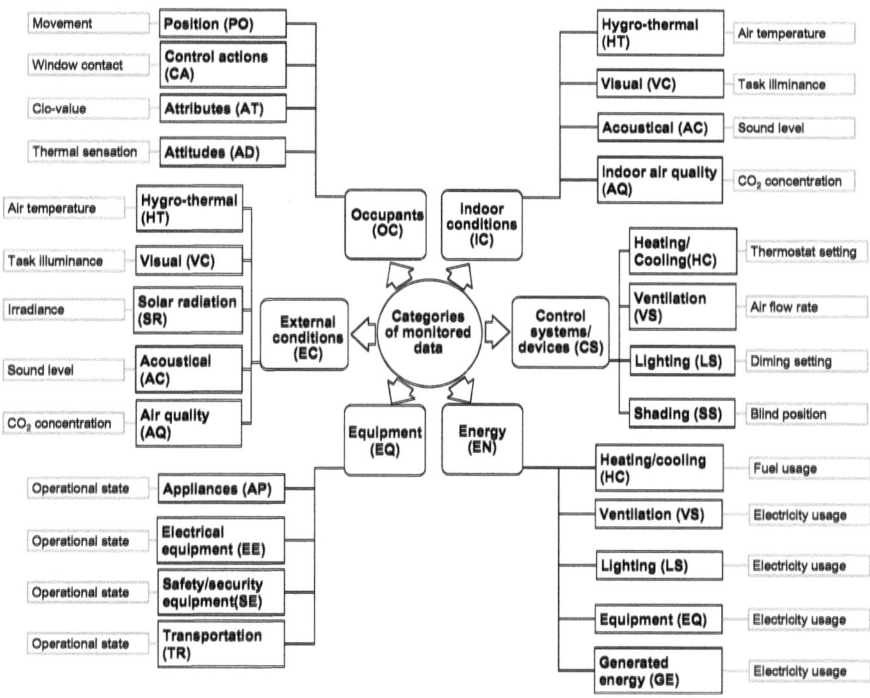

Fig. 10.1 Fundamental elements of a building monitoring ontology: data categories, sub-categories, and examples of corresponding monitored variables

10.2.1.1 Occupants

People's presence and actions in buildings need to be captured in a systematic and versatile manner (see Chap. 2). It is suggested to structure information in the category of buildings' occupants in terms of four sub-categories, namely (a) position (PO), (b) control actions (CA), (c) attributes (AT), and (d) attitudes (AD).

Time series data of occupants' presence in building is a prerequisite for most use cases pertaining to building operation. Likewise, occupants' actions (operation of indoor environmental control devices, as well as office equipment and household appliances) are essential for use cases such as building operation and building performance assessment. Such actions must be either directly monitored, or—in case of exclusively user-driven device and equipment actuators—extracted from corresponding device/equipment state change data (Mahdavi and Taheri 2016). Moreover, depending on the building systems' type and configuration, occupants may have the possibility to control pertinent set-points for heating, cooling, ventilation, lighting, etc. Hence, occupant-driven changes to the values of such set-points (e.g., via operating a thermostat) need to be registered as well. Depending on the resolution and coverage of intended applications, additional data concerning occupants may be required. This includes occupants' attributes or state data (clothing, activity, physiology) as well as perceptual and evaluative (attitudinal) information such as (both short-term and long-term) subjective characterizations of the of indoor environmental conditions, for example, via thermal sensation and thermal comfort scales. Note that the expression attitudinal is used here to denote a broad range of information generated by occupants, including subjective sensation, perception, and evaluation. In these instances, the human agent may be arguably considered to be the sensor. Note that, aside from technical feasibility issues, data collection campaigns addressing occupants may be also considerably constrained due to privacy issues (see also Chap. 11).

10.2.1.2 Indoor Environmental Conditions

Building performance assessment processes typically require indoor environmental data. In fact, most theories regarding subjective evaluation processes of indoor environmental conditions, as well as causal theories of occupants' control-oriented behavior involve one or more indoor environmental parameters as independent variables (e.g., air temperature and illuminance levels). High-resolution spatial and temporal data from multiple domains (hygro-thermal, visual, acoustical, air quality) would be obviously most preferable. However, practical and economic constraints may limit the extent of respective monitoring campaigns and infrastructures.

10.2.1.3 External Environmental Conditions

The objective assessment of the degree to which a building's energy and indoor climate performance has been satisfied requires the consideration of the building's contextual circumstances. Likewise, user-related behavioral models frequently require information regarding external conditions. For instance, prediction of adaptive actions (e.g., operation of windows) may need information concerning the prevailing external conditions (e.g., air temperature, solar radiation, precipitation, sound levels). While standard weather stations can provide a good part of required data, special applications (e.g., building performance simulation model calibration) may require additional sensory equipment.

10.2.1.4 Control Systems and Devices

The performance of buildings obviously depends on the quality of the installed control systems (for heating, cooling, ventilation, lighting, etc.). Reliable information on the state of such systems is thus of essential interest, not only with regard to immediate technical control functionalities, but also in view of their environmental performance and deployment patterns. The latter aspect concerns—amongst other things—occupants' interactions with buildings' control systems and devices. The state information regarding devices (windows, luminaires, radiators, fans, shades, etc.) and associated actuators is thus of critical importance. Monitored data might include state information from devices that can be controlled: (a) only automatically; (b) only by occupants; or (c) both automatically and by occupants (e.g., via user override of automated control routines).

In many buildings, occupants have the possibility to manipulate the values of the control parameters of the buildings' environmental systems (e.g., set-point temperatures for room heating and cooling). Thus, adjustment of the control parameter values must be monitored as well.

In the case of device states and control set-points, changes in observed values point to control events. Such events/actions are implicitly captured in the monitored state data, as they can be extracted from device state data and associated control set-points. However, information regarding events/actions should be ideally accompanied with information on actors (e.g., human initiators or agents, or control software). To address this matter in the ontology, one simple approach would be to assign an agent or actor ID to every monitored device or set-point state at time t_i, if it displays a change with respect to the previous observation at time t_{i-1}.

10.2.1.5 Equipment

Aside from environmental control systems, buildings also house various technical components typically deployed by occupants for different purposes. Roughly speaking, this includes electrical equipment (e.g., computers and associated peripherals such as printers and scanners), appliances (e.g., clothes washers and dryers, ovens, refrigerators), safety and security equipment (e.g., smoke detectors), and transportation equipment (e.g., elevators and escalators). Monitored data on equipment operation can benefit multiple applications (e.g., energy optimization, smart grids, performance simulation tools, and behavioral models). Such equipment are not subsumed into the previous category (systems and devices), as the existing functional difference justifies a logical differentiation. Control systems and devices fulfil the explicit functionality of influencing the indoor environmental conditions via heating, cooling, ventilation, lighting, etc. Appliances and electrical equipment can also influence indoor climate—for instance, as sources of internal heat gains, air pollutants, or noise. But such influences represent side effects of the equipment's operation, not its designated functionality (Mahdavi and Taheri 2016).

10.2.1.6 Energy

Evidence-based building design and energy performance verification requires high-resolution energy use monitoring (energy metering). Here, resolution can be understood: (a) in spatial terms (e.g., micro-zones, rooms, floors, whole buildings), (b) across multiple systems (e.g., heating, lighting, equipment), and (c) in temporal terms (e.g., sub-hourly, hourly, daily, monthly, annual). Metering is typically associated with energy use documentation. However, buildings might be equipped with energy-harvesting systems such as solar-thermal collectors or photovoltaic panels. In such cases, the magnitude of energy produced is metered. Note that occupants' presence and actions not only influence indoor environmental conditions, but also have considerable implications for the energy performance of buildings and their systems. This underlines the critical importance of continuous energy use monitoring in view of related behavioral models.

10.2.2 The Structure of Monitored Data

Sensor, meters, and other sources of data (e.g., simulation-powered virtual sensors, human agents) in the above six categories generate streams of data to be captured, stored, and processed. A suitable ontology for the monitored information must thus clearly define the nature of the monitored variables. Toward this end, it is possible to demonstrate that all monitored data can be captured in terms of the profile (structure) shown in Figs. 10.2 and 10.3. Given each data category and the respective sub-categories, monitored variables are specified in terms of their values, associated sources, and possible actors.

10.2.2.1 Values

Variables pertain to properties that are subject to change. Observational data are typically in the category of measured (quantitative) data. Measured values of scalar nature, such as temperature, have a magnitude. Physical phenomena represented as vectors (e.g., air flow velocity or sound intensity) can be expressed in terms of two linked variables (i.e., a magnitude and a direction) with respective numeric values.

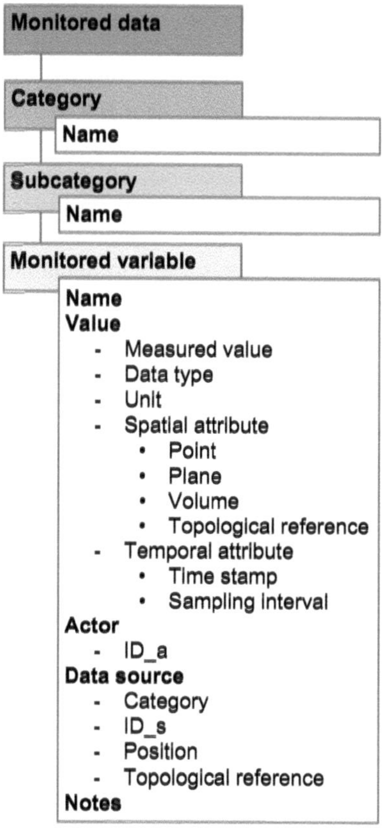

Fig. 10.2 The general structure of monitored data

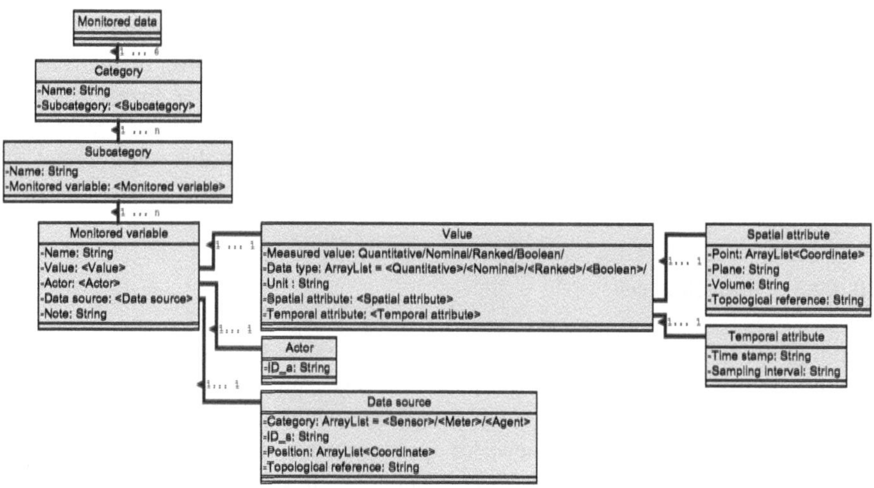

Fig. 10.3 Representation of the structure of monitored data (adopted from Mahdavi and Taheri 2016) using UML (Unified Modelling Language)

A data type should be assigned to a monitored variable value. Most measured variables in building monitoring have values that can be expressed in terms of real numbers. However, there are some variables whose values are not quantitative. A class of such variables is typically characterized as nominal data. For instance, user-based thermal comfort evaluation may be expressed in terms of classifications and categories. Another class of data, namely ranked data, refer to variables whose values display a certain order (e.g., successive positions of a valve). Applied to actuators, device states may be captured in terms of Boolean variables. Both nominal and ranked data can be made subject to quantitative operations when, for instance, variable values are mapped into a set of ordinal numbers and treated via statistical operations.

Typically, a unit must be specified for the recorded values of a variable (e.g., degrees Celsius or Kelvin for air temperature) in order to correctly interpret the variables' numeric values.

Spatial and temporal attributes (or extensions) can be assigned to variable values. The spatial attribute could involve a one-dimensional point (with x, y, and z coordinates), a two-dimensional plane such as a polygon, or a three-dimensional volume such as a polyhedron. In case individual sensor readings of different points are aggregated (e.g., in the course of post-processing) for a plane or a volume, the mode of aggregation (e.g., arithmetic averaging) should be noted (Mahdavi 2011b). Typically, a variable value has also a temporal attribute. Specifically, the point in time when the reading occurs must be recorded in terms of a time stamp. Moreover, sensor readings may be assigned to a discrete time interval (sampling interval).

10.2.2.2 Actors

As discussed before, changes in the state of control devices, equipment, and associated settings may be triggered by different agents (or actors). For instance, windows may be operated by human agents, and motorized shades may be operated based on programmed rules in the building automation systems. In many cases buildings accommodate both automated control actions and user override opportunities. Ideally, the monitoring system should identify for each change of state instance the responsible agent. The ontology accommodates this requirement via associated labels (ID_a).

10.2.2.3 Data Sources

Building monitoring can integrate not only common technical sensors (e.g., temperature sensors) and meters (e.g., power meters), but also human agents. For instance, attitudinal information (e.g., subjective evaluation of indoor climate) is customarily assessed via interviews or questionnaires. Given the extensive and heterogeneous characteristics of the corresponding information, the ontology can include a reference (ID_s) to an external document with details on the nature of the source of monitored data (sensors, meters, or human agents) and relevant technical specifications, such as measuring range and sensor precision.

Data sources must be also specified in terms of their location. In ontology, this is accommodated via position information as well as through mapping to the building's topology. The spatial hierarchy of a building can be captured, for instance, in a BIM (Building Information Modeling) environment. Sensors and their sensing targets can be associated with whole building, floors, sections, rooms, zones, workstations, etc. The spatial association of data sources and buildings' geometric and functional units is essential for a seamless representation of monitoring hardware, networks, topology, and architecture.

10.2.3 Expressions of the Ontology for Multiple Data Categories

Given the established data categories (Fig. 10.1) and monitored data's basic structure (Figs. 10.2 and 10.3), one can examine the aptitude of the proposed building monitoring ontology in capturing multiple categories and instances of data. As Figs. 10.4, 10.5, 10.6, 10.7, 10.8 and 10.9 demonstrate, the proposed framework can accommodate multiple instances of data (window contact, temperature, relative humidity, valve position, operational state, electricity usage) in selected sub-categories (CA, HT, HC, EE) of the aforementioned six data streams of building monitoring (OC, IC, EC, CS, EQ, EN) (see Sect. 10.2.1 and Fig. 10.1). Furthermore, Fig. 10.10 provides for each of these six data streams an illustrative example of a monitored variable specification in selected sub-categories (CA, VC, HT, SS, AP, HC).

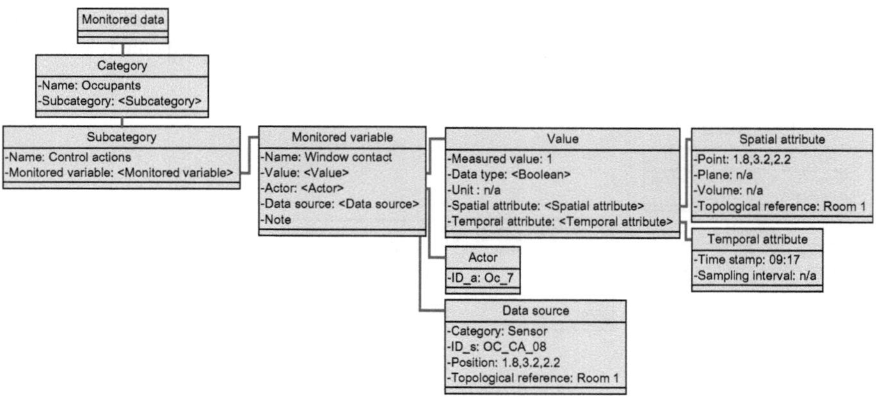

Fig. 10.4 Expression of the monitored variable window contact (*data category* occupants; *sub-category* control actions) in the proposed building monitoring ontology

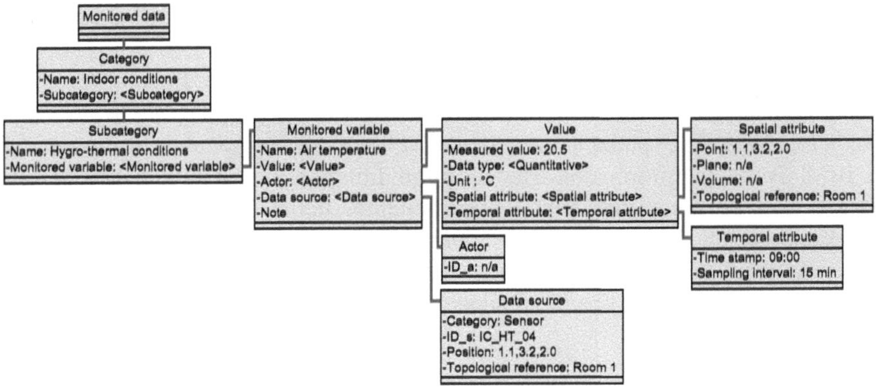

Fig. 10.5 Expression of the monitored variable temperature (*data category* indoor conditions; *sub-category* hygro-thermal conditions) in the proposed building monitoring ontology

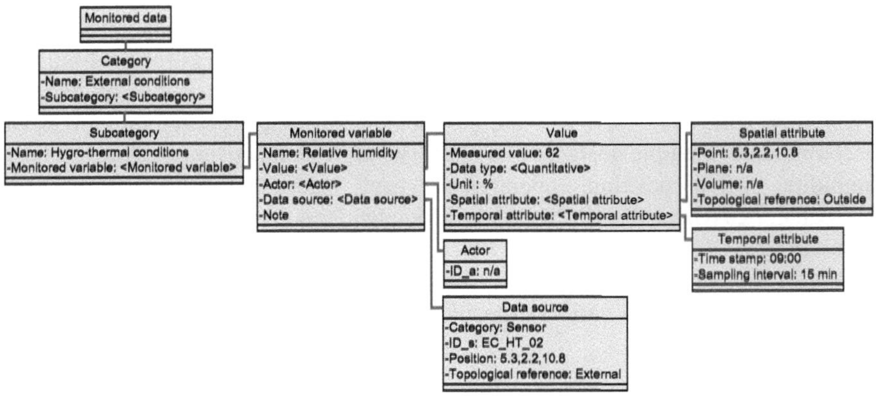

Fig. 10.6 Expression of the monitored variable relative humidity (*data category* external conditions; *sub-category* hygro-thermal conditions) in the proposed building monitoring ontology

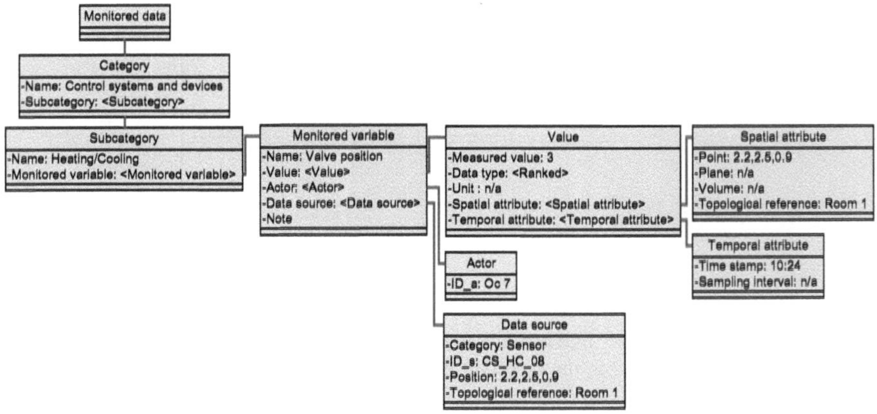

Fig. 10.7 Expression of the monitored variable valve position (*data category* control systems and devices; *sub-category* heating/cooling) in the proposed building monitoring ontology

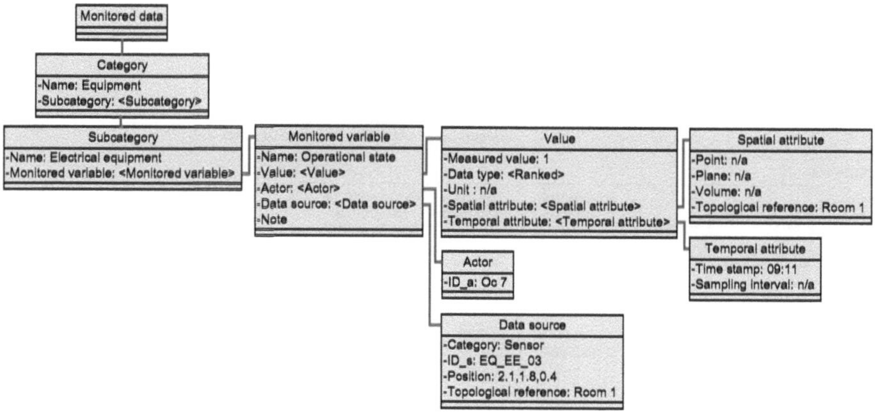

Fig. 10.8 Expression of the monitored variable operational state (*data category* equipment; *sub-category* electrical equipment) in the proposed building monitoring ontology

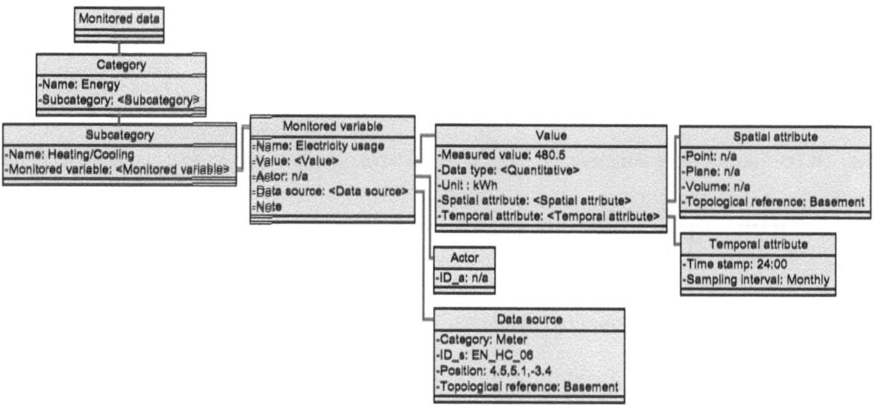

Fig. 10.9 Expression of the monitored variable electricity usage (*data category* energy; *sub-category* heating/cooling) in the proposed building monitoring ontology

10.3 Data Processing and Typical Queries

The preparation of monitored data can involve very different data processing paths and options. The necessary steps of the related processing routines are strongly dependent on the specific attributes and behavior of the data collection sequence (see Fig. 10.11) entailing the sensor, the signal convertor, data pre-processing, storage, retrieval, and post processing.

Generally, data post processing could be separated into two main categories, one for periodic data and the other for event-triggered or event-related data. The result for most typical data processing routines consists of periodic data streams with fixed

Monitored data						
Category						
Name	Occupants	Indoor conditions	External conditions	Control systems and devices	Equipment	Energy
Subcategory						
Name	Control actions	Visual conditions	Hygro-thermal conditions	Shading	Appliances	Heating/ cooling
Monitored variable						
Name	Window contact	Illuminance	Relative humidity	Blind position	Operational state	Electricity usage
Value						
- Measured value	1	280	62	0	1	480.5
- Data type	Boolean	Quantitative	Quantitative	Boolean	Ranked	Quantitative
- Unit	-	lx	%	-	-	kWh
- Spatial attribute						
• Point	1.8/3.2/2.2	2.0/1.7/0.9	5.7/12.2/9.8	-	-	-
• Plane	-	Desk R1	-	Window R1	-	-
• Volume	-	-	-	-	-	-
• Topological reference	Room 1	Room 1	External	Room 1	Room 1	Basement
- Temporal attribute						
• Time stamp	09:17	11:00	09:00	10:45	09:00	24:00
• Sampling interval	-	15 min	15 min	-	-	Monthly
Actor						
- ID_a	Oc_7	-	-	BMS	Oc_7	-
Data source						
- Category	Sensor	Sensor	Sensor	Sensor	Sensor	Meter
- ID_s	OC_CA_08	IC_VC_08	EC_HT_02	CS_SS_07	EQ_AP_03	EN_HC_06
- Position	1.8/3.2/2.2	2.0/1.7/0.9	5.7/12.2/10	1.8/2.2/2.2	2.0/1.7/0.3	4.3/6.5/-3.2
- Topological reference	Room 1	Room 1	External	Room 1	Room 1	Basement
Notes						

Fig. 10.10 Illustrative examples of a monitored variable specifications in selected sub-categories

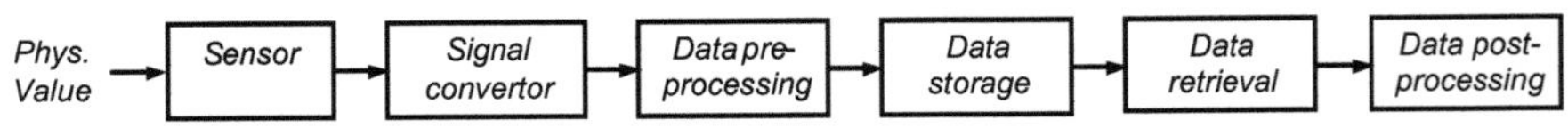

Fig. 10.11 Data measurement, pre-processing, storage, retrieval and post-processing

intervals, whereby the time stamps are synchronized. In case periodically measured data are exported into building simulation tools, the interval values are typically generated such that they represent the averaged value of the preceding interval.

10.3.1 Periodic Raw Data

Periodic data is provided by systems that store measurements at regular time intervals based on an internal cycle timer. Corresponding typical systems are Building Automation Systems (BASs) and measurement systems or data loggers. The interval is usually defined by an internal setup value of such systems. A cycle timer triggers the execution of an internal polling algorithm and the data storage routine. Data processing for this type of data is mainly a simple averaging or an interpolation of the raw data as illustrated in Fig. 10.12. The dots show the periodically generated data as weighted averages of the raw data in the preceding

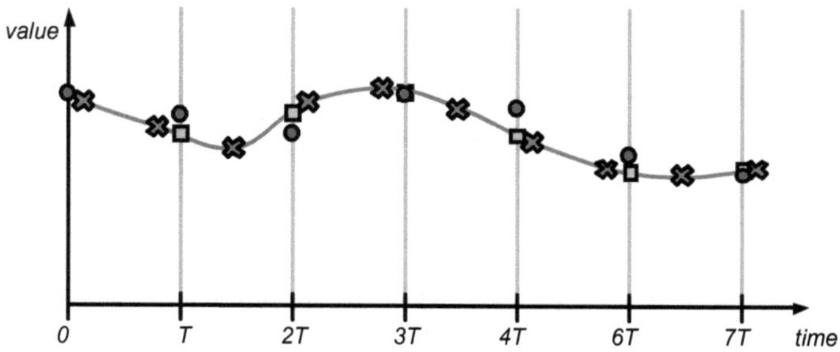

Fig. 10.12 Illustration of measured data (*crosses*), snap shots or instantaneous data (*squares*), and generated periodic data as the time-weighted mean value of the measurements in the preceding interval (*dots*)

interval. The squares represent generated periodic values for the exact periodic timestamp based on interpolations. As mentioned earlier, when data is used for building simulation mostly the first method is used.

10.3.2 Event Related Raw Data

Data monitoring systems that are triggered by events (e.g., detection of a movement, opening of a door or window, activation of devices, alarms or warnings) tend to store the raw data with corresponding—typically irregular—time stamps. Usually this data has to be post-processed to generate periodic synchronized data for the subsequent analysis, evaluation or export into other applications (e.g., building simulation tools). Figure 10.13 shows a typical trend of event based raw data together with an example of generated periodic data from a data processing algorithm. Commonly, event-based data is only stored when a change occurs. The periodic data generation process works in terms of a sample and hold process and repeats the last value as long as no new event is recorded. If more than one value was measured during an interval, different post processing options may be relevant. For instance, periodic instantaneous data may be generated using the last recorded value at each interval (dots in Fig. 10.13). However, in certain use cases (e.g., building energy simulation), multiple measurements within an interval are aggregated (for instance via time-weighted averaging) and assigned as the periodic interval value.

10.3.3 Interval Data from BAS (Building Automation System)-Integrated Sensors or Data Loggers

Raw data from BAS logging routines usually contain time stamps that are not synchronized, hence a data processing with an interpolation and subsequent weighted averaging is necessary. An interpolation of values is always needed when

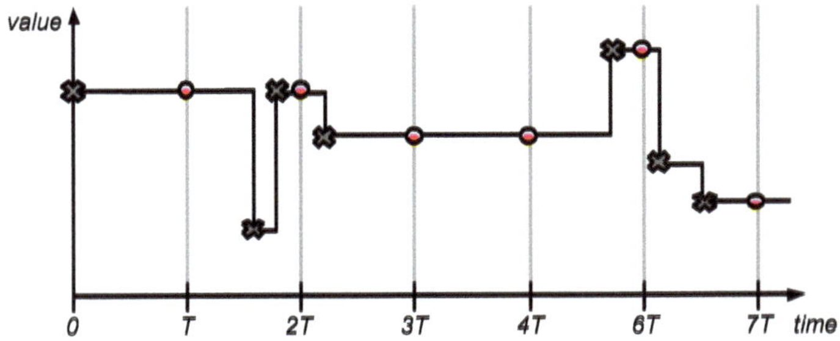

Fig. 10.13 Illustration of event based data (*crosses*) and generated periodic data (*dots*)

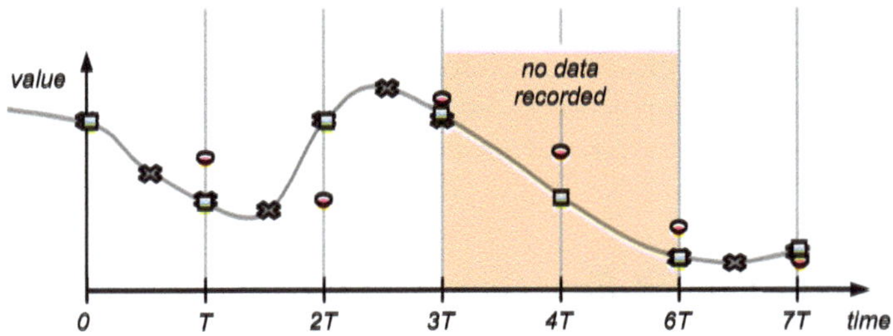

Fig. 10.14 Illustration of data post-processing for periodic interval BAS data (*crosses*) and the generated snap shots or instantaneous data (*squares*), and generated periodic data as the time-weighted mean value of the measurements in the preceding interval (*dots*)

the polling interval is in a similar range as or smaller than the needed periodic data. Likewise, data interpolation may be necessary to account for a gap in measurements (see Fig. 10.14). If the process involves an interpolation or other data processing steps, additional qualification is needed to indicate whether the output is a raw data or the result of an interpolation or averaging.

10.3.4 Practical Examples of Building Monitored Data Processing

10.3.4.1 Generation of Occupancy Data from PIR-Motion Raw Sensor Data

Passive infrared (PIR) motion sensors usually report a change of state (i.e., from occupied to vacant or vice versa). After reporting the occupied state, these sensors delay switching back to the vacancy state with an internal timer. This value could be fixed or variable depending on the specific product. This internal function is necessary to avoid unreasonably rapid state fluctuations. Depending on the time interval of the desired periodic data, even with this sensor-integrated filtering, the

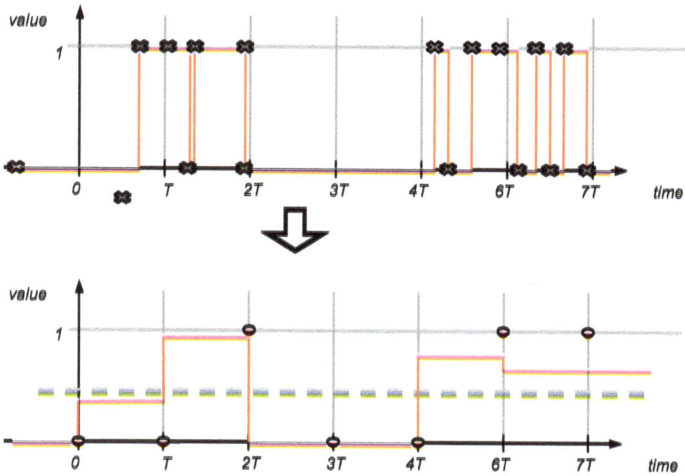

Fig. 10.15 Data pre-processing for event related occupancy raw data from PIR sensors. The *dashed line* represents the threshold used to determine binary occupancy states based on the weighted average of measured values in each interval

raw data may contain multiple fluctuations within each interval. For the generation of typical binary occupancy schedules the raw data must to be processed as follows (see Fig. 10.15):

1. Generation of weighted average values for each interval resulting in the fraction of time interval in which the occupant is present.
2. A sample-hold pattern (with the last actual measurement) to generate values for the intervals with no measurements.
3. Generation of binary occupancy data using a threshold value. The designation of this threshold value requires some experience with the sensor's behavior (Gilani and O'Brien 2016). Moreover, it is necessary to distinguish between transitional events (e.g., a co-worker passing by a non-occupied workstation in an open-plan office) and actual occupancy.

Event related data from sensors that report non-binary values can be processed in a similar way as explained above, however excluding the last step.

10.3.4.2 Generation of Presence Probability Profiles

The occupants' presence patterns are commonly included in the building performance models as typical schedules (Yan et al. 2015). Thereby, in the absence of on-site occupancy data, typical occupancy diversity profiles from standards such as ASHRAE Standard 90.1 (ASHRAE 2013) are widely used. However, previous studies (e.g., Duarte et al. 2013) have shown that such default schedules can differ significantly from actual occupancy patterns. Moreover, use of stochastic occupancy models to generate random non-repeating daily occupancy profiles based on standard-based schedules cannot compensate for lack of reliable on-site occupancy information (Tahmasebi and Mahdavi 2015).

In this context, the use of monitored data to generate presence probability profiles improves the reliability of building performance simulations. Using the generated Boolean occupancy interval data (as described in Sect. 10.3.4.1), the average occupancy profiles can be simply generated as the statistically aggregated daily probability profile of past presence data.

10.3.4.3 Generation of Boolean Daily Occupancy Profiles

For specific building performance simulation scenarios (e.g., predictive building systems control, or integration of stochastic occupant behavior models), Boolean daily occupancy profiles are required. Stochastic occupancy models (e.g., Page et al. 2008) can generate non-repeating daily occupancy profiles based on average presence probability profiles. However, more recent studies (Mahdavi and Tahmasebi 2016) have outlined that for deployment scenarios pertaining to building operation, in which one-to-one agreement between predicted and actual daily occupancy profiles is desired, simple non-stochastic approaches yield more reliable results.

In principle, non-stochastic approaches to derive Boolean daily occupancy profiles from presence probability profiles use a probability threshold to determine the state of occupancy at each interval (vacant or occupied). While this threshold can be defined in a heuristic process, Mahdavi and Tahmasebi (2015) suggested setting the threshold such that the area under the resulting binary occupancy profile is as close as possible to the area under the profile of presence probability.

10.3.4.4 Use of Electric Energy Meter Data to Determine Usage Profiles

The metering of electric energy use for plug loads or lighting is challenging in terms of equipment and data processing. Building and energy management systems usually integrate accurate meters on the whole building or floor level. This metering approach, however, is not promising for energy data collection with a high spatial resolution. Energy meters with wireless data transmission allows for new solutions toward more affordable and less invasive implementations.

Amongst different approaches to collect electric lights usage data, deployment of wireless energy meters represents a relatively simple one. Thereby, however, certain challenges have to be addressed. Specifically, battery-less wireless meters cannot report data when the lights are switched off. When the power supply is switched off, the meter keeps the value of the internal counter, but no message is communicated until the power supply is turned on again. This circumstance needs to be addressed in data processing. If the meter provides only a series of discrete

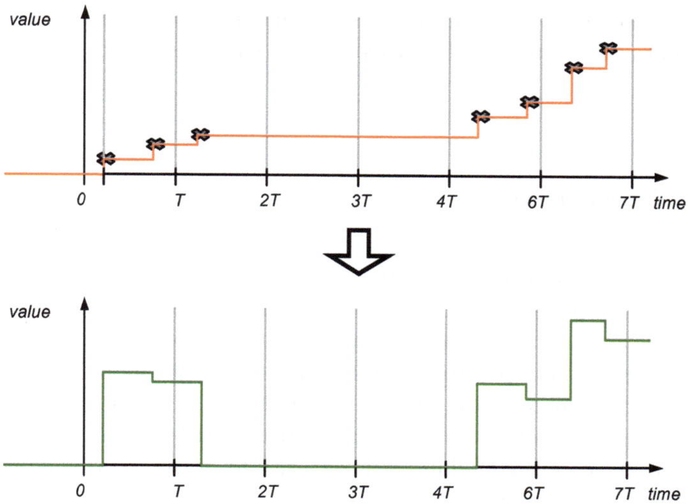

Fig. 10.16 Data pre-processing for light usage based on raw data from energy meters with wireless communication

energy values (i.e., no explicit power data), the raw data must be processed as follows (see Fig. 10.16):

- Calculation of the used energy for each interval based on the difference between successive measured values (crosses in the figure).
- Average interval loads derived by dividing used energy by interval duration.

For the aforementioned case of light usage monitoring with electrical energy meters, similar steps are taken. Typically, wireless energy meters report the values on a regular basis, e.g., every half hour. This can be used in further data processing. If the time difference between two measurements is larger than the sensor's default reporting interval, the last metered value can be taken to represent intervals without measurements.

10.4 Building Monitoring Repositories and Prototypical Implementations

Building monitoring data can support a multitude of use cases, from performance simulation of individual buildings to urban energy computing (Mahdavi et al. 2016a, b). This implies the need for distributed monitoring systems that are easy to maintain, offer the possibility to efficiently accommodate resource demand peaks, and can be integrated with Internet of Things (IoT) solutions (Schuss et al. 2016). The following sections describe related formal specifications that are applicable to (1) real-time monitoring applications, and (2) historic data repositories. Note that the subject matter of this section has also a number of implications for data security, privacy, and associated ethical considerations. The latter issues are discussed in Chap. 11.

10.4.1 System Design

A variety of monitoring systems (open source and proprietary) are available that serve different purposes and thus follow different system designs. Considering the pertinent use cases, one can conclude that modular monitoring applications are best suited to handle multi-purpose systems: compared to monolithic application designs, they offer more flexibility, maintainability, and optimized resource distribution. Independent software modules support the realization of a scalable architecture. Loosely coupled modules can be used to create distributed monitoring systems that fit various use cases, such as:

- Data processing
- Data retrieval
- Data persistence
- Data access
- Data presentation

To realize scalability and to increase reliability on the application level, modules must be deployable and usable on different machines. This practice requires the implementation of stateless core components, which allow new instances to be added during runtime. This is necessary to accommodate load peaks (i.e., monitoring occupancy during the morning rush hour). However, this approach also allows to remove instances in cooling-down periods to free resources that might be needed elsewhere.

Such a concept implementation requires a central distribution mechanism that routes requests between modules between physical machines that could be distributed across buildings within a city. For instance, a Java-based implementation could bundle the components with a Message Oriented Middleware (MOM) that can be accessed via a Java Message Service (JMS) Application Programming Interface (API). The communication process is then established by dynamically created queues (point-to-point) and topics (publish-subscribe). On the binary protocol level, there are various protocols that can be used, for instance the Advanced Message Queuing Protocol (AMQP). With this technique, it is possible to develop a robust system core that consists of variously deployable modules that can reside on different physical machines but use one centralized communication mechanism. The system core consists of at least a data access layer that implements the necessary web services to communicate building data via standard industry protocol implementations, such as OPC Unified Architecture (OPC UA), and Open Building Information Exchange (oBIX), or custom RESTful (Representational state transfer) APIs. Sensor data can either be requested from distributed sensor webs in real-time via sensor observation services (e.g., IoT networks) or from the application's data stores (e.g., historic data) via the persistence layer. The system core enriches the raw sensor data with further semantic information from the sensor ontology and builds a sensor data result set that is communicated to client applications or other application services via the MOM (internal) or web services (external).

10.4.2 Data Repositories

Creating high-performance data repositories implies a thorough requirement analysis. The stability of the data repository does not only depend on the amount of data that has to be stored, but also on the queries to be supported, necessary pre- and post-processing, number of requests, desired response time (real-time vs. historic data access), amount of data per request, distribution channels, caching, indexing and partitioning techniques, and many more. Depending on the data store concept to be adopted, the requirements will change. Most monitoring applications either store sensor data in files (e.g., CSV), relational databases (e.g., MySQL), NoSQL databases (e.g., MongoDB, Cassandra), embedded databases, in-memory databases, or NewSQL databases. Based on considerations pertaining to availability and market distribution, the following section discusses the implementation of data repositories with MySQL and Cassandra.

10.4.2.1 MySQL Data Repository

MySQL is a well-established RDMS that is used by a large number of applications. Relational databases use strict schema definitions that cannot be altered at runtime. Due to this fact, the data model has to be well planned. The proposed ER model shown in Fig. 10.17 focuses on a generic representation to support various building data sources. All measurement points are expressed in the datapoint table and hold information about

- The location of the measurement (by referencing a zone);
- If it is a virtual or physical sensor/actor (by referencing a data source); and
- The unit, accuracy, value range, deadband, sample interval, etc.

As it can be seen from Fig. 10.17, a generic repository abstracts physical and virtual sensors to datapoints (Zach et al. 2012a). Multiple datapoints can be aggregated into a physical device. Measurements are stored within the data table by adding a new row with datapoint name, timestamp, and value. Furthermore, additional zonal information can be stored inside the datapoint. The row-locking feature of MySQL default storage engine (InnoDB) enhances multi-user concurrency and performance.

Various performance optimization techniques can be used to minimize the required computer resources and to improve the database performance. Most commonly, access speed is optimized by indexing certain tables. As the data table

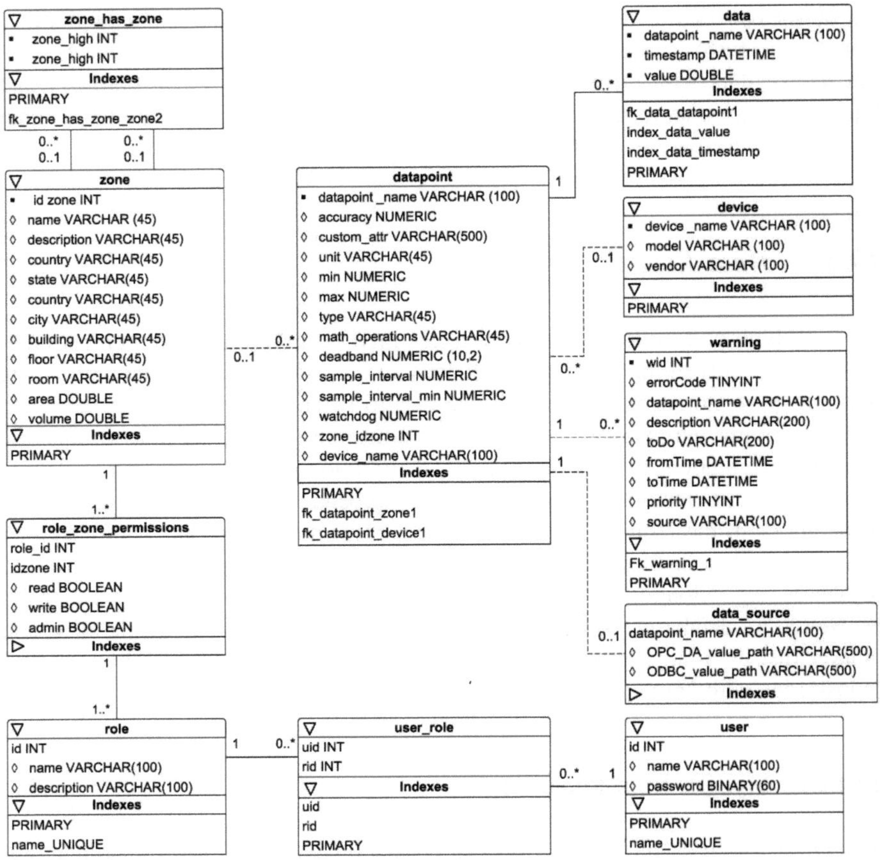

Fig. 10.17 Proposed generic relational repository ER-model

contains a majority of the systems data and is frequently queried it should be highly indexed. Depending on the Entity-Relationship (ER) model, a multi-column hash-index based on certain attributes (in this case *datapoint_name, timestamp, value*) is used to pre-sort measurements. The index resides in memory and enables the database to process datapoint- and timeframe-specific requests without accessing the hard disk.

To further improve access performance, partitions facilitate the organization of measurements into different virtual tables as illustrated in Fig. 10.18. The measurements are separated depending on the timestamps. However, the user only accesses one data table. This strategy allows using diverse indexing approaches on different partitions. Different partitions can thus be indexed to serve certain use cases (e.g., historic data analysis with large datasets, real-time data access with little datasets). The differing indexing techniques allow to reduce the overall memory usage and furthermore improves performance for highly loaded databases. Besides partitioning, specific use cases (large batch-processing, long-term data archiving,

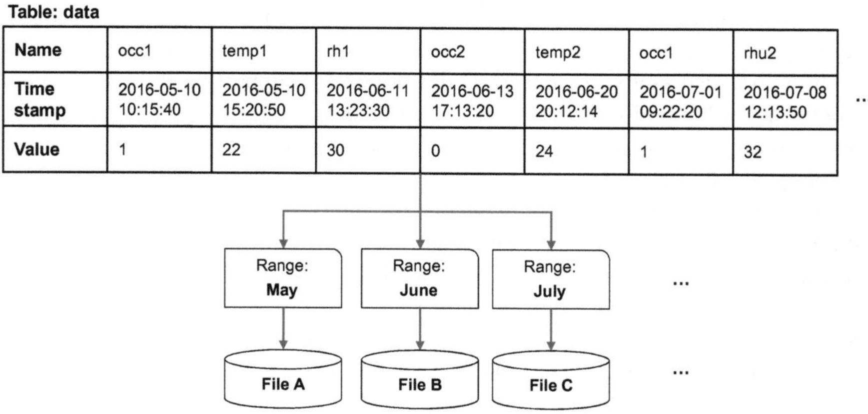

Fig. 10.18 Structuring measurements in different partitions according to their time stamps

etc.) can be optimized by using replication with adapted partition and index layouts. This setup is useful for large data repositories with diverse usage patterns.

10.4.2.2 Cassandra Data Repository

To test the introduced repository in a NoSQL environment, the model shown in Fig. 10.17 was migrated to a Cassandra cluster. Besides the optimized MySQL environment, managing a growing number of buildings proved to be a challenging task, as scaling partitioned relational databases is hard to accomplish and complex to maintain (Leavitt 2010; Hecht and Jablonski 2011). Relational data stores often sacrifice access speed to assure transaction security. However, transaction security is not a crucial condition for storing sensor measurements. In practice, sensors send a value every minute or even every second. Considering the nature and availability of building sensor measurements [intervals i = (1, 3600) sec], it can be concluded that the effect of one lost measurement on an entire day's data collection is negligible. Due to the reduced number of transactional checks, NoSQL stores provide better access speeds than relational databases. Due to the schemaless data handling, NoSQL databases are predestined for scaling applications. Tudorica and Bucur (2011) compared various databases and found that NoSQL data stores (specifically Cassandra) provide a better write and read latency in a write intensive environment than relational products (MySQL). At approximately 7000 read or write operations MySQL became unresponsive.

NoSQL databases typically distribute incoming data randomly across nodes in a cluster. This is why query design is a crucial part in developing NoSQL depending solutions. In the worst case, not supported queries have to be implemented in the client application at the cost of performance. The generic design of the monitoring application defines a column family for every *datapoint* in the database. To improve access speed, keys are indexed. Sensor data are stored in a temporally linear matter.

The introduced key space design is developed for supporting historic data collections, as well as real-time access. If a sensor stores a value every second, a search algorithm must access one month of data, which equals a maximum of 2.6×10^6 randomly distributed values.

To improve real-time access, a secondary index on the unique measurement timestamp is applied, which removes the need for sorting in query procedures. Cassandra defines a maximum of 2 billion cells (rows x columns) by partition. By considering the proposed example of a sensor recording a measurement every second, one sensor executes 3.15×10^7 operations per year. To optimize data access, the data is divided into 12 monthly shards. This means that one partition can hold 1.6×10^8 measurements, which equals 5.3 years of data before new partitions are necessary. If sensors commit a measurement every minute, one partition can hold approximately 300 years of data per sensor. The maximum file size of 5 TB is never reached before the maximum number of cells is exceeded.

10.4.3 Prototypical Implementation—Monitoring System Toolkit (MOST)

The introduced monitoring system design concepts were prototypically implemented in the monitoring system MOST (2012). It was optimized to handle multiple building data on an urban level (Glawischnig 2016). Thus, the discussed implementation of redundant, stateless core components was a vital step. Figure 10.19 shows the latest application design (Glawischnig et al. 2014). As can be seen, the application consists of four layers that communicate internally via the MOM. The persistence layer offers multiple repository implementations.

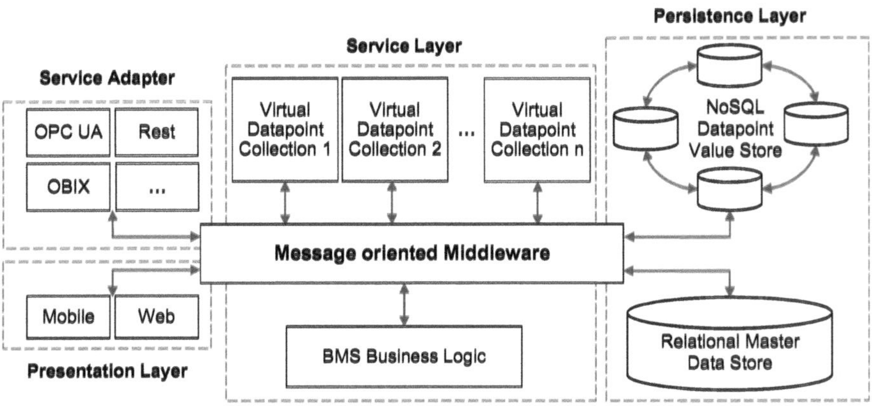

Fig. 10.19 A modular monitoring framework architecture

Depending on the use case, either the introduced MySQL repository or Cassandra repository can be used as a sensor data store. The BAS business logic and virtual datapoint implementations, which are written in the MOST-domain specific language, reside in the service layer. Virtual sensors are extensively used to support the integration of third party simulation software and to offer automatized interfaces to frequently applied mathematical procedures. Furthermore, the ontology that is used to enrich sensor data resides in the BAS business logic. The service adapter holds implementations of various standard industry protocols, such as OPC UA and oBIX, as well as a custom RESTful interface to offer access to client applications. Finally, the presentation layer currently consists of a web-application and a mobile app. All modules are loosely coupled and can thus be redundantly deployed on different physical machines, but still share the same application context.

10.4.4 Module Overview

Currently, the implemented MOST modules include (Zach et al. 2014):

- Connector: a driver for the data-source (sensor/actor) to MOST;
- MySQL: handles database access for meta-data like datapoints and zones;
- Neo4j: handles database requests for datapoint measurements stored in Neo4j;
- Cassandra: handles database requests for datapoint measurements stored in Cassandra;
- Calibration: calibrates a building simulation model—currently only EnergyPlus (EnergyPlus 2016)—in an automated and periodic manner;
- Virtual Datapoint: contains several implementations of the Datapoint interface to provide common access for data that is not directly measureable;
- MOST Server: routes requests between different MOST modules (Only required for distributed deployments);
- REST: exposes MOST data as RESTful web service;
- OPC UA: exposes MOST data through OPC Unified Architecture;
- oBIX: exposes MOST data through oBIX (oBIX 2016); and
- Web: provides an out of the box web application to query, visualize and export monitored data.

10.4.5 Virtual Sensor Implementation

As mentioned before, virtual sensors are a vital concept in the proposed monitoring structure. Virtual sensors can provide data about phenomena that cannot directly be measured with physical sensors (e.g., average temperature across multiple zones). Currently, MOST includes the following virtual sensor implementations in the most-vdp module, which can easily be extended, as the modules can be redundantly deployed:

- A virtual datapoint, which takes the mean surface temperature of a radiator and calculates its heating power based on geometric information of the radiator and the surrounding room temperature.
- A virtual datapoint wrapping the MOST domain specific language (most-DSL) implemented in Scala. It enables users to weave datapoint values into mathematical expressions where particular values are evaluated at runtime based on the requested timeframe for evaluation. An expression computing the average temperature in °C of two datapoints "tem1" and "tem2" would be written as follows:

```
(dp("tem1") + dp("tem2")) / 2.
```

- Integrating most-DSL as a VDP allows nesting an arbitrary graph of most-DSL expressions, whereas loops are not allowed. Assuming the last expression would be accessible as the VDP "avgTem", a new VDP can be built to convert the result to °F:

```
dp("avgTem") * 1.8 + 32.
```

10.5 Conclusion

This chapter focused on a number of necessary high-level efforts and developments to better realize the multi-faceted potential of building monitoring in supporting the building design, delivery, and operation processes. Toward this end, the chapter addressed the need for richly structured approaches to the collection, storage, sharing, and analyses of monitored data, particularly as relevant to the occupants' presence/activities in (and their impact on) the buildings. Specifically, the present chapter introduced an ontology for the representation and incorporation of multiple layers of data (occupants' presence, control actions, indoor climate, outdoor conditions, devices, and equipment) in pertinent computational applications, such as building performance simulation tools and building automation systems. Moreover, a number of typical data processing requirements relevant to building monitoring data repositories were described. Finally, to illustrate modular and scalable monitoring system architectures, the requirements, characteristics, and specific implementations of data repositories for structured collection, storage, processing, and multi-user exchange of monitored data were explained.

References

ASHRAE (2013) ASHRAE 90.1-2013: Energy standard for buildings except low-rise residential buildings. American Society of Heating, Refrigerating and Air-Conditioning Engineers, Atlanta

Böhms M, Rieswijk T (2015) Linked energy data: enabling monitoring and decision support for improved energy management. In: IEEE International Conference on Engineering, Technology and Innovation/International Technology Management Conference (ICE/ITMC), Belfast, Northern Ireland, pp 22–24 June 2015

Duarte C, Wymelenberg KVD, Rieger C (2013) Revealing occupancy patterns in an office building through the use of occupancy sensor data. Energy Build 67:587–595

EnergyPlus (2016) EnergyPlus engineering reference. http://apps1.eere.energy.gov/buildings/energyplus/. Accessed 20 Feb 2016

Gilani S, O'Brien W (2016) Review of current methods, opportunities, and challenges for in-situ monitoring to support occupant modelling in office spaces. J Build Perform Simul. doi:10.1080/19401493.2016.1255258

Glawischnig S (2016) An urban monitoring system for large-scale building energy assessment. Dissertation, TU Wien

Glawischnig S, Hofstätter H, Mahdavi A (2014) Applications of web-technologies in building information systems: the case of the Monitoring System Toolkit (MOST). In: Proceedings of 2nd ICAUD, Tirana, 2014

Guerra-Santin O, Tweed CA (2015) In-use monitoring of buildings: an overview of data collection methods. Energy Build 93:189–207

Hecht R, Jablonski S (2011) NoSQL evaluation: a use case oriented survey. In: Proceedings of International Conference on Cloud and Service Computing, IEEE, Hong Kong, pp 336–341

IEA (2016) IEA-EBC Annex 66. http://www.annex66.org/. Accessed 30 Mar 2017

Leavitt N (2010) Will NoSQL databases live up to their promise? IEEE Comput 43(2):12–14

Mahdavi A (2011a) People in building performance simulation. In: Hensen J, Lamberts R (eds) Building performance simulation for design and operation, 1st edn. Taylor & Francis Group, New York, pp 56–83

Mahdavi A (2011b) The human dimension of building performance simulation. In: Proceedings of the 12th International IBPSA Conference, Sydney, 14–16 November 2011, pp K16–33

Mahdavi A, Bachinger J, Suter G (2005) Toward a unified information space for the specification of building performance simulation results. In: Proceedings of the 9th International IBPSA Conference, Montreal, 15–18 August 2005, pp 671–676

Mahdavi A, Tahmasebi F (2015) Predicting people's presence in buildings: an empirically based model performance analysis. Energy Build 86:349–355

Mahdavi A, Glawischnig S, Ghiassi N (2016a) Urban energy computing: a multi-layered approach. In: Proceedings of CEUR Workshop, pp 130–135

Mahdavi A, Glawischnig S, Schuss M, Tahmasebi F, Heiderer A (2016b) Structured building monitoring: ontologies and platform. In: Mahdavi A, Martens B, Scherer R (eds) eWork and eBusiness in Architecture, Engineering and Construction: ECPPM 2016. CRC Press, pp 651–658

Mahdavi A, Taheri M (2016) An ontology for building monitoring. J Build Perform Simul. doi:10.1080/19401493.2016.1243730

Mahdavi A, Tahmasebi F (2016) The deployment-dependence of occupancy-related models in building performance simulation. Energy Build 117:313–320

MOST (2012) Monitoring system toolkit. http://most.bpi.tuwien.ac.at. Accessed 10 March 2016

oBIX (2016) Open Building Information Xchange. http://www.obix.org. Accessed 20 Jan 2016

Page J, Robinson D, Morel N, Scartezzini JL (2008) A generalized stochastic model for the simulation of occupant presence. Energy Build 40(2):83–98

Roda F, Musulina E (2014) An ontology-based framework to support intelligent data analysis of sensor measurements. Expert Syst Appl 41(17):7914–7926

Schuss M, Glawischnig S, Mahdavi A (2016) A flexible and scalable approach to building monitoring and diagnostics. In: Christodoulou S, Scherer R (eds) eWork and eBusiness in Architecture, Engineering and Construction: ECPPM 2016. CRC Press, pp 235–242

Tahmasebi F, Mahdavi A (2015) The sensitivity of building performance simulation results to the choice of occupants' presence models: a case study. J Build Perform Simul. doi:10.1080/19401493.2015.1117528

Tudorica BG, Bucur C (2011) A comparison between several NoSQL databases with comments and notes. In: Proceedings of ProEduNet 2011, Iasi, pp 1–5

Yan D, O'Brien W, Hong T, Feng X, Gunay HB, Tahmasebi F, Mahdavi A (2015) Occupant behavior modeling for building performance simulation: current state and future challenges. Energy Build 107:264–278

Zach R, Hofstätter H, Mahdavi A (2014) A distributed and scalable approach to building monitoring. In: Mahdavi A, Martens B, Scherer R (eds) eWork and eBusiness in Architecture, Engineering and Construction: ECPPM 2014. CRC Press, pp 231–236

Zach R, Schuss M, Bräuer R, Mahdavi A (2012a) Improving building monitoring using a data pre-processing storage engine based on MySQL. In: Gudnason G, Scherer R (eds) eWork and eBusiness in Architecture, Engineering and Construction: ECPPM 2012. CRC Press, pp 151–157

Zach R, Glawischnig S, Hönisch M, Appel R, Mahdavi A (2012b) MOST: an open-source, vendor and technology independent toolkit for building monitoring, data pre-processing, and visualization. In: Gudnason G, Scherer R (eds) eWork and eBusiness in Architecture, Engineering and Construction: ECPPM 2012. CRC Press, pp 97–103

Chapter 11
Ethics and Privacy

Chien-fei Chen, Marcel Schweiker and Julia K. Day

Abstract When conducting research, one of the primary considerations should be to maintain high scientific and ethical standards, including protecting the rights and benefits of all participants. Researchers should take great care to ensure scientific validity during the design of a study; at the same time, ethical conduct should not be considered a researcher's burden, but rather an important consideration for any type of research. This chapter provides guidelines for ethics approval by discussing common types of ethics applications, the concepts of informed consent, privacy, and confidentiality, and additional ethical considerations particular to occupant research. While ethical review processes differ across countries and institutions, this chapter provides basic guidance to researchers in the field of occupant behavior to (a) improve their interactions with ethics review boards, (b) help them meet crucial requirements, and (c) ensure their studies are conducted ethically.

11.1 Introduction

In previous chapters, the importance of studying occupants' behavior was highlighted, and a variety of research methods were introduced. The underlying objectives of occupant research include efforts to (a) improve occupants' experi-

C. Chen (✉)
NSF Center for Ultra-Wide-Area Resilient Electrical Energy Transmission Networks (CURENT), Department of Electrical Engineering and Computer Science, University of Tennessee, 508 Min H. Kao Building, 1520 Middle Dr., Knoxville, TN, USA
e-mail: cchen26@utk.edu

M. Schweiker
Building Science Group, Faculty of Architecture, Karlsruhe Institute of Technology, Englerstr. 7, 76131 Karlsruhe, Germany
e-mail: Marcel.Schweiker@kit.edu

J.K. Day
School of Design and Construction, Washington State University, PO Box 645815, Pullman, WA 99164, USA
e-mail: julia_day@wsu.edu

A. Wagner et al. (eds.), *Exploring Occupant Behavior in Buildings*,
DOI 10.1007/978-3-319-61464-9_11

ences within built environments, (b) minimize occupants' efforts to satisfy their needs, e.g., in terms of thermal, visual, or aural requirements, and (c) reduce the impact of occupancy on natural resources. While occupant behavior researchers conduct important research and enjoy freedom of inquiry and expression, they must also hold their work to high ethical standards, including protecting the rights and benefits of participants (Canadian Institutes of Health Research et al. 2014). Primarily, these efforts need to consider the protection of an individual's privacy and physical and mental safety. Moreover, conducting research that involves human participants also needs to ensure that participants' time and efforts are not wasted due to a poorly designed study, e.g., one that is unable to answer a given research question. Therefore, part of a researcher's ethical conduct is to ensure scientific validity during the design of a study (see Chap. 3 for an introduction to research designs). Ethical conduct should not be considered as a burden to a researcher, but rather as an important consideration for any type of research to minimize potential harm to participants, especially when considering the potentially high level of personal interaction that accompanies occupant behavior studies or experiments. In addition, occupant researchers have the responsibility, as members of the broader research community, to build the trust and confidence with the public by conducting research ethically (Canadian Institutes of Health Research et al. 2014).

The European Commission defines ethics as "the systematic reflection on, and development of standards for, right and wrong conduct and their application to situations in which such standards may be violated" (Shelley-Egan et al. 2015, p. 18). To ensure that relevant standards are met and that the behavior of a single researcher (or team) does not harm the research community or participants, research involving human participants most likely requires ethics approval by some kind of formal board or body. Ethics approval requirements and the processes for obtaining ethics approval differ from country to country, and even from university to university; nevertheless, internationally, research ethics codes of conduct consistently evoke principles such as beneficence, justice, and autonomy.

This chapter first provides a brief explanation of the institutions involved in the process of ethical review and review categories (e.g., exempt, expedited, and full board review), and then discusses ethical considerations of participant recruitment, including risks and anticipated benefits. Next, aspects of privacy, confidentiality, and informed consent are introduced, followed by submission procedures for ethics clearance and a brief introduction of the research debriefing process. Then, implications for multiple-site and cross-country studies are discussed. Later, the chapter introduces typical written policies and procedures and offers tips for obtaining ethics approval for a project. Wherever possible, both the importance of an element of ethical conduct and the mechanisms for compliance are discussed. Finally, the chapter considers potential changes in human subjects' protection programs relating to publicly available Internet data, e.g., from social media platforms. Overall, this chapter's discussion of ethics applies to many occupant behavior research methods, including in situ, laboratory experiments, surveys, and interviews. Interested readers can find more details about each of these in Chaps. 6 (in situ), 7 (laboratory), and 8 (surveys and interviews).

11.2 Institutions Involved in Ethical Review Processes

When a researcher—whether student or professional—collects data by interacting or intervening with an individual or their confidential information, this individual is known as a “participant” (Protection of Human Subjects 2009a). In most cases, research activities involving human participants are reviewed, monitored, and approved by a group of individuals independent from the researchers conducting the studies, i.e., an independent committee. Depending on the country or organization, there are different names for these review boards, including “ethical review board,” “ethics committee,” “research ethics board” (“REB”), or “research ethics committee” (“REC”). This chapter’s authors use the term “institutional review board” (or “IRB”), borrowed from the USA context, to refer to ethics review boards more broadly.

Typically, an IRB reviews and oversees all research activities involving human participants (including human biological samples, e.g., blood or tissue), in addition to studies involving animals or those potentially related to military applications or for malevolent, criminal, or terrorist abuse in an institution (e.g., involving hazardous materials). Ethics committees are in place to (a) ensure the rights, safety, and welfare of human research participants, and (b) enforce compliance with all applicable federal and state laws/regulations.

Specific organizations may have committees or groups dedicated to ensuring research of the highest ethical standard is conducted—for example, the European Network of Research Ethics Committees (EUREC) or the World Health Organization (WHO) Research Ethics Review Committee [for more information on the international compilation of human research standards see the USA Department of Health and Human Services (2017)]. Not every institution; however, has its own ethics committee or ethics application process, and the specific regulations may be different. For example, the United Kingdom (UK) has 104 Research Ethics Committees (RECs) that differ by region and purpose, although they all focus on reviewing and approving various forms of research. The aforementioned European Network of Research Ethics Committees (EUREC) works specifically with existing European RECs from various countries to coordinate the process of international ethical review. Because similar entities exist in many countries, researchers organizing international studies should follow the ethical principles presented in the United Nations Declaration of Human Rights, the Nuremberg Code, the Declaration of Helsinki, and the Belmont Report, depending on the ethics of researchers’ institutions (Markham and Buchanan 2012).

11.3 Review Categories

Before starting an ethics application, investigators should determine whether they need to submit their research to an ethics board. Some institutions do not require an ethics approval for class research projects [e.g., the IRB at the University of Michigan (2017)]; others require an ethics approval to present research results outside of their own institution [e.g., the IRB at the University of Tennessee (2017)]. Therefore, researchers should thoroughly review their respective institution's requirements. If any doubts exist regarding planned procedures, some ethics boards will accept voluntarily (non-required) submissions for review.

When planning to submit an ethics application, it is wise for investigators to first consult with their ethics board to determine whether there are different types or levels of review and to identify the review category that matches the proposed project. The review type determines which questions need to be answered in the application process. For example, there are three major types of review in the USA: exempt, expedited, and full board review. Exempt and expedited reviews apply to projects considered minimal risk to participants—that is, the risk or discomfort involved in participation is no greater than that ordinarily encountered in daily life. In the field of occupant research, these are the most commonly used review types and will be further discussed in the paragraphs that follow. Full board review is not common in occupant studies because many of them use non-intrusive behavioral observation without identifying personal information (e.g., anonymous survey or sensor data), though some studies may include participants' identification (e.g., names). Still, some studies in occupant research may involve above minimal risk and thus require full board review. Likewise, any research involving vulnerable participant groups (e.g., children, prisoners, institutionalized individuals) is subject to full board review.

11.3.1 Exemption

In some situations, a research study may be exempt from review. Ethics exemption does not translate to ethics being abandoned and that researchers have no responsibilities to the participants; rather, it solely means that it is not mandatory to go through the ethics or IRB review and approval process (Penslar 1993). Exempt status may include, for example, studies using secondary data analysis (Protection of Human Subjects 2009b). Research studies may also qualify for exempt status if there is very minimal or no risk involved. For example, exempt studies must not include identifiable information or disclosure of responses that could risk the participants' employability or reputation (Oakes 2002, p. 458). Within the context of occupant research, a web-based survey distributed to occupants of a building would likely fall into this category if it does not collect or store Internet Protocol

(IP) addresses, personal identifiers, or other information allowing one to identify the specific building location (as that could be linked to individuals in the building).

11.3.2 Expedited Review

Two kinds of ethics applications may undergo an expedited review procedure: new applications posing minimal risk (or other met requirements), and previously authorized applications where minor changes to the protocol are being requested (Oakes 2002). Data collection methods relevant to social science and occupant behavior studies that may pose a minimal risk and thus receive expedited review include: the compilation of data from voice, video, digital, or image recordings made for research purposes, or research employing a survey, interview, focus group, or oral history. Most institutions offer an expedited review for anonymous surveys, but this is not always the case, and so it is important to be familiar with the ethical guidelines at each home institution.

11.4 Recruitment of Participants

The following discussion focuses on procedures that researchers should consider when selecting participants. Readers can refer to Chaps. 3 and 8 for more details on sampling and sample size.

11.4.1 Selection of Participants

The selection of participants should consider fairness and equity. In other words, researchers should be concerned with the issue of inclusiveness in that neither a particular individual, group, nor community should "bear an unfair share of the direct burdens of participating in research, nor should they be unfairly excluded from the potential benefits of research participation" (Canadian Institutes of Health Research et al. 2014, p. 47). Therefore, it is important to decide the criteria for including or excluding participants. There are three primary considerations:

(1) Equitable selection regarding gender, race, ethnicity, etc., without personal bias, unless the use of one particular group has significance to the purposes of the study, which needs to be specified—for example, limiting the number of independent variables by choosing only young females.
(2) Fair distribution of benefits among the populations (e.g., findings would serve not only high-income people who can afford the particular technology being investigated, but also low-income people).

(3) Provision of additional safeguards for vulnerable populations, as defined above (Collaborative Institutional Training Institute (CITI) 2016).

In all cases, it is important that participants not be pressured to participate during the selection process. Such pressure can unintentionally be applied—for example, if a group of people is asked for their participation at the same time, some people may push others to participate.

11.4.2 Vulnerable Populations

Researchers should pay special attention to certain types of participants who have the potential to be more vulnerable than the rest of the community to risks, including children, prisoners, pregnant women, individuals with mental or physical disabilities, economically disadvantaged persons, educationally disadvantaged persons, terminally ill persons, students and employees affiliated with the organization conducting the study, and so on (University of Tennessee IRB 2012). Research involving vulnerable groups sometimes requires full board review, and so researchers should leave ample time to consult with the review board, researchers and personnel frequently working with those groups, as well as people from those groups, if possible. Although special attention needs to be paid to the requirements of vulnerable participants, this population should not be excluded from consideration as potential participants. New regulations enforced by the EU in 2014 focus on allowing vulnerable participants to participate in research when the research is potentially beneficial to the particular group being researched (Diaz et al. 2015).

11.5 Risks and Anticipated Benefits

In the case of research studies, "risk" can be defined as "the probability of harm or injury (physical, psychological, social, or economic) occurring as a result of participation in a study. Both the probability and magnitude of possible harm may vary from minimal to significant" (Penslar 1993). Researchers should reflect on the probability and magnitude of each identified potential type of risk.

11.5.1 Identification of Risks

Selected study design features, as well as a particular intervention program, may pose potential risks to research participants. In social and occupant behavior studies, for example, the risk of invasion of privacy and possibly compromising confidentiality are among the most important risks that need to be minimized by the research design and data collection methods.

11.5.2 Minimal Risk

Minimal risk is defined in USA federal regulations 45 CFR 46.102(i) as follows: "the probability and magnitude of harm or discomfort anticipated in the proposed research are not greater, in and of themselves, than those ordinarily encountered in daily life or during the performance of routine physical or psychological examinations or tests". Minimal risks are referenced to a normal healthy adult person; adjustment for illness or disability is not permitted (Protection of Human Subjects 2009c). Maintaining minimal risk is usually considered a priority in research applications because it leads to the possibility of an expedited review (Protection of Human Subjects 2009d). For example, participants exposed to thermal conditions in laboratory studies can be regarded as participating in a minimal risk study only when the thermal conditions are within the observed conditions of a regular office and/or residential space. Note that one of the primary responsibilities of ethics committees is to determine if the research presents greater than minimal risk only after the risks have been identified (Penslar 1993).

11.5.3 Hard and Soft Impacts

The ethical risks that accompany research can be broken down further into the potential level of impact the research methods might have on a participant. Two levels have been identified in the literature: hard impacts and soft impacts. The former involves research that affects health or physical well-being, while the latter involves research that impacts a person's social goals or individual identity (Shelley-Egan et al. 2015). While the effects of hard impacts on research participants may be measured more easily than soft impacts, any impact on participants should be considered serious and an explanation of the ways in which these two impacts vary should be discussed.

11.5.4 Risk Issues Specific to Occupant Behavior Research

Common risks associated with occupant behavior research and procedures to reduce them are as follows:

1. *Experimental study design*: Ethics and data protection in laboratory studies can be more challenging compared to anonymous survey-based investigations, in part because of the nature of the spaces used for experiments. Often, the number of

people in the laboratory at a particular time is limited due to the restrictions of the corresponding facility, e.g., a maximum of four persons can participate at the same time in a small laboratory. As a result, it may be relatively easy for researchers or others to identify specific participants. In field studies, the same problem could arise in the case that several office characteristics (e.g., orientation, floor level, location) are provided.

Another type of risk arises in studies with between- and within-subject designs, where specific data are linked to each participant in order for researchers to track the results of pre- and post-experiments and compare results among participants. For example, participants may be asked their birth date, weight, height, gender, and so on. Collecting these data from participants would not be anonymous since these are often required variables for certain thermal comfort models (e.g., to calculate the body surface area). However, in this example, the combination of these sensitive four variables (i.e., age, weight, height, and gender) is considered "health-related data" because they can be combined to identify participants' health status (e.g., obesity). There is thus a risk of identifying private information and of this information leaking.

The issue of privacy and leaking participants' personal information may remain even after preliminary precautions are taken. For example, deleting the date stamps from physical and questionnaire data would lower the chances that someone outside the research team could match the date of participation to the unique combination of personal characteristics. However, when outdoor environmental data are also logged (as is commonly done), it is possible to use an algorithm to compare the logged data with publicly available data about the weather during the experimental period to identify a specific date. Thus, deleting date stamps is not a sufficient risk mitigation measure in this case.

An acceptable risk mitigation procedure for the above scenario is for data to be pseudonymised and for any information linking a person's name to these data to be deleted—whether it is in the form of a (digital or non-digital) list connecting the name of the participant with the experimental days they were assigned to, or an email sent to the participant informing them about their appointments. As long as such an email or other trail persists (e.g., in backups of the email server), collected data cannot be regarded as anonymous.

2. *Video data collection*: Surveillance video data presents notable potential ethics risks. For example, employees in a building may fear the loss of promotion opportunities—or worse, their job—if footage captured via video recording reveals any suspicious or undesirable behavior, including a lack of presence. This uneasiness can lead to a loss of productivity if the employee being recorded fears performing as he or she normally would.

To enhance the confidentiality of taped materials in workplaces, the following guidelines should be followed:

(1) The surveillance should be done with equipment not owned by the workplace to prevent it from being intercepted or stored beyond the parameters of the

occupant behavior study (and this matter should be communicated to employees);
(2) Only the researchers and authorized study personnel can obtain access to the recordings;
(3) The data are analyzed at a group level, where individual-level data are used only when necessary. An example of when individual-level data may be necessary is a study evaluating individual behavioral changes in response to a newly implemented energy conservation program in the workplace;
(4) If possible, the equipment should also be clearly marked and visible to protect employees' feelings of security and autonomy; and
(5) Alternative procedures are used to obtain consent if a signed consent form is the only link between the recording and the individual's identity.

Some researchers have used the "implicit consent" scenario by posting a sign allowing participants to know that "videotaping" is taking place in a particular public area for research purposes (Gutwill 2002, 2003). It is important to note that privacy is a subjective matter and therefore researchers need to constantly remind themselves of the procedures to protect participants' privacy.

3. *Sensor data collection*: Motion sensors and remote sensing are other tools that can pose privacy risks in occupant studies and make previous restrictions on data largely irrelevant. This is because changes in spatial monitoring technologies have increased opportunities for misuse—fears similar to the ones provoked by digital recording are likely (Slonecker et al. 1998). The same applies to any type of sensory data which would allow identifying a single occupant, as the observed occupancy could be in contrast to the stated occupancy of an employee towards the employer. Until a comprehensive legal framework is in place for the use of sensing technology, the procedures to ensure ethical research must be similar to the steps described to reduce the privacy risks of video surveillance.

4. *Smart phone applications* (*Apps*): A unique concern is directed towards the utilization of smart phone applications ("apps") to gather data. The users must be made aware of the data being collected and the intended use when accepting the terms of agreement and downloading the application. However, because people do not always read the terms closely, a separate consent form should also be incorporated to explicitly reiterate the purpose of the collected data and how it will be managed (Miluzzo et al. 2010). Anticipated benefits should always outweigh potential risks. Furthermore, as the use of phone applications in occupant behavior studies would facilitate knowledge of individuals' identity, identities should be coded and data encrypted to remain confidential. This way, both privacy and confidentiality are ensured, and users will not risk being reprimanded for their individual behaviors. Again, data should only be available to the researchers and must immediately be destroyed once the study has concluded.

5. *Secondary data*: Many researchers do not collect their own (primary) data, and instead use data collected from another sources, i.e., secondary data. Secondary data use may include cases where the original researcher uses their own previously collected data for further analysis by returning to the same participants after an

initial analysis has been completed (Grinyer 2004). It may also include the secondary use of some datasets, e.g., the American Time Use Survey (Bureau of Labor Statistics at the U.S. Department of Labor 2003) or utility companies' smart meters. The use of these latter datasets may be relatively uncontroversial, but there are still ethics issues relating to data linkage and data security, for example. In general, for studies using secondary data, an IRB ethics approval including procedures for the protection of privacy and data is still required; however, informed consent from original participants might not be required.

11.5.5 Anticipated Benefits

Broadly speaking, benefits may include: "increased knowledge, improved safety, technological advances, and better health" (Penslar 1993, p. 2). It is good practice to disclose to potential participants any anticipated benefits stemming from their involvement in the research, not only for ethical purposes, but also because a person's decision to participate might depend on the perceived benefits associated with their participation. Direct benefits to participants may be easier to identify in other fields, such as medicine (e.g., receiving a medical treatment), than for occupant studies; nonetheless, depending on the type of study, direct and indirect benefits likely exist at the individual and/or societal level—for example, fixing faulty controls for increased occupant comfort. More broadly, a potential benefit is circulating aggregated survey results among the occupants so that they learn new things about their building and controls. A better understanding of behaviors in a particular building or community may, in turn, influence personal thermal or visual comfort, provide useful insights for building designers and operators to better design new buildings, and lead to better building operation/maintenance systems (O'Brien et al. 2013). Anticipated benefits may also be even more broad, such as promoting knowledge about the relationships between energy use and job satisfaction or occupant productivity, energy concerns, reduction of greenhouse gas emissions, and so on (O'Brien and Gunay 2014).

It should be noted that any monetary payment offered to participants for their participation in a study should not be considered a benefit; rather, it is considered compensation for their time and effort. Although participation in research has the ability to elicit personal reward or be considered a humanitarian contribution, these subjective benefits should not be a factor in an ethics committee's analysis of the potential benefits and risks that the study entails (Penslar 1993).

11.6 Privacy and Confidentiality

Privacy is defined as the amount of control an individual has over the degree of physical or intellectual information that is shared about them with another person (U. S. National Commission for the Protection of Human Subjects of Biomedical and Behavioral Research 1979). In other words, privacy is about individuals having a sense of control regarding the access that others have to their information (Penslar 1993). For research purposes, privacy should be considered from the point of view of the participant, not the researchers or the ethics committee. Importantly, informed consent does not guarantee privacy (U.S. National Commission for the Protection of Human Subjects of Biomedical and Behavioral Research 1979).

Confidentiality, an extension of privacy, refers to the maintenance and accessibility of identifiable data to prevent inappropriate disclosure of protected information (Office of Research. University of California Irvine 2017). Specifically, confidentiality is applied when an individual discloses private information to someone of authority (e.g., a researcher) with the mutual understanding that the privacy of the information will not be compromised without permission (Penslar 1993). For example, the two main principles of the Belmont Report (U.S. National Commission for the Protection of Human Subjects of Biomedical and Behavioral Research 1979) support the ideas of privacy and confidentiality. The first principle, respect for individuals, argues that an individual, as an autonomous agent, should be able to exercise their autonomy freely. An individual has a fundamental right to privacy, and thus, a right to keep confidential information private with the only exception being special circumstances where legal requirements may require disclosure. Note that there may be exceptional circumstances in which there are legal requirements to disclose individual data to authorities, or when data may not remain confidential to the project staff (Office of Research. University of California Irvine 2017). The second principle, beneficence, entails the obligation to maintain participants' privacy and confidentiality, and to protect participants from potential harms, including psychological and social harms and criminal or civil liability. This is a particularly important issue for occupant behavior researchers because individuals' participation in some surveys and experimental studies could be identified as mentioned earlier.

11.6.1 Anonymous Data

For data to be considered anonymous, those data must be collected without any use of personal or identifiable information. Through collecting anonymous data from participants, potential ethical and legal concerns about confidentiality can be addressed, but concerns about confidentiality and security of the data still remain (Carroll et al. 2004). Note that anonymity and confidentiality are not interchangeable terms when applying for ethics approval or writing up study results.

11.6.2 Privacy in the Use of Personally Identifiable Records

Researchers should always respect the privacy of research participants. No personally identifiable records should be disclosed or published without the consent of other researchers or ethics organizations. Sufficient protections should be present to prevent any potential harms resulting from an invasion of privacy or a violation of confidentiality.

11.6.3 Potential Steps to Protect Participants' Privacy

In sum, the following list explains potential steps a researcher could take to protect participants' privacy. Note that the list is in no particular order.

(1) For any study, researchers are expected to justify the purpose of asking participants' personal information or information linked to personal identifiers.

(2) Data should be securely stored to eliminate the risk of a third party getting access to the dataset by making it as limited as possible (see more discussion in Sect. 11.6.4).

(3) Any potential links between a participant and their identifiers or pseudonym data are to be avoided. For example, data protection requirements by some universities (e.g., Karlsruhe Institute of Technology, Germany) stipulate that researchers are not allowed to have an electronic version of the list with participants' real names and the times that they participated. The existence of such a list is often unavoidable when participants participate once or multiple times in studies during non-consecutive days, and so such a list should exist only in a non-digital paper version protected by researchers and locked in a secure location (e.g., the researcher's private office).

(4) Researchers should assign each participant a random and unique identification (ID) for data collection instead of using their real name. The same applies to office numbers in a field study. This specific ID will be linked to survey responses and other data. Researchers are expected to eliminate any evidence providing the participants' (or offices') real identities or contact information immediately following their number assignment. According to the UK Data Archive, "researchers can create an anonymization log of all replacements, aggregations or removals made; such a log should be carefully stored separately from the anonymized data file" (University of Essex 2012).

(5) Researchers should ensure that participants' personal information (as well as survey, sensor, or other laboratory-related data) is kept completely confidential. Only the investigators can access these data.

(6) Researchers should remove all the apps or online devices/programs for tracking participants immediately after the study is completed.

(7) Researchers should erase data collected through computers from each computer used after the data are downloaded so that no one can access this information.

(8) In quantitative research, the results of the study should be reported only as statistical averages and never in terms of individuals. In qualitative or mixed methodology research, reporting at the individual level is permissible, but names and identifying information should be removed to ensure privacy and confidentiality of the participants.

11.6.4 Data Storage, Processing, and Sharing

In addition to privacy protection, an explanation of a researchers' plan for data collection, processing, analysis, storage, and data sharing procedures needs to be addressed in an ethics application. Penslar (1993) offers many important points to consider regarding data storage, processing, and sharing, including:

- How will the collected data be documented and stored?
- Has a plan been developed to reduce the likelihood of risk during data collection, processing, and sharing?
- Can information be supplied to the ethics committee if unforeseen results are found?
- Does the researcher's institution have a board to monitor data and its safety?
  - If so, should the board recognize projects that are still under evaluation by the ethics committee?
  - If not, should the ethics committee suggest one be established?

In terms of cross-institutional studies, researchers should pay attention to data transmission and sharing. In some cases, researchers might use cloud-based file sharing services to share data (e.g., Dropbox); however, such methods are not recommended for storing and sharing confidential data due to data security and privacy protection issues (University of Essex 2012). Likewise, data should not be transmitted through emails. Rather, to safely transmit sensitive or personal data, researchers should adopt an appropriate encrypted procedure. In addition, procedures need to be implemented to protect against data leakage—for example, password-protected access to certain areas of the storage space and/or encrypted data files. The procedures for data storage and protection need to be addressed through the architecture of the computer server structure and/or by encrypting sensible data. It is important to note that some data protection officers consider any encryption temporary until a method is found to decrypt the data.

The following are examples of a researcher's steps in the IRB ethics application process regarding data storage and protection:

- All electronic data will be encrypted and stored on computer servers at the researcher's university, as well as on the researcher's strongly password-protected computer(s) and a backup hard drive that will be in a locked space in the researcher's office.
- Computer and network administrators in the researchers' departments will work to ensure data safety by monitoring computer and network use, controlling user access, and preventing intrusions and failures.
- These data are expected to be maintained for as long as needed, or at least 5 years. Note that in some countries, such as Germany, this may be up to 10 years minimum.
- In order to recover data from potential intrusions, failures, or unexpected situations, emergency backup systems have been actively put in place.

The following statement—or parts thereof—might be used in the ethics application to communicate how privacy would be protected with regard to data sharing:

> No one other than the researchers involved in the research will have access to these data. Personal identifiers will be removed from the original data sets before sharing research data. For participants to remain anonymous, researchers must agree not to disclose, publicly announce, or make known to any persons who are unauthorized, the information collected throughout the course of this research project that could potentially identify any participants in the study.

11.7 Informed Consent

Informed consent addresses anonymity and consent issues and is one of the primary ethical requirements underpinning research with human participants (U.S. National Commission for the Protection of Human Subjects of Biomedical and Behavioral Research 1979). Initial consent is the first requirement, but is not sufficient by itself; consent must be a continuous process because changes may be experienced throughout the course of a study (Penslar 1993).

The purpose of informed consent is to ensure prospective participants understand (a) the nature of the research, (b) that they can voluntarily decide whether or not to participate, and (c) that they can cease participation at any point. In some countries, the following information must be provided to each participant (Protection of human subjects 2009e):

(1) A statement explaining the purposes of the research, the estimated duration of the subject's participation, a description of the procedures, and identification of any experimental procedures;
(2) A description of any anticipated risks or discomforts;
(3) A description of benefits towards the participant or others;
(4) Disclosure of alternative procedures, if any, that might be beneficial to the participant;
(5) A statement describing the protection of participant-identifying records to maintain privacy, if applicable;

(6) Participants subject to research posing greater than minimal risk receive an explanation regarding potential compensation and possible medical treatments in the case of incurred injury, what these treatments consist of, or where they may find additional information;
(7) Instructions regarding whom to contact with important questions about participants' rights and
(8) A statement of voluntary participation, and that refusal to participate or stop participation at any point would not result in any penalty or loss of benefits granted to the participant.

11.8 Submission Procedures for Ethics

Typical ethics submissions may include the contents listed below (Protection of Human Subjects 2009a; University of Tennessee Institutional Review Board 2012; Carleton University 2016; University of Illinois at Champaign-Urbana 2017):

(1) Summary of research: brief explanation of research purpose and objectives, significance of research, and targeted participants.
(2) Recruiting procedures: reasons for selecting specific participants, participants' background, and the method of recruitment (e.g., flyer, email, letter, phone call).
(3) Inclusion and exclusion criteria: details and justification for selecting and excluding participants.
(4) Withheld information (i.e., deception): justification for withholding a study's true purpose and the procedure for describing the study's true purpose to participants following the study's completion.
(5) Research procedures: methods of contacting and informing participants, length and location of the study, usage of foreign language(s), details of explaining the research procedures to the participants, consent procedures, incentive method, and so on.
(6) Equipment: use of a system or equipment, including how each one interfaces with each other and with participants (attach manufacturer's printed material, as appropriate). This includes specifications, industry ratings, and industry standards for all stimulation, amplification, transduction, and data acquisition equipment, as applicable.
(7) Data collection procedures: description of (a) how the data will be collected, anonymously or not, along with specific procedures for collecting data if the study is not anonymous, and (b) the method for informing participants of research results.
(8) Data security and retention: procedures for ensuring all kinds of collected data (including sensor data, audiotapes, videos, coded transcripts, signed consent forms, survey, or written interview notes) will be kept in the principal investigator's or main researcher's office, or in another secure location where only the main researcher(s) will have access to the data. Videos or audiotapes will immediately be destroyed following transcription. Generally,

data will be locked in a file cabinet in the researchers' locked office for at least five years.

(9) Staff training: qualifications of the researchers relating to the proposed study.

(10) Dissemination results: method of distributing the results via websites, academic conferences, social media, journal publications, etc.

(11) Consent process: details of consent information, distribution of consent forms (online, phone, face-to-face, etc.), and use of language.

(12) Risks: description of any physical and mental risks or discomfort. One of the issues in occupant behavioral research relates to the privacy of statements that might put the current job of the employees at risk. A good example of describing risks involved in the research can be found in in the IRB guidelines at the University of Tennessee Knoxville (2012).

(13) Benefits: explanation of how research results will benefit all participants, as well as the research community.

(14) Risk/benefit assessment: description detailing how the benefits of the study outweigh the risks.

11.9 Debriefing

Debriefing is the procedure where in researchers explain the purpose of the study after the study is conducted. Debriefing can take the form of a conversation, an interview, or a written document informing study participants of the purpose, design, and sometimes the results of the study. This is especially important in occupant-related experiments where deception was used as an essential part of the study design. Ethical importance must be centered on the debriefing process to guarantee all participants are fully informed about the purpose of the experiment and that they will not sustain any type of physical or mental injury over the course of the study. Informed consent and debriefing are the basic building blocks of an ethical safeguard within research that involves the participation of human beings.

11.10 Multiple-Site and Cross-Country Studies

When working with researchers from several institutions, whether nationally or internationally, it is important to remember that each institution may have specific requirements for research and ethical considerations. Although many nations have well-established regulations concerning the ethics, it is important to consider differences between one country and another's positions on ethical practices (Shelley-Egan et al. 2015). For instance, research occurring in both the U.S. and Germany may require a separate ethics application and approval from the institutions or universities in these two countries. Specifically, cooperative research involving more than one institution generally requires that each institution is responsible for protecting the rights of participants. Each institution must gain approval from the department or agency head, and cooperative projects may rely on the review of another qualified IRB or ethics review committee as part of a joint review arrangement or some similar arrangement (Protection of human subjects 2009c).

Any multiple-site or international agreement must usually be formalized before the lead university will accept research proposals from the other institution or rely on their review. Instances will also arise where each institution needs to process its own ethics application. In any case, researchers will likely be required to identify all institutes involved in the project, the respective ethics applications sought, and the method in which protocol information is diffused among all participating establishments. In addition, researchers are responsible for coordination with outside regulatory agencies and other participating facilities, as well as for all aspects of internal review and oversight procedures. Importantly, the involved researchers must ensure that any participating facility obtain the review and approval of their specific ethics applications and adopt all protocol revisions in a concise manner. The researchers' responsibility is to guarantee that, prior to recruiting participants, participating facilities and any other applicable committees have reviewed and approved the study (Brown University 2016).

11.11 Tips for Improving Interactions with Ethics Committee

The following tips are useful when researchers interact with ethics committee staff (Oakes 2002, p. 469):

- Draw attention to human participant interactions through an attached cover letter along with your ethics application;
- Intrinsically pose the question of whether you would enjoy having someone you care about engage in your study;
- Establish recruitment materials that procure impartial and just results;
- Write consent forms in a manner that middle-school-aged students could read and understand;
- Emphasize risks and benefits;
- Make sure procedures are established that delink any distinguishable material from primary data sets and origins;
- Encrypt any identifying information through previously set procedures and destroy it at the earliest opportunity;
- Make sure to read and understand the statute if you do not agree with a decision made by the IRB, and then schedule a face-to-face meeting to further discuss the conclusion that was reached;
- Keep in mind that research is not a right, but a privilege, and IRBs are strictly enforced peer review groups;

- Be aware that the purpose of ethics/IRBs is not to irritate scientists; rather, they stem from a multitude of ethics violations from previous studies.

11.12 Internet Research Ethics

Recent growth in the popularity of analyzing "big data" through the Internet has raised new ethical considerations for researchers from all fields of study. The human subject ethics review process and procedures will likely evolve to reflect advances in using social media platforms (e.g., Facebook, Twitter, blogs, etc.) as a data collection site. The definition of Internet research includes research that (a) uses the Internet to collect data or information, (b) observes people's activities or participation on social networking sites; (c) involves data processing, analysis, or storage of datasets; (d) studies software, code, and Internet technologies, (e) analyzes the design or structure of computer systems, interfaces, and other elements, (f) conducts visual and textual analysis, and (g) investigates large-scale production and regulations of the Internet by governments or industry (Markham and Buchanan 2012).

Recently in the USA, changes have been made regarding ethical decisions about Internet research; in particular, studies conducted with data from public social media platforms, which provide almost no protection for participants beyond that of user agreements. It was deemed that because individuals have no reasonable expectation of privacy in these platforms, even if the information is identifiable, the use of the data may be exempted from human subject review (Fiske and Hauser 2014). For example, analyses of posts to a public forum would likely not require a human subject review. This logic extends to other types of data collection, such as observing, coding, and recording behavior in public places, as well as using certain other digital data where an individual has no reasonable expectation of privacy.

Markham and Buchanan (2012) point out several key principles that are fundamental to ethical Internet research. First and foremost, ethical decision-making is best approached using practical judgment attentive to the specific context. Digital information may involve individual participants and their personal information, and so considering the ethical criteria for its use is necessary. Second, when making ethical decisions, researchers must balance the rights of participants with researchers' rights to conduct research. Finally, ethical decision-making is a deliberative process, and researchers should consult the IRB or ethics committee and experienced experts.

11.13 Conclusion

Researchers have a responsibility to conduct quality research while also protecting participants' privacy and rights (Canadian Institutes of Health Research et al. 2014). Ethical conduct is an important consideration for any type of research to minimize potential harm to participants. The approval of ethical conduct will not be granted by an IRB/ethics board unless all ethical considerations are carefully and thoroughly deliberated and expressed; as a result, the ethics application process often involves multiple rounds of revisions requested on behalf of the committee.

This chapter provided a guide to navigating the process of ethics approval by explaining the common types of review, and then discussing the various ethical considerations of conducting research with human subjects. These considerations included participant recruitment, potential risks specific to occupant behavior studies, and guidance to ensure that such risks are significantly outweighed by anticipated benefits. Informed consent is a necessary tool and an ongoing process of protecting participants' anonymity and confidentiality. This chapter also provided basic information on research projects involving multiple sites, nationally or internationally, where each institution has its own set of ethics requirements that must be met. The tips provided in this chapter are designed to help researchers from any country or institution improve their interactions with ethics review boards, meet crucial requirements, ensure an ethically conducted study, and most of all, protect research participants.

References

Brown University (2016) Human research protection program policy and procedure manual–Section 1

Bureau of Labor Statistics at the U.S. Department of Labor (2003) American time use survey. https://www.bls.gov/tus/. Accessed 14 Apr 2017

Canadian Institutes of Health Research, Natural Science and Engineering Research Council of Canada, Social Sciences and Humanities Research Council of Canada (2014) Tri-council policy statement: ethical conduct for research involving humans

Carleton University (2016) Forms and templates. http://carleton.ca/researchethics/human-ethics/forms/. Accessed 14 Apr 2017

Carroll AM, Perez M, Toy P (2004) Appendix D: ethical considerations in human subjects research. In: UCLA Center Health Policy Research. Health DATA Train the Trainer Project Performing a Community Assessment Curriculum. Appendix

Collaborative Institutional Training Institute (CITI) (2016) Training materials on informed consent. https://www.citiprogram.org/. Accessed 6 Jul 2016

Diaz JA, Martín-Arribas MC, Olivera LH, et al (2015) Ethics assessment and guidance in different types of organisations—Research Ethics Committees. http://satoriproject.eu/media/3.a-Research-ethics-committees.pdf. Accessed 15 Jan 2017

Fiske ST, Hauser RM (2014) Protecting human research participants in the age of big data. Proc Natl Acad Sci USA 111:13675–13676. doi:10.1073/pnas.1414626111

Grinyer A (2004) The narrative correspondence method: what a follow-up study can tell us about the longer term effect on participants in emotionally demanding research. Qual Health Res 14:1326–1341. doi:10.1177/1049732304269674

Gutwill JP (2002) Gaining visitor consent for research: testing the posted-sign method. Curator Mus J 45:232–238

Gutwill JP (2003) Gaining visitor consent for research II: Improving the posted-sign method. Curator Mus J 46:228–235

Institutional Review Board at the University of Tennessee (2017) Frequently asked questions. http://irb.utk.edu/faq/. Accessed 14 Apr 2017
Markham A, Buchanan E (2012) Ethical decision-making and internet research: recommendations from the AoIR ethics working committee (Version 2.0). https://aoir.org/reports/ethics2.pdf. Accessed 19 Apr 2017
Miluzzo E, Lane ND, Lu H, Campbell AT (2010) Research in the app store era: experiences from the cenceme app deployment on the iphone. In: First Workshop on Research in the Large at UbiComp
O'Brien W, Gunay HB (2014) The contextual factors contributing to occupants' adaptive comfort behaviors in offices–a review and proposed modeling framework. Build Environ 77:77–87
O'Brien W, Kapsis K, Athienitis AK (2013) Manually-operated window shade patterns in office buildings: a critical review. Build Environ 60:319–338. doi:10.1016/j.buildenv.2012.10.003
Oakes JM (2002) Risks and wrongs in social science research: an evaluator's guide to the IRB. Eval Rev 26:443–479
Office of Research. University of California Irvine (2017) Privacy and Confidentiality. http://www.research.uci.edu/compliance/human-research-protections/researchers/privacy-and-confidentiality.html. Accessed 15 Jan 2017
Penslar RL (1993) Protecting human subjects: institutional review board guidebook. Office for Human Research Protections, U.S. Department of Health and Human Services, Rockville, MD
Protection of human subjects (2009a) 45 C.F.R §46.111(a)(7). https://www.hhs.gov/ohrp/regulations-and-policy/regulations/45-cfr-46/index.html#46.111. Accessed 14 Apr 2017
Protection of human subjects (2009b) 45 C.F.R §46.101(b). https://www.hhs.gov/ohrp/regulations-and-policy/regulations/45-cfr-46/index.html%7B#%7D46.101. Accessed 14 Apr 2017
Protection of human subjects (2009c) 45 C.F.R §46.102(i). https://www.hhs.gov/ohrp/regulations-and-policy/regulations/45-cfr-46/index.html%7B#%7D46.102. Accessed 14 Apr 2017
Protection of human subjects (2009d) 45 C.F.R §46. https://www.hhs.gov/ohrp/regulations-and-policy/regulations/45-cfr-46/index.html. Accessed 14 Apr 2017
Protection of human subjects (2009e) 45 C.F.R §46.116. https://www.hhs.gov/ohrp/regulations-and-policy/regulations/45-cfr-46/index.html#46.116. Accessed 14 Apr 2017
Shelley-Egan C, Brey P, Rodrigues R, et al (2015) Ethical assessment of research and innovation: a comparative analysis of practices and institutions in the EU and selected other countries. http://satoriproject.eu/media/D1.1_Ethical-assessment-of-RI_a-comparative-analysis.pdf. Accessed 21 Mar 2017
Slonecker ET, Shaw DM, Lillesand TM (1998) Emerging legal and ethical issues in advanced remote sensing technology. Photogramm Eng Remote Sens 64:589–595
U.S. Department of Health and Human Services (2017) International compilation of human research standards. https://www.hhs.gov/ohrp/international/compilation-human-research-standards/index.html. Accessed 14 Apr 2017
U.S. National Commission for the Protection of Human Subjects of Biomedical and Behavioral Research (1979) The belmont report: ethical principles and guidelines for the protection of human subjects of research. https://www.hhs.gov/ohrp/regulations-and-policy/belmont-report/. Accessed 14 Apr 2017
University of Essex (2012) United Kingdom Data Archive
University of Illinois at Champaign-Urbana (2017) Tip sheet examples. http://oprs.research.illinois.edu/?q=irb-admin/tipsheet_example.html. Accessed 14 Feb 2017
University of Michigan (2017) Class assignments & IRB approval. http://research-compliance.umich.edu/human-subjects/human-research-protection-program-hrpp/hrpp-policies/class-assignments-irb-approval. Accessed 14 Apr 2017
University of Tennessee Institutional Review Board (2012) Institutional Review Board Guide for Faculty, Staff, and Students. http://irb.utdev4.wpengine.com/wp-content/uploads/sites/29/2013/05/handbook-2012-20.pdf. Accessed 20 Sep 2016

Chapter 12
Concluding Remarks and Future Outlook

William O'Brien, Andreas Wagner and Bing Dong

Abstract This book has defined relevant terms in the field of building occupant research and provided a comprehensive overview of the steps required to study occupants' behavior in buildings, whether in situ or through laboratory experiments or surveys. It has offered both broad and specific guidance about research design and methodological approaches, including data collection, storage, and processing, and presented relevant discussions of ground truth and ethics. At the time of this book's publication, the field of occupant research is relatively new, but with rapidly increasing activity. Therefore, the motivation was to significantly improve the state of the art of occupant behavior research methodologies, considering the multidisciplinarity of the field by including authors from the broad backgrounds of engineering, architecture, interior design, information technology, and social sciences. Readers of this book will realize that the field of occupant behavior research still holds a large number of unanswered fundamental questions to be tackled. Thus, this concluding chapter provides the editors' perspectives on research needs future outlook for the field of occupant behavior.

A. Wagner (✉)
Building Science Group, Faculty of Architecture, Karlsruhe Institute of Technology, Englerstr. 7, 76131 Karlsruhe, Germany
e-mail: wagner@kit.edu

W. O'Brien
Department of Civil and Environmental Engineering, Carleton University, 3432 Mackenzie Building, 1125 Colonel By Drive, Ottawa, ON K1S 5B6, Canada
e-mail: Liam.obrien@carleton.ca

B. Dong
Department of Mechanical Engineering, University of Texas at San Antonio, Engineering Building, One UTSA Circle, San Antonio, TX 78249-0670, USA
e-mail: Bing.dong@utsa.edu

A. Wagner et al. (eds.), *Exploring Occupant Behavior in Buildings*,
DOI 10.1007/978-3-319-61464-9_12

12.1 Research Needs

As evidenced throughout the book, there have been numerous notable insights into building occupants' behavior; nonetheless, there is considerable potential for advancement in the field. To begin, there are few established precedents for deciding whether to measure certain predictors in occupant studies—in fact, many occupant action domains (e.g., window blinds) have only a handful of existing studies. Additionally, the sensors used for occupant studies are often borrowed from other applications and thus not optimized for the current studies. For instance, daylight illuminance sensors for occupant studies are often designed for building controls, where they serve a different purpose (e.g., measure relative illuminance on the ceiling). The slim selection of sensors aimed at measuring occupants directly is relatively immature and lacks the robustness and accuracy required for rigorous research. Furthermore, for large-scale and long-term studies, self-configured and "peel and stick" occupant sensors would be highly desirable, but do not yet exist in a readily available and affordable format. The four major components of a sensor—sensing element, power consumption, processing, and communication—need further research and innovations.

It is essential to ensure the validity of measurements based on occupant sensing technologies and the reliability of the collected data. Unfortunately, there is not a standard procedure to collect, verify, and validate ground truth data. In addition, any sensing technology introduces measurement errors. Hence, it is challenging to construct ground truth occupancy datasets and this requires further research.

The cost, effort, and technological and practical limitations of building occupant research have resulted in most existing occupant studies involving no more than tens of participants (unless a survey approach is used). As of yet, it is unclear whether such small-scale studies can be extended to a broader population or different contexts—this is an issue of generalizability that suggests poor practice. Moreover, studies' methods largely vary from one to the next and therefore cannot be compared or cross-validated. Paradoxically, the remedy is more studies, yet the validity of those would also be in question. This is why the current book is so timely: consistency and quality of occupant research methods is critical to the advancement of this field.

While in situ occupant studies have been the dominant approach for informing statistical occupant models in building performance simulation, laboratory studies and surveys are emerging as promising alternative methods. A major research need for the future is to establish the validity of these latter methods for simulation. In the broader sense, beyond simulation applications, valuable qualitative and quantitative findings can emerge from all occupant studies. For instance, the usability of building systems can be assessed and occupants' building and energy literacy—as well as cause and effect relationships—can be established. However, currently both the studies and dissemination of these results to building designers and operators are lacking.

Ethics and privacy remain a major consideration for occupant researchers. Despite the low risk of most occupant studies, ethical requirements often mandate that researchers obtain informed consent from occupants. In order to avoid biases, e.g. Hawthorne effect, resulting from participants' knowledge about the study, it has to be negotiated with ethical boards to keep the information about the scientific goals and purposes of the study as general as possible. Only information about an overall aim, the method and probable interactions with participants should be given. In addition, sensor and meter data may be readily available to researchers, however this does not mean that they can necessarily be used without clearance from an ethics board. It is expected that occupant researchers will learn to better navigate the ethics process over time and that ethics procedures will better accommodate occupant research needs.

Despite—or perhaps as a result of—the cost of conducting occupant studies, occupant researchers are generally quite conservative about sharing data and no widely used data repository currently exists for this purpose. At best, data are shared via personal communication, which is in great contrast to more mature scientific fields where datasets are often required to accompany publications. Before the opportunity to create a data repository can be seriously pursued, significant efforts are still required to develop a scientifically sound approach for providing generally usable and comparable data sets. Notably, Chaps. 6 and 10 provide significant guidance on data structures and documentation.

One topic not covered in this book is modeling occupant behavior, where data are needed to train and validate these models. Modeling spans a huge field from physically based (i.e., white box) to pure statistical approaches (i.e., gray or black box) and calls researchers from different disciplines onto the scene. In line with experimental work on occupant behavior, numerous activities can be seen in the scientific community on model development and implementation in building simulation software. As with the results of occupant studies, the quality and transferability of models strongly depends on a sound scientific basis and knowledge about possibilities and constraints of the different modeling techniques. Interested readers will find more in-depth information on this topic in the book "Statistical Modeling of Occupant Behaviour" that also emerged from the activities of the International Energy Agency (IEA) Energy in Buildings and Communities (EBC) Annex 66 Definition and Simulation of Occupant Behavior in Buildings.

12.2 Future Outlook

The fundamental drivers for increased interest and research in occupant behavior are expected to remain for many decades to come. Heightened environmental awareness and policy-driven demands for better building performance will continue into the future, as well. Contrary to expectations, new technologies and deeper automation systems have highlighted—not made obsolete—the importance of understanding occupants' needs, preferences, and interactions with buildings. As such, occupant behavior studies and their methods are predicted to remain center stage.

At the same time, the field of occupant behavior is at the peak of inflated expectations of the hype cycle. This suggests that the coming years are likely to be met with challenges as researchers discover that the field is far more complex than the physical relationships that can be used to describe most other building performance-related phenomena. Nevertheless, researchers must pass to the next stage of the hype cycle in order for widespread adoption of the valuable research outputs to occur. This book aims to guide the field through the anticipated turbulence and keep the research high-quality, ethical, and impactful.

In the future, we expect many outstanding fundamental questions to be answered, such as: What is the best systematic way to collect ground truth data? What sample time and size of data are required to capture occupant behavior? Do laboratory and survey studies have sufficient ecological validity? Is the Hawthorne effect significant in this context? What are the critical contextual factors (e.g., climate, culture, presence of air conditioning, etc.) to be identified and what impact do they have on occupant behavior?

Meanwhile, we expect that technological advances will improve the accuracy and cost effectiveness of studying occupant behavior. Occupant sensors will be able to count the number of occupants, comfort level, location, direction of movement, and even posture of occupants, while processing these phenomena locally to maintain data security. Already, sensors and sensor infrastructure (e.g., communication and data acquisition and storage) have advanced considerably to the extent that many of them are now applicable to studying occupants. Wireless communication for sensors will become commonplace. Sensors will routinely generate sufficient power to process signals and communicate to hubs. Industry will better cooperate in developing and adhering to communication protocols that will greatly improve accessibility of sensor data to researchers. In the future, we will see sensors that are better able to capture occupant comfort and indoor air quality, rather than measuring fairly indirect proxies. Weather stations and their integration into building automation systems will be commonplace as costs drop and communications protocols standardize. Finally, the spatial resolution of electricity and water meters will greatly improve as the ratio of the cost of the measurement equipment to the measured substance continues to decrease.

Finally, and central to any research field, we expect a growing critical mass of worldwide occupant behavior researchers to remain strong and integrated. While a repository of occupant data will be a key ingredient for this thriving community, this book has contributed to standardizing and elevating the quality of future research in the field. Research programs like IEA EBC Annex 66 are one avenue for advancing a field and developing international standards and formalisms. Documentation of methods and results of occupant studies need to be targeted at all stakeholders, including building design practitioners, building technologies developers, building operators, and, of course, researchers.

FSC
www.fsc.org
MIX
Papier aus verantwortungsvollen Quellen
Paper from responsible sources
FSC® C105338

If you have any concerns about our products,
you can contact us on
ProductSafety@springernature.com

In case Publisher is established outside the EU,
the EU authorized representative is:
Springer Nature Customer Service Center GmbH
Europaplatz 3, 69115 Heidelberg, Germany

Printed by Libri Plureos GmbH
in Hamburg, Germany